INTERMEDIATE MECHANICS

SI EDITION

D. HUMPHREY, B.A., B.Sc.
Late Director of Education, The Polytechnic, London

Edited by

J.TOPPING, M.Sc., Ph.D., D.I.C., F.INST.P.
Vice-Chancellor and Principal, Brunel University, Uxbridge

VOLUME TWO

STATICS

LONGMAN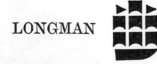

LONGMAN GROUP LIMITED
LONDON
*Associated companies and representatives
throughout the world*

New edition © Longman Group Ltd 1964, 1971

SI edition © Longman Group Ltd 1971

First edition 1930
Second edition 1964
SI edition 1971

ISBN 0 582 32238 3

Printed in Great Britain by
Richard Clay (The Chaucer Press) Ltd,
Bungay, Suffolk

INTERMEDIATE MECHANICS
SI EDITION

STATICS

PREFACE TO THE SI EDITION

In this new edition the special features and format of the Second Edition have been maintained, and the content is essentially unchanged. The justification and indeed need for a new edition is the agreement by Great Britain to adopt an international system of units, known as SI.

The text and the examples throughout the book have been accordingly modified. In many of the examples the changes have been made to keep the arithmetic as simple as possible and in others to bring the problem up to date. The newton has been used exclusively as the unit of force; it has been thought best not to introduce any other, and consequently the unit equal to the weight of a body of mass 1 kilogramme, sometimes called the kilogramme-force and denoted by 1 kgf, has not been used at all.

This volume, and the companion volume on Dynamics, are designed for students carrying their study of the subject beyond the General Certificate of Education at Advanced Level, while the *Shorter Intermediate Mechanics* in one volume should suit the purposes of those for whom Advanced Level is an adequate attainment. It is hoped that the books may continue to find wide acceptance in schools, whatever changes may be made in the next few years in school examinations and curricula, and to prove useful in first year University courses.

It is a pleasure to express my thanks again to the teachers and students who have written to me from time to time. It is also a pleasure to thank the various examining bodies for permission to use questions.

J. TOPPING

Brunel University, Uxbridge

v

ACKNOWLEDGEMENTS

WE are indebted to the following for permission to reproduce questions from past examination papers:

The Joint Matriculation Board of the Northern Universities; Oxford and Cambridge Schools Examination Board; University of Cambridge Local Examinations Syndicate, and the University of London.

The following abbreviations are used to denote the source from which examples are taken:

B.Sc.	London University. B.Sc. Examination.
C.A.	G.C.E. Advanced Level. Cambridge University.
C.S.	College Scholarships. Cambridge University.
C.W.B.	Central Welsh Board. Higher Certificate.
Ex.	Exhibitions. London University.
H.C.	Oxford and Cambridge Higher Certificate.
H.S.C. & H.S.D.	Higher Schools. London University.
I.A.	Intermediate Arts. London University.
I.C.	Imperial College. Entrance and Scholarship Examination.
I.E.	Intermediate Engineering. London University.
I.S.	Intermediate Science. London University.
L.A.	G.C.E. Advanced Level. London University.
N.U.	Joint Matriculation Board. Northern Universities. Higher Certificates and G.C.E. Advanced Level.
O.C.	G.C.E. Advanced Level. Oxford & Cambridge.
Q.E.	Qualifying Examination. Mechanical Sciences Tripos, Cambridge.
S.	Scholarship Examination. London University.

CONTENTS

CHAPTER ONE

STATICS OF A PARTICLE

1.1. We now consider that part of mechanics, known as statics, which deals with solids at rest under the influence of forces. Again we shall assume to start with that the dimensions of the solid body may be neglected. We shall, as earlier, refer to such a body as a *particle* and represent it by a point.

1.2. Force

We may define *force* as any cause which alters or tends to alter a body's state of rest or of uniform motion in a straight line (see Dynamics, Chapter Three). Force accelerates a body, and only if there is no resultant force acting on the body will it continue at rest, relative to some assigned system of reference, or continue to move with uniform velocity. In statics we are concerned to examine the relations that must exist between the forces acting on a body when the resultant force is zero.

1.3. To specify completely a force which acts on a particle, we require: (i) its magnitude, and (ii) its direction in space.

In the case of a rigid body the point of application of the force must *also* be specified.

The magnitude of a force can be measured by its effect. In dynamics this is usually done by the motion it will produce in a given body in a given time, or more strictly, by the acceleration it will produce in the body at any instant. Thus the unit of force, *the newton* (N), is defined as the force which, acting on a mass of 1 kg, produces an acceleration of 1 ms^{-2}.

We can also measure a force by the weight it will just support. In this case the value will depend on the force of gravity at the place where the force is measured.

Such a unit of force has a different value at different places on the earth's surface, because of the variation in the gravitational force.

However, in Statics we are usually concerned only with the *relative* values of forces at the same place. We can therefore conveniently use any unit of force, for the ratios of forces all expressed in terms of any chosen unit will not vary from place to place.

1.4. Vectors

We can represent a given force completely by means of a straight line AB. For one extremity A can represent the point of application, the direction of the line in space can represent the direction of the force, and the length of the line can be made to represent the magnitude of the force to some convenient scale.

Any quantity which, like a force, possesses both magnitude and direction is called a *Vector*.

Other examples of vectors which occur in dynamics are velocity, momentum, and acceleration, and they can all be represented completely by straight lines.

When a force is represented by a line AB the direction of the force is indicated by the order of the letters, i.e. AB represents a force acting from A to B, and BA represents a force acting from B to A. These are denoted by the vectors **AB** and **BA**, and we note that **AB** = − **BA**.

The fundamental law satisfied by all vectors is the *law of vector addition*. This states that the sum of any two vector quantities of the same kind represented by **AB** and **BC** is a vector represented by **AC**.

If **AB** and **BC** are given, **AC** can be found by drawing the triangle ABC.

1.5. Basis of statics

To build up the science of Mechanics we start with some laws or postulates based on, or supported by, experiment or observation.

Now Statics is really a particular case of Dynamics in which the bodies are at rest relative to some system of reference. If we choose the laws relating to the action of forces on bodies by considering the more general case of bodies in motion (Dynamics) they will hold for the particular case of when the bodies are at rest (Statics).

It is, of course, possible to base the science of Statics on laws which are independent on the idea of motion. Historically, in fact, Statics was developed before Dynamics. For the latter we had to wait until the seventeenth century, the foundations being laid by Galileo and Newton, whereas Statics goes back at least to the early Greeks, some two thousand years earlier. Although Aristotle (384–322 B.C.) made some contribution to mechanics, it was Archimedes of Syracuse (287–212 B.C.) who developed the principle of the lever and effectively constituted a science of Statics (cf. 11.17).

Now we can take as our starting-point in Statics the fundamental theorem known as the *Parallelogram of Forces*. This merely states that the law of vector addition is true for forces, but it can be deduced from the fundamental laws of Dynamics, known as *Newton's Laws of Motion*. This is explained in Chapter Three of the volume on Dynamics, 3.12. Thus Statics and Dynamics comprise one science based on one set of fundamental postulates.

1.6. The parallelogram of forces

If two forces, acting on a particle at O, be represented in magnitude and direction by the two straight lines OA, OB drawn from O they are equivalent to a single force which is represented in magnitude and direction by the diagonal OC of the parallelogram OACB.

This is the fundamental theorem of Statics, and it can be verified by experiment. As for any other scientific law, this is its basic justification. **OC** (Fig. 1.1) is the vector sum of **OA** and **OB**, so that the theorem simply states that forces are added (or compounded) like vectors.

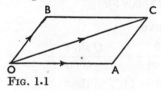

FIG. 1.1

The single force represented by OC is known as the *resultant* of the two forces represented by OA and OB. Conversely, the forces represented by OA and OB are known as the *components* of the force represented by OC.

In the simple case when the two forces are in the same direction their resultant is clearly equal to their sum, and when they act in opposite directions their resultant is equal to their difference and acts in the direction of the greater.

The resultant is zero only if the two forces are equal in magnitude and act in opposite directions.

The magnitude and direction of the resultant of any two forces acting at O can be found by drawing a parallelogram OACB on an appropriate scale, or by calculation, as is now explained.

1.7. Let two forces P and Q acting at O be represented by OA and OB (Fig. 1.2). Their resultant, R, is represented by the diagonal OC of the parallelogram OACB.

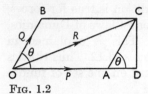

FIG. 1.2

Let θ be the angle between the directions of P and Q, i.e. angle AOB = θ. Draw CD perpendicular to OA, produced if necessary.

Then
$$OC^2 = OD^2 + DC^2$$
$$= (OA + AD)^2 + DC^2$$
$$= (OA + AC \cos \theta)^2 + (AC \sin \theta)^2.$$
$\therefore$
$$R^2 = (P + Q \cos \theta)^2 + (Q \sin \theta)^2$$
$$= P^2 + Q^2 (\cos^2 \theta + \sin^2 \theta) + 2PQ \cos \theta$$
$$= P^2 + Q^2 + 2PQ \cos \theta.$$

Therefore $R = +\sqrt{(P^2 + Q^2 + 2PQ \cos \theta)}$.

To find the direction of OC we have
$$\tan COD = \frac{CD}{OD} = \frac{AC \sin \theta}{OA + AC \cos \theta} = \frac{Q \sin \theta}{P + Q \cos \theta}.$$

If the forces are at right angles, $\theta = 90°$, so that
$$R = \sqrt{(P^2 + Q^2)}, \text{ and } \tan COA = Q/P.$$

If the forces are equal, say each equal to P, then
$$R = \sqrt{[P^2(1 + 1 + 2 \cos \theta)]} = P\sqrt{[2(1 + \cos \theta)]}$$
$$= P\sqrt{(\cos^2 \tfrac{1}{2}\theta)} = 2P \cos \tfrac{1}{2}\theta.$$

Also
$$\tan COA = \frac{P \sin \theta}{P + P \cos \theta} = \frac{2 \sin \tfrac{1}{2}\theta \cos \tfrac{1}{2}\theta}{2 \cos^2 \tfrac{1}{2}\theta} = \tan \tfrac{1}{2}\theta.$$

Hence in this case the resultant bisects the angle between the forces. This can also be deduced from the fact that when the forces are equal the parallelogram is a rhombus.

1.8. EXAMPLE (i)

If the resultant of forces $3P$, $5P$ *is equal to* $7P$, *find the angle between the forces.*

If θ is the angle between the forces $3P$ and $5P$, and R is the resultant,

$$R^2 = 9P^2 + 25P^2 + 30P^2 \cos \theta$$

$\therefore \qquad 34P^2 + 30P^2 \cos \theta = 49P^2$, since $R = 7P$

$\therefore \qquad 30P^2 \cos \theta = 15P^2$

$\therefore \qquad \cos \theta = \tfrac{1}{2}$

$\therefore \qquad \theta = 60°.$

EXAMPLE (ii)

If two forces P *and* Q *act at such an angle that* $R = P$, *show that if* P *is doubled the new resultant is at right angles to* Q.

Let OA, OB (Fig. 1.3) represent P and Q. Then the diagonal OC of the parallelogram OACB, which represents R, is equal to OA.

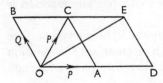

FIG. 1.3

Produce OA to D, making AD = OA, then OD represents $2P$, and the resultant of Q and $2P$ is represented by the diagonal OE of the parallelogram ODEB.

Now $\qquad\qquad$ BC = OA = P

and $\qquad\qquad$ CE = AD = P.

$\therefore \qquad\qquad$ CB = CE = CO.

Therefore BE is the diameter of a semicircle with centre C passing through O,

$\therefore \qquad\qquad \angle$BOE is a right angle.

Or $\qquad\qquad \angle$CEO = $\angle$COE, since CE = CO,

and $\qquad\qquad \angle$CBO = $\angle$COB, since CB = CO,

$\therefore \qquad\qquad \angle$CEO + $\angle$CBO = $\angle$BOE,

$\therefore \qquad\qquad \angle$BOE is a right angle.

1.9. Smooth bodies

There is, of course, no such thing as a perfectly smooth body. In practice, when one body is pressing against another there is always some force acting along their common surface tending to prevent the slipping of one over the other. This is known as *friction*.

With highly polished surfaces this force may be very small, and in many problems the bodies are supposed to be perfectly smooth. In such cases the only force between the bodies is perpendicular to their common surface. This is called the *normal reaction*.

The *direction* of the normal reaction acting on any smooth body, where it is in contact with another body or surface, is always perpendicular to the direction in which the body is capable of moving.

When a rod is resting against a smooth plane the reaction is normal to the plane.

When a rod is resting against a smooth peg the reaction is perpendicular to the rod.

When the end of a rod is resting against the curved surface of a smooth sphere or against a smooth circular arc the reaction is normal to the sphere or circle, and therefore passes through its centre.

These cases should be noted particularly, as they occur frequently.

1.10. Tension of a string

When a string is used to suspend a weight or move a body the string is in a state of tension.

If we consider a string of negligible weight supporting a weight W vertically the tension in the string is approximately the same throughout its length and may be taken equal to W. If, however, the string is heavy the tension in the string varies from point to point; the tension at any point A, if the string is at rest, equals W plus the weight of the string below A. If the weight of the string is to be neglected we shall refer to it as a *light* string.

If a string supporting a weight is passed over a small, smooth pulley as at A (Fig. 1.4) it is found that the force required to keep the weight in position is unaltered, and it makes no difference whether the string is held in the position AB, AC, or AD. This is *not* true if the pulley is rough (see 8.19).

We note that the tension in a *light* string passing round a smooth peg or pulley is the same throughout its length.

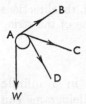

FIG. 1.4

We shall usually assume that the string is inextensible, but we shall consider a few problems in which the string is elastic and satisfies Hooke's law. In the latter case the tension varies with the extension of the string (see Dynamics, 4.19).

1.11. EXAMPLE

Two equal weights of mass 10 kg are attached to the ends of a thin string which passes over three smooth pegs in a wall arranged in the form of an equilateral triangle with one side horizontal. Find the thrust on each peg.

Let A, B, C (Fig. 1.5) be the positions of the pegs.

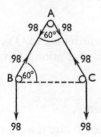

FIG. 1.5

Since the pegs are smooth, the tension is the same throughout the string and equal to 10g N = 98 N.

The thrust on A is the resultant of the two tensions of 98 N inclined at an angle of 60°. If R N is the magnitude of this resultant,

$$R^2 = 98^2 + 98^2 + 2 \times 98^2 \cos 60° = 3 \times 98^2$$

$$\therefore \qquad R = 98\sqrt{3} = 170.$$

The thrust on B or C is the resultant of the two tensions of 98 N inclined at an angle of 150°. If S N is the magnitude of this resultant,

$$S^2 = 98^2 + 98^2 + 2 \times 98^2 \cos 150°$$
$$= 98^2(2 - \sqrt{3}) = 98^2 \times 0.268$$
$$\therefore \qquad S = 98 \times 0.518 = 50.8.$$

Since the component forces in each case are equal, the directions of the thrusts bisect the angles between the portions of the string on each side of the peg.

1.12. Resolution of a force

A force may be resolved into two components in an infinite number of ways; for an infinite number of parallelograms can be constructed on a given line as diagonal.

In practice, the directions of the components are known, and the most important case is when these directions are at right angles to each other.

Let OC (Fig. 1.6) represent a given force F, and suppose we wish to resolve it into two components, one along OX, and the other along a perpendicular direction OY.

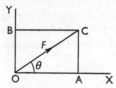

FIG. 1.6

Draw CA perpendicular to OX and CB perpendicular to OY.

Then OA and OB represent the components of F along OX and OY respectively.

If the angle COX $= \theta$ we have

$$OA = F \cos \theta \text{ and } OB = F \sin \theta.$$

Hence a force F is equivalent to a force $F \cos \theta$ along a line making an angle θ with its own direction together with a force $F \sin \theta$ perpendicular to the direction of the first component.

When a force is resolved in this manner into two forces whose directions are at right angles these forces are sometimes called the *Resolved Parts* or *Resolutes* of the given force in these two directions.

The resolved part of a force F in a direction making an angle θ with it is $F \cos \theta$.

The expression 'horizontal component' of a force, which is often used, must be taken to mean the resolved part in the horizontal direction, i.e. it is understood that the other component is vertical.

Generally, we shall use the term component (and not the term resolved part), it being understood, unless the contrary is indicated, that we are concerned with resolving the force (or the vector) into two perpendicular components.

1.13. If we do require the components of a force F in directions making angles α and β with it they can be found as follows:

Let OC (Fig. 1.7) represent F. Draw OA and OB, making

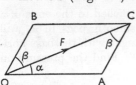

Fig. 1.7

angles of α and β with OC, and through C draw parallels to complete the parallelogram OACB.

Then OA and OB, or OA and AC represent the required components.

Hence, from the triangle OAC

$$\frac{OA}{\sin \beta} = \frac{AC}{\sin \alpha} = \frac{OC}{\sin (\alpha+\beta)}$$

$$\therefore \qquad OA = \frac{F \sin \beta}{\sin (\alpha+\beta)}.$$

Similarly, $\qquad OB = \frac{F \sin \alpha}{\sin (\alpha+\beta)}.$

EXAMPLES 1.1

1. In the following cases P and Q denote two forces, θ the angle between them, and R is their resultant:

      (i) If $P = 9, Q = 12, \theta = 90°$,   find R.

      (ii) If $R = 13, Q = 5, \theta = 90°$,   find P.

      (iii) If $Q = 7, Q = 8, \theta = 60°$,   find R.

      (iv) If $P = 10, Q = 10, \theta = 120°$,   find R.

      (v) If $P = 12, Q = 5, R = 11$,   find θ.

      (vi) If $P = 3, Q = 5, \theta = \sin^{-1} \frac{3}{5}$,   find R.

2. Show that the resultant of two forces each equal to P, and inclined at any angle of $120°$, is also equal to P.

3. Find the angle between two equal forces P when their resultant is: (i) equal to P; (ii) equal to $P\sqrt{3}$.

4. The resultant of two forces P and Q is equal to P in magnitude, and that of two forces $2P$, Q (acting in the same directions as before) is also equal to P. Find the magnitude of Q, and prove that the direction of Q makes an angle of $150°$ with that of P.

(H.C.)

5. The sum of two forces is 24 N, and their resultant, which is at right, angles to the smaller of the two forces, is 12 N. Find the magnitudes of the two forces, and the angle between them.

6. Two equal weights of mass 10 kg are attached to the ends of a light string which passes over three smooth pegs in a wall arranged in the form of an isosceles triangle, with the base horizontal and with a vertical angle of $120°$. Find the pressure on each peg.

7. Two equal weights of mass 5 kg are attached to the ends of a light string which passes over two smooth pegs in a wall, one of which is higher than the other, and such that the line joining them makes an angle of $30°$ with the horizontal. Find the pressure on each peg.

8. At what angle must two forces of 5 and 12 N be inclined if they are balanced by a force of 15 N?

9. A, B are two fixed points on the circumference of a circle ABC. P is a point which moves on the arc ACB. Forces of constant magnitude act along PA and PB. Prove that the line of action of their resultant passes through a fixed point.

10. The resultant of two forces P and $2P$, acting at a point, is perpendicular to P. Find the angle between P and $2P$.

11. The resultant of forces of 3 N and 5 N is at right angles to the smaller force. Find the magnitude of their resultant and the angle between the forces.

12. Forces of 4 N and 6 N act at an angle of $60°$. Find their resultant graphically, and check by calculation.

13. Find graphically the resultant of forces of 5 N and 6 N, inclined at an angle of $40°$.

14. The resultant of two forces is 8 N, and its direction is inclined at $60°$ to one of the forces whose magnitude is 4 N. Find graphically the magnitude and direction of the other force.

15. When two equal forces are inclined at an angle 2α their resultant is twice as great as when they are inclined at an angle 2β. Show that $\cos \alpha = 2 \cos \beta$.

16. A force of 25 N is inclined at $\theta°$ to the horizontal. If the vertical component is 15 N, find the horizontal component and the value of θ.

17. Resolve a force of 10 N into two perpendicular components such that: (i) the components are equal, and (ii) one component equals 3 times the other.

18. A force P N is resolved into two components each of magnitude $2P$ N; find their directions.

19. One component of a force of 12 N is 4 N at an angle of 30°. Find the magnitude and direction of the other component.

20. In pulling a garden roller of mass 60 kg across a horizontal lawn a man exerts a force of 500 N at an angle of 45° with the ground. Find: (i) the forward pull exerted on the roller, and (ii) the vertical reaction between the roller and the lawn.

21. A body of mass 25 kg is supported by two strings, one inclined at 60° and the other at 20° to the vertical. Calculate the tension in each string, and check by means of a graphical construction.

1.14. Particle subject to forces

In dynamics, we have shown that if a particle is subject to *one* force F, the particle will move in the direction of F with an acceleration a given by $F = ma$, where m is the mass of the particle. Again, if the particle is subject to *two* forces P and Q the particle will move in the direction of the resultant of P and Q. This resultant is zero only if P and Q are equal in magnitude and opposite in direction. In this case the particle will remain at rest (or if in motion, continue to move with uniform velocity). For example, if a weight W is suspended from a fixed point by a string it rests in equilibrium under the action of two forces viz. its weight W acting vertically downwards and a force T vertically upwards due to the string equal to the tension in the string. In this case $T = W$.

1.15. Three forces acting on a particle

Let P, Q, R (Fig. 1.8) be three forces acting on a particle at O.

If two of these, say P and Q, be represented by OA and OB their resultant is represented by the diagonal OC of the parallelogram OACB.

Hence if R can be represented in magnitude and direction by CO it will balance the resultant of P and Q, and the three forces P, Q, R will be in equilibrium.

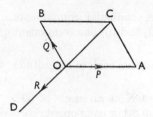

Fig. 1.8

Hence if the three forces P, Q, R can be represented in magnitude and direction by OA, AC, and CO they will be in equilibrium.

(It should be noticed that AC represents Q in magnitude and direction but not in line of action.)

This result is known as the **Triangle of Forces**, which is usually stated as follows:

If three forces, acting at a point, can be represented in magnitude and direction by the sides of a triangle taken in order the forces will be in equilibrium.

The converse of this is also true:

If three forces acting at a point are in equilibrium they can be represented in magnitude and direction by the three sides of a triangle taken in order.

This can be verified experimentally.

1.16. *Graphical method*

The converse of the triangle of forces enables us to obtain simple graphical solutions of problems on the equilibrium of three forces.

Suppose we know that three forces P, Q, R acting at a point O in given directions OX, OY, OZ (Fig. 1.9) are in equilibrium.

Then if we know the magnitude of *one* of the three forces, say
P, we can obtain the magnitudes of the other two as follows:

Since the forces are in equilibrium, we know that they can
be represented in magnitude and direction by the sides of a
triangle taken in order.

Draw AB parallel to OX and make its length represent the
magnitude of P to some convenient scale.

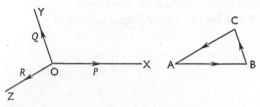

FIG. 1.9

From A draw AC parallel to OZ, and from B draw BC
parallel to OY. Since AB represents P, the other two sides of
the triangle ABC must represent Q and R. BC represents Q, and
CA represents R, on the same scale that AB represents P. The
order of the letters represents in each case the direction of the
force, and this order is obtained by continuing round the
triangle in the same direction.

1.17. In Fig. 1.9 we saw that CA represents the force R which
is in equilibrium with the resultant of P and Q. Hence AC must
represent the resultant of P and Q in magnitude and direction.
We may therefore obtain the magnitude and direction of the
resultant of P and Q from the triangle ABC without using the
parallelogram.

*If two forces acting at some point are represented in magnitude
and direction by the sides* AB, BC *of a triangle* ABC *the third side*
AC *will represent their resultant in magnitude and direction.*

In vector notation **AC** is the vector sum of **AB** and **BC**, that
is,

$$\mathbf{AC} = \mathbf{AB} + \mathbf{BC}$$

or $$\mathbf{AB} + \mathbf{BC} + \mathbf{CA} = 0.$$

It must be clearly understood that AC does not represent
the *position* of the resultant which will act at the point where
the forces represented by AB and BC act.

If AB and BC are actually the lines of action of the forces their resultant will act at B, but it will be equal and parallel to AC.

1.18. The converse of the triangle of forces can also be put in a trigonometrical form, usually called *Lami's Theorem*.

If three forces acting at a point are in equilibrium each is proportional to the sine of the angle between the other two.

Let P, Q, R (Fig. 1.10) be the three forces acting at O.

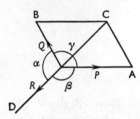

FIG. 1.10

Let OA represent P, and OB represent Q, then CO represents R in magnitude and direction.

Let α, β, γ be the angles between Q and R, R and P, P and Q respectively.

In the triangle OAC we have

$$\angle \text{ACO} = \angle \text{BOC} = 180° - \alpha, \quad \therefore \sin \text{ACO} = \sin \alpha.$$
$$\angle \text{COA} = 180° - \beta, \qquad\qquad \therefore \sin \text{COA} = \sin \beta.$$
$$\angle \text{CAO} = 180° - \gamma, \qquad\qquad \therefore \sin \text{OAC} = \sin \gamma.$$

Also
$$\frac{\text{OA}}{\sin \text{ACO}} = \frac{\text{AC}}{\sin \text{COA}} = \frac{\text{CO}}{\sin \text{OAC}}$$

$$\therefore \qquad \frac{P}{\sin \alpha} = \frac{Q}{\sin \beta} = \frac{R}{\sin \gamma}.$$

1.19. EXAMPLE (i)

Forces equal to $7P$, $8P$ and $5P$ acting on a particle are in equilibrium; find by drawing and by calculation the angle between the forces $8P$ and $5P$.

Since the forces are in equilibrium, they can be represented by the sides of a triangle taken in order. As in Fig. 1.11A, we draw the triangle ABC with AB = 7 units, BC = 8 units and CA = 5 units. The forces must therefore be parallel to the sides of this triangle.

By measurement $\angle ACB = 60°$. By calculation, if $\angle ACB = \theta$ we have

$$7^2 = 8^2 + 5^2 - 2 \times 8 \times 5 \times \cos \theta$$

$\therefore$ $$49 = 64 + 25 - 80 \cos \theta$$

$\therefore$ $$\cos \theta = 40/80 = \tfrac{1}{2}$$

$\therefore$ $$\theta = 60°.$$

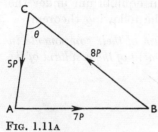

Fig. 1.11a

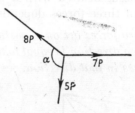

Fig. 1.11b

The forces must be directed as shown in Fig. 10.11B, where the vectors representing the forces have been drawn parallel to the corresponding sides of the triangle ABC.

The angle between the forces $8P$ and $5P$ is $180° - \theta = 120°$.

EXAMPLE (ii)

A weight of mass 5 kg hangs from a fixed point O by a light inextensible string. It is pulled aside by a horizontal force P N and rests in equilibrium with the string inclined at an angle of 30° to the vertical. Find P.

In Fig. 1.12 A is the equilibrium position of the weight.

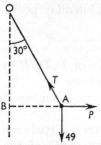

Fig. 1.12

The weight is acted upon by three forces: (i) its weight 49 N; (ii) the horizontal force P N; and (iii) a force T N equal to the tension in the string.

If AB is drawn horizontally the sides of the triangle OAB are parallel to the three forces, and therefore represent them on a certain scale.

$$\therefore \qquad \frac{T}{AO} = \frac{P}{BA} = \frac{49}{OB}$$

$$\therefore \qquad P = 49\frac{BA}{OB} = 49 \tan 30° = 49/\sqrt{3} = 28{\cdot}3.$$

1.20. Another method which may be used to solve many problems associated with a particle in equilibrium under the action of three forces depends upon the following theorem.

If two forces act on a particle the sum of their components in any given direction is equal to the component of the resultant of the two forces in that direction.

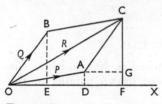

FIG. 1.13

Let OA and OB represent the two forces P and Q and OC their resultant R, so that OACB is a parallelogram (Fig. 1.13).

Let OX be the given direction.

Draw AD, BE, CF perpendicular to OX, and AG perpendicular to CF.

The triangles OBE, ACG have their sides OB and AC equal, and their sides are parallel, so that they are equal in all respects,

$$\therefore \qquad OE = AG = DF$$

$$\therefore \qquad OF = OD + DF = OD + OE.$$

But OD, OE, OF represent the components of P, Q, R in the direction of OX, and hence the theorem follows.

This theorem may obviously be extended to any number of forces.

It follows that if three or more forces acting on a particle are in equilibrium the algebraic sum of the components of the forces in *any* direction must be zero.

It can be shown, conversely, that if the algebraic sums of the components of the forces in *any two* directions are zero, then the forces must be in equilibrium (see 1.26).

1.21. EXAMPLE (i)

A particle of mass 50 kg is suspended by two strings 3 m and 4 m attached to two points at the same level, whose distance apart is 5 m. Find the tensions in the strings.

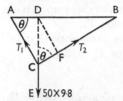

FIG. 1.14

Let AC, BC (Fig. 1.14) be the two strings, and let T_1, T_2 N be the tensions in the strings.

AC = 3 m, BC = 4 m, AB = 5 m, and ACB is a right angle.

Let CD be perpendicular to AB; since AB is horizontal, DC is vertical, and the weight of the particle at C acts along DC produced.

Let $\angle DCB = \theta$, then $\angle BAC = \theta$, and

$$\cos \theta = \tfrac{3}{5}, \sin \theta = \tfrac{4}{5}.$$

Method (a): Resolving horizontally, the component towards the left must equal that towards the right,

$$\therefore \qquad\qquad T_1 \cos \theta = T_2 \sin \theta. \qquad\qquad (i)$$

Resolving vertically, the sum of the upward components must equal the downward one, and hence

$$T_1 \sin \theta + T_2 \cos \theta = 50 \times 9.8 = 490. \qquad (ii)$$

From (i) $\tfrac{3}{5}T_1 = \tfrac{4}{5}T_2$ or $T_2 = \tfrac{3}{4}T_1$
and from (ii) $\tfrac{4}{5}T_1 + \tfrac{3}{5} \times \tfrac{3}{4}T_1 = 490$

$$\therefore \qquad\qquad\qquad \tfrac{25}{20}T_1 = 490$$

$$\therefore \qquad\qquad T_1 = 392 \text{ and } T_2 = 294.$$

Otherwise, resolving along AC and CB we have immediately

$$T_1 = 490 \sin \theta = 392$$
and $$T_2 = 490 \cos \theta = 294.$$

Method (b): Produce DC to E, then by Lami's Theorem,

$$\frac{T_1}{\sin BCE} = \frac{T_2}{\sin ACE} = \frac{50}{\sin ACB}.$$

But sin BCE = sin θ, sin ACE = sin $(90-\theta)$ = cos θ, and $\angle ACB = 90°$.

Hence
$$\frac{T_1}{\frac{4}{5}} = \frac{T_2}{\frac{3}{5}} = \frac{490}{1}$$

$$\therefore \qquad T_1 = 392 \text{ and } T_2 = 294.$$

Method (c): If the figure be drawn to scale, and DF be drawn parallel to AC, the sides of the triangle DCF are parallel respectively to the weight 490, the tension T_2, and the tension T_1.

Hence CF and FD represent T_2 and T_1 to the same scale that DC represents 490. But $CF/CD = \cos \theta = \frac{3}{5}$. Hence $T_2 = 294$.

EXAMPLE (ii)

A particle of mass 5 kg is placed on a smooth plane inclined to the horizontal at an angle of 30°. Find the magnitude of the force (a) acting parallel to the plane, (b) acting horizontally, required to keep the particle in equilibrium.

Let AB (Fig. 1.15) be the surface of the plane, AC horizontal, P the particle.

Since the plane is smooth, it cannot exert any force parallel to its surface tending to prevent the particle sliding down. There is only a normal reaction equal to: (a) R_1 N, and (b) R_2 N.

(a) the force F_1 acting parallel to the plane (Fig. 1.15A) necessary to prevent the particle sliding down is evidently 49 sin 30° = 24·5 N, acting upwards. The component of the weight perpendicular to AB is balanced by the reaction R_1 of the plane.

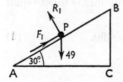   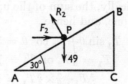

FIG. 1.15A   FIG. 1.15B

(b) If a force F_2 acts horizontally as shown (Fig. 1.15B), its component F_2 sin 30 perpendicular to AB merely presses the particle against the plane. Its component $F_2 \cos 30°$ parallel to the plane must therefore be sufficient to balance the component of the weight, 24·5 N, acting down the plane.

$$\therefore \qquad F_2 \cos 30° = 24·5$$

$$\therefore \qquad F_2 = \frac{49}{\sqrt{3}} = \frac{49}{3}\sqrt{3} \text{ N}.$$

We note that $R_2 = F_2 \sin 30° + 49 \cos 30° = 98\sqrt{3}/3$.

EXAMPLE (iii)

A string ABCD is suspended from two points A and D at the same level. A weight of 10 N is hung from a string knotted to ABCD at B, and a weight w N from a string knotted to ABCD at C. If AB and CD make angles of 45° and 60° respectively with the horizontal, and if BC is horizontal, find the tension in the string BC and the weight w.

Let the tensions in the strings AB, BC, CD be T_1, T_2, T_3 N respectively (Fig. 1.16).

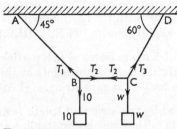

FIG. 1.16

Considering the three forces acting at B, and resolving horizontally and vertically, we get

$$T_1 \cos 45° = T_2$$

and $\qquad\qquad T_1 \sin 45° = 10.$

Hence, $\qquad\qquad T_1 = 10\sqrt{2}$ and $T_2 = 10.$

Similarly, considering the three forces acting at C, and resolving horizontally and vertically, we get

$$T_3 \cos 60° = T_2$$

and $\qquad\qquad T_3 \sin 60° = w.$

Hence, $\qquad\qquad T_3 = 2T_2 = 20$ and $w = 10\sqrt{3}.$

Alternatively, we can construct a triangle of forces for the forces acting at B by drawing BM perpendicular to AD. It follows immediately that $T_1 = 10\sqrt{2}$ and $T_2 = 10.$ Similarly, if we draw CN perpendicular to AD the triangle CND will serve as a triangle of forces for the forces acting at C.

EXAMPLES 1.2

1. A weight of mass 20 kg is suspended by two light strings of lengths 6 m and 8 m from two points at the same level and 10 m apart. Find the tensions in the strings.

2. A mass of 90 kg is suspended by two light strings of lengths 9 m and 12 m from two points at the same level and 15 m apart. Find the tensions in the strings.

3. A weight of 26 kg is suspended by two light strings of lengths 5 m and 12 m from two points at the same level, and 13 m apart. Find the tension in the strings.

4. A smooth, weightless pulley carries a weight of mass 15 kg, and can slide freely up and down a smooth vertical groove. It is held up by a string passing round the pulley so that the two parts of the string make angles of 30° and 60° with the horizontal; show that the tension in the string is slightly under 108 N.   (H.C.)

5. Two strings of lengths 1·5 m and 1·8 m are fastened to a particle of mass 10 kg, their other ends being fastened to points at the same level 2·4 m apart. Find the tension in each string.   (I.A.)

6. A weight of mass 5 kg is suspended by two strings of length 7 cm and 24 cm attached to two points in the same horizontal line at a distance 25 cm apart. Find the tensions of the strings.   (I.S.)

7. A particle of mass 10 kg is placed on a smooth plane of inclination 60°. What force applied (i) parallel to the surface of the plane, (ii) horizontally, will keep the particle at rest?

8. A small body of mass 10 kg is suspended from two points A, B, 12 m apart, and in the same horizontal line, by strings of lengths 7 m and 10 m attached to the same point in the body. Find the tension in each string.   (I.S.)

9. A particle of mass 6 kg, resting on a smooth inclined plane of slope 30°, is connected by a light string passing over a smooth pulley at the top of the plane to a weight of mass M kg hanging vertically. Find the value of M so that the weights may be in equilibrium and find also the pressure on the pulley when this condition is satisfied.

10. A 10 kg mass C is hung up by two cords AC, BC of respective lengths 2 m and 3 m, the ends A, B of the cords being attached to pegs 4 m apart on the same level. Find the tensions in the cords.

11. Three forces, P, Q, R, meet at a point, and the resultants of Q and R, of R and P, and of P and Q are each known in magnitude and direction. Show how to determine each of the forces, P, Q, R.   (I.S.)

12. Find (i) the horizontal force, and (ii) the force up the plane required to support particles of the following masses on planes of the inclinations given:

  (i) 10 kg on an incline of length 10 m and height 6 m.

  (ii) 45 kg on an incline of length 25 m and height 20 m.

  (iii) 5 Mg on an incline of 30°.

13. A body of mass 10 kg rests on a smooth plane inclined at 30° to the horizontal. Find the least value of the force required to keep it in equilibrium and the resultant reaction of the plane.

14. A particle of mass 5 kg is supported by means of two strings attached to it. If the direction of one string be at 60° to the horizontal, find the direction of the other in order that its tension may be as small as possible, and the values of the tensions in the two strings in this case.

15. A weightless wire is stretched between two points A and B on the same level 1·2 m apart. A mass of 5 kg is hung at the mid-point of the wire, and causes it to drop 5 cm below the line AB. Find the tension of the wire.

16. Three strings are attached to a light particle P which is in equilibrium. Two of them pass over pulleys and then hang vertically, carrying weights, while the third supports a weight of mass M kg. The inclinations of the upper strings are respectively 30° and 45° to the upward vertical through P. An additional weight of mass 10 kg is now attached to W. Find what additional weights must be added to the other two strings to ensure that P shall remain in equilibrium in the same position.

(N.U. 3 and 4)

17. Two strings, 40 cm and 30 cm long, are tied to a mass of 6 kg and have their other ends fastened to two nails, 50 cm apart in a horizontal line. By drawing or otherwise, find the tension in the longer string.

  The nails are now replaced by smooth pulleys, over which the strings are passed with mass of 4·5 kg and 6 kg tied respectively to their free ends and hanging freely. In the new position of equilibrium find, by drawing or otherwise, the angles which the sloping portions of the strings make with the vertical.

18. A weight of mass 10 kg hangs from a fixed point O by a light inextensible string. It is pulled aside by a force P N which always acts perpendicular to the string. Find P, if in the equilibrium position the string makes an angle of (i) 45°, (ii) 30° with the vertical.

19. A string, which breaks at a tension of 100 N, hangs freely from a fixed point, and has a weight of mass 6 kg attached to it at B. A horizontal pull P now acts at B and drags it sideways. In the position of equilibrium when P is 25 N, find the tension in the string and the slope of the string to the vertical.

In the position when the string is about to break, find the magnitude of the horizontal pull P and the slope of the string.

20. The figure (Fig. 1.17) represents a weight W attached to two strings which pass over two smooth pegs at the same level and support weights P, Q at their free ends. Prove that, in the position of equilibrium,

$$\sin \theta = (W^2 + P^2 - Q^2)/2WP.$$

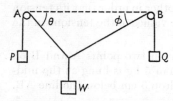

Fig. 1.17

21. A light string ABCD is supported at A and D, which are points in the same horizontal plane. At B hangs a weight of mass 10 kg and at C a weight of mass M kg. The string AB is at 45° to the vertical, DC is at 30° to the vertical and BC slopes upward at 45° to the horizontal. Find the tensions in AB, BC and CD, and the value of M.

22. ABCD is a light string attached to two fixed points A and D, has two equal weights attached at B and C and rests with the portions AB and CD inclined at 30° and 60° respectively to the vertical. Prove that the tension in the portion BC is equal to either weight, and that BC is inclined at 60° to the vertical.

23. A and D are points in a horizontal line 48 cm apart; the ends of a light cord 66 cm long are attached to A and D; B is a point in the cord 25 cm from A and C a point in the cord 29 cm from D. A mass of 14 g is suspended from B. Find the mass of Q which must be suspended from C, so that the portion BC of the cord shall be horizontal.

24. Weights of 2, 4, and 3 kg are attached to points B, C and D of a string of which the ends A and E are attached to fixed supports. If each of the portions of the string BC and CD makes an angle of 25° with the horizontal, prove that AB and DE make angles of approximately 43° and 49° 23′ with the horizontal.   (Q.E.)

1.22. Particle subject to more than three forces

If a particle is acted upon by several forces we can show that the forces are in general equivalent to a single force, the resultant of the forces. The particle will be in equilibrium if the resultant of the forces is zero.

Two important problems arise; firstly, how can the resultant of any number of forces be found, and secondly, what are the relations between the forces when they are in equilibrium?

1.23. The polygon of forces

If any number of forces, acting on a particle, can be represented in magnitude and direction by the sides of a polygon taken in order, the forces will be in equilibrium.

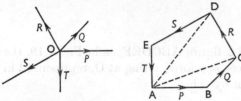

Fig. 1.18

Let the sides AB, BC, CD, DE, EA of the polygon ABCDE (Fig. 1.18) represent the forces acting on a particle at a point O.

The resultant of forces represented by AB and BC is represented in magnitude and direction by AC.

The resultant of forces represented by AC and CD is represented in magnitude and direction by AD, and similarly the resultant of forces represented by AD and DE is represented in magnitude and direction by AE. Hence the resultant of P, Q, R, S is equal and opposite to T and, since all the forces act at a point, this resultant and T balance and the system of forces is in equilibrium.

This method of proof will evidently apply whatever the number of forces.

It is also clear that the forces need not be in one plane.

Resultant of any number of forces

1.24. If we have a number of forces acting on a particle at O, as in the last paragraph, and we draw a polygon with its sides proportional and parallel to these forces (always drawing the next side from the point where the previous side ended), then if the polygon closes we know that the system is in equilibrium, but if it does *not* close the resultant of the forces is represented by the straight line necessary to close the figure, the direction of this resultant being opposite to that in which the figure has been drawn, that is, from the first vertex to the last vertex.

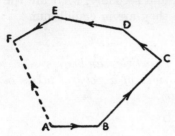

FIG. 1.19

Thus, if we obtained a figure ABCDEF, as in Fig. 1.19, the resultant of the five forces is a force acting at O, represented in magnitude and direction by AF.

In vector notation,

$$\mathbf{AB+BC+CD+DE+EF = AF.}$$

The vectors representing the forces can be drawn in *any* order, that is, the resultant is independent of the sequence in which the forces are added. In Fig. 1.19A the force-polygon

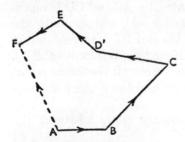

FIG. 1.19A

differs from that in Fig. 1.19, the order of the third and fourth forces having been interchanged, but AF is the same in each case.

This method of obtaining the resultant of a number of

forces graphically is obviously very convenient and much easier than applying the parallelogram of forces to pairs of the forces in turn.

It is often used, as will be seen later, when the forces are acting on a rigid body.

However, it is more accurate and often quicker to calculate the resultant of a number of forces by resolving the forces in two perpendicular directions in the manner illustrated in the following paragraph.

1.25. *To find the resultant of any number of forces in one plane acting on a particle.*

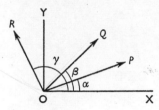

FIG. 1.20

Let the forces, P, Q, R, etc., act upon a particle at O (Fig. 1.20). Through O draw any two axes OX, OY, at right angles to each other. We assume all the forces act in the XOY plane.

Let the forces, P, Q, R . . . make angles α, β, γ . . . with OX. The components of P in the directions OX and OY are $P \cos \alpha$ and $P \sin \alpha$ respectively; similarly the components of Q are $Q \cos \beta$ and $Q \sin \beta$, and so on.

Hence the forces are equivalent to a component

$$P \cos \alpha + Q \cos \beta + R \cos \gamma \text{ . . . along OX,}$$

and a component

$$P \sin \alpha + Q \sin \beta + R \sin \gamma \text{ . . . along OY.}$$

Let these components be X and Y respectively, and let F be their resultant and θ its inclination to OX.

Then $\qquad F \cos \theta = \text{X}$

and $\qquad F \sin \theta = \text{Y}$

∴ $\qquad F^2 = \text{X}^2 + \text{Y}^2$

and $\qquad \tan \theta = \dfrac{\text{Y}}{\text{X}}.$

EXAMPLE (i)

A particle is acted on by forces of 1, 2, 3, and 4 N, the angles between them being 60°, 30°, and 60° respectively. Find the magnitude and direction of the resultant.

It is convenient to take as the axis of x the line of action of one of the forces.

Let O (Fig. 1.21) be the position of the particle, and OX the line of action of the 1 N force. The 3 N force then acts along OY.

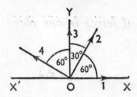

FIG. 1.21

The component of the resultant along OX is

$$1+2 \cos 60-4 \cos 30 = 1+1-2\sqrt{3}$$
$$= 2-3 \cdot 464 = -1 \cdot 464,$$

i.e. a force of 1·464 N in the direction OX'.

The component of the resultant along OY is

$$2 \cos 30+3+4 \cos 60 = \sqrt{3}+3+2 = \sqrt{3}+5 = 6 \cdot 732.$$

The resultant is

$$\sqrt{[(1 \cdot 464)^2+(6 \cdot 732)^2]} = 6 \cdot 9 \text{ N.}$$

If θ is the angle made by this resultant with OX,

$$\tan \theta = -\frac{6 \cdot 732}{1 \cdot 464} = -4 \cdot 597$$

∴ $\theta = 102° 16'$ about.

This problem may also be solved graphically as follows. As in Fig. 1.21A, draw AB, BC, CD, DE parallel to the given forces and of magnitudes 1, 2, 3, 4 respectively.

Then **AE** represents the resultant of the forces in magnitude and direction.

By measurement, AE = 6·9 and angle EAB = 102°. If these

quantities are calculated, using Fig. 1.21A, the equivalence of the
two methods is obvious.

It is instructive to draw the sides of the force-polygon ABCDE in
a different order.

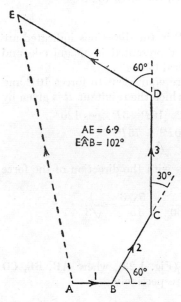

$AE = 6.9$
$E\hat{A}B = 102°$

FIG. 1.21A

EXAMPLE (ii)

Three forces of magnitude $15P$, $10P$, $5P$ *act on a particle in directions
which make* $120°$ *with one another. Find their resultant.*

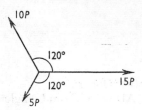

FIG. 1.22

Method (i)

Sum of the components of the forces parallel to the force $15P$

$$= 15P - 10P \cos 60° - 5P \cos 60°$$
$$= 15P/2.$$

Also, sum of the components of the forces perpendicular to the force $15P$

$$= 10P \sin 60° - 5P \sin 60° = 5\sqrt{3}P/2.$$

Hence, the resultant $R = \frac{1}{2}P\sqrt{[15^2 + (5\sqrt{3})^2]} = 5\sqrt{3}P$ at an angle $\theta = \tan^{-1}(5\sqrt{3}/15) = 30°$ with the force $15P$.

Method (ii)

We note that forces $5P$, $5P$, $5P$ in the directions indicated are in equilibrium, since they can be represented in magnitude and direction by the sides of an equilateral triangle.

Hence, the three given forces are equivalent to forces $10P$ and $5P$ inclined at an angle of $120°$, of which the resultant R is given by

$$R^2 = (10P)^2 + (5P)^2 + 2 \times 10P \times 5P \times \cos 120°$$
$$= 100P^2 + 25P^2 - 50P^2 = 75P^2$$

$$\therefore \qquad R = 5\sqrt{3}P.$$

The angle θ the resultant makes with the direction of the force $15P$ is given by

$$\tan \theta = \frac{5 \sin 60°}{10 - 5 \cos 60°} = \frac{5\sqrt{3}}{15} = \frac{1}{\sqrt{3}}$$

so that $\theta = 30°$.

Method (iii)

Draw a vector-polygon ABCD (Fig. 1.23), where **AB**, **BC**, **CD** represents the forces $15P$, $10P$, $5P$ respectively.

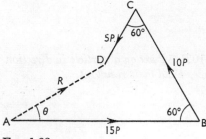

FIG. 1.23

Then **AD** represents the resultant of the three forces.

By measurement, or by calculation, $AD = 8\cdot7P$ and angle $DAB = 30°$.

EXAMPLES 1.3

1. Forces of 4, 3, 2, and 1 N act as a point A in directions AB, AC, AD, and AE, where $\angle BAC = 30°$; $\angle CAD = 30°$, $\angle DAE = 90°$. Find the magnitude of their resultant, and the

inclination of its direction to AB. Verify by drawing a vector-diagram.

2. A particle is acted on by forces of 2 and 4 N at right angles, and also by a force of $4\sqrt{2}$N, whose direction bisects the angle between the other two. Find the resultant force on the particle.

3. Three forces of 5, 10, and 13 N act in one plane at a point, the angle between any two of their directions being 120°. Find the magnitude and direction of their resultant.

4. ABCD is a square, and forces of 2, 4, and 5 N act at A in the directions AB, AC, AD respectively. Find the magnitude of their resultant.

5. Find the magnitude and direction of the resultant of the following forces acting at a point: 20 N east, 42 N north-west, 60 N 30° south of west, 15 N south.   (H.C.)

6. ABCDEF is a regular hexagon. Forces of 2, $4\sqrt{3}$, 8, $2\sqrt{3}$, and 4 N act at A in the directions AB, AC, AD, AE and AF respectively. Find the magnitude of their resultant and the inclination of its direction to AB. Verify by drawing a vector-diagram.

7. Four horizontal wires are attached to a telephone post and exert the following tensions on it: 80 N north, 120 N east, 160 N south-west, and 200 N south-east. Calculate the resultant pull on the post and find its direction.   (H.C.)

8. Four smooth pegs, A, B, C, D are fixed in the same vertical plane so that they form the four lower corners of a regular hexagon with the side BC horizontal. A string is tied to A, passes under B and C, over D and A, and has a weight of mass 10 kg attached to the free end so that the weight hangs vertically. Find the pressures on the four pegs.

9. Three forces, each of equal magnitude F, act at the point O along the lines OA, OB, OC, which are in a plane. The angle BOA is $+45°$, and the angle BOC is $-90°$. Find the magnitude of the resultant force, and determine its direction by finding the tangent of the angle it makes with OB.

10. Three forces act upon a point in directions north-west, north-east, and south, the forces being respectively of the value 100, 200, 300 N. Find and state accurately the direction and value of the resultant equivalent single force of these three forces.

11. The sides BC, CA, AB of a triangle are 7, 5, and 3 cm respectively. Find graphically, or otherwise, the magnitude and the inclination to BC of the resultant of the following forces *acting at a point*: 5 N in direction BC, 9 N in direction AC, and 3 N in direction AB.   (I.A.)

12. Forces of 2, 3, and 4 N act at a point in directions parallel to the sides AB, AC, BC respectively of an equilateral triangle. Find their resultant.

13. Forces of 5, 4, 12, and 4 N act at a point A along AB, AC, AD, AE respectively. $\angle$BAC = 30°, $\angle$BAD = 90°, $\angle$BAE = 150°. Find the magnitude of the resultant and the angle its line of action makes with AB.

14. ABC is an equilateral triangle, and G the point of intersection of its medians. Forces of 8, 8, and 16 N act at G along GB, GC, GA respectively. Find the magnitude and direction of their resultant.

15. Forces of 11, 20, and 5 N act along the lines OA, OB, OC respectively, where tan AOB = cot AOC = $\frac{3}{4}$, and OB lies between OA and OC. Find the resulting force in the direction OA, and also the magnitude of the total resultant.

16. Four forces, 10, 9, 4, and 1 N, act at a point O in directions north, east, south, and west respectively. Find the magnitude of their resultant, and its inclination to the north.

    Find also the magnitude and direction of the new resultant if a force of $\sqrt{8}$ N be added at O in a north-west direction.

17. The following coplanar forces act at a point, the unit of force being 1 N, and the directions being reckoned counterclockwise from a given line OX:

    2 at 0°, 3 at 30°, 3 at 150°, 2 at 240°, 4 at 270°.

    Find the magnitude and direction of their resultant: (i) graphically by the polygon of forces; (ii) by calculating its components parallel and perpendicular to OX.

18. ABCD is a square of side 30 cm. Forces of 4, 2, and 2 N act along the sides DA, AB and BC respectively. Show that the resultant of the forces along AB and BC cuts DA produced at E such that AE = 30 cm, and hence find the magnitude, direction and line of action of the resultant of the three forces.

19. Forces of 1, 2, and 3 N act along the sides AB, BC and CA of an equilateral triangle of side 5 cm. Find by drawing where the resultant of the forces along AB and BC cuts AC produced, and hence find the magnitude and direction of the resultant of the three forces and where its line of action cuts BC produced.

20. If k is a constant and ABC is a triangle with sides of length a, b, c, find: (i) the resultant of a force ka along BC and a force kc along AB; (ii) the resultant of a force ka along BC and a force kc along BA. Give the magnitude and line of action in each case.

Prove that a force $2ka$ along BC and a force kc along BA are equivalent to a force $3k$BD along BD, where D is a point on CA such that AD = 2DC.

1.26. *Conditions of equilibrium of any number of forces acting on a particle.*

If we resolve the forces in any two directions at right angles and the sums of the components in these directions be X and Y, the resultant F is given by

$$F^2 = X^2 + Y^2.$$

But if the forces are in equilibrium F must be zero.

Now the sum of the squares of two real quantities cannot be zero, unless each quantity is separately zero,

$$\therefore \qquad\qquad X = 0 \text{ and } Y = 0.$$

Hence, *if any number of forces acting on a particle are in equilibrium the algebraic sums of their components in any two directions at right angles must separately vanish.*

Conversely, *if the sums of their components in two directions at right angles are both zero the forces are in equilibrium.*

For then both X and Y are zero, and therefore F is zero.

1.27. In problems where forces acting at a point are in equilibrium the result of the last paragraph gives us the most general method of obtaining a solution by calculation.

To apply this result we resolve the given forces in two directions at right angles (usually horizontally and vertically) and equate the sum of the components in each of these directions to zero.

The graphical methods involving the triangle and polygon of forces can be used for any number of forces, but they usually take longer and are less accurate than calculation.

A vector method is often very convenient.

EXAMPLE (i)

A string is tied to two points at the same level, and a smooth ring of weight W which can slide freely along the string is pulled by a horizontal force P. If in the position of equilibrium the portions of the string are inclined at angles 60° and 30° to the vertical, find the value of P and the tension in the string.

Let A, B (Fig. 1.24) be the points to which the string is tied, C the position of the ring, and CD perpendicular to AB, so that $\angle ACD = 60°$ and $\angle BCD = 30°$. Since the ring is smooth, the tension T is the same throughout the string.

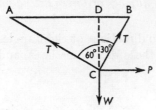

Fig. 1.24

Resolving vertically,

$$T \cos 30° + T \cos 60° = W$$

$$\therefore \quad \left(\frac{\sqrt{3}}{2} + \frac{1}{2} \right) T = W$$

$$\therefore \quad T = \frac{2W}{\sqrt{3}+1} = W(\sqrt{3}-1).$$

Resolving horizontally,

$$P + T \sin 30° = T \sin 60°$$

$$\therefore \quad P = \frac{\sqrt{3}}{2} T - \tfrac{1}{2} T = \frac{T}{2} (\sqrt{3}-1)$$

$$\therefore \quad P = \frac{W}{2} (\sqrt{3}-1)^2 = W(2-\sqrt{3}).$$

EXAMPLE (ii)

ABCD *is a parallelogram and* P *is any point. Prove that the system of forces represented by* PA, BP, PC, DP *is in equilibrium.*

In Fig. 1.25 it is clear that the resultant of the forces represented by BP, PA is represented in magnitude and direction by BA. This resultant acts, of course, at P.

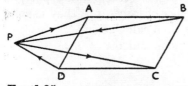

Fig. 1.25

The resultant of the forces represented by DP, PC is represented in magnitude and direction by DC, and acts at P.

Now AB is equal and parallel to DC, so that the resultants are equal in magnitude, and as they act at the same point P they are in the same straight line. Since their directions are opposite, they will balance and the system is in equilibrium.

Vectorially, we can write:

The vector-sum of the forces

$$= (\mathbf{BP}+\mathbf{PA})+(\mathbf{DP}+\mathbf{PC})$$
$$= \mathbf{BA}+\mathbf{DC}$$

which is zero, since AB and DC are equal and parallel and the two vectors are oppositely directed.

Since the forces act at a point and their vector-sum is zero, they are in equilibrium.

In this case resolving the forces in two perpendicular directions would not be a good method to apply.

EXAMPLES 1.4

1. Two forces, 9 and 10 N, act at a point and are inclined to each other at an angle whose tangent is $\frac{4}{3}$.

   Two other forces P and Q are introduced at the point so that the four forces are in equilibrium; Q acts in a direction opposite to that of the force of 9 N, and P is perpendicular to Q. Find their magnitudes.

2. E is the mid-point of the side CD of a square ABCD. Forces 16, 20, P, Q N act along AB, AD, EA, CA in the directions indicated by the order of the letters. Find P and Q, if the forces are in equilibrium.

3. A string of length 0·6 m is attached to two points A and B at the same level and at a distance of 0·3 m apart. A ring of weight 50 N, slung on to the string, is acted on by a horizontal force P which holds it in equilibrium vertically below B. Find the tension in the string and the magnitude of P.   (N.U. 3 and 4)

4. A string of length 31 cm has its ends tied to two points in a horizontal line at a distance 25 cm apart. A small ring from which is suspended a weight of mass 90 g can slide on the string, and is acted on by a horizontal force of such a magnitude that in the position of equilibrium the ring is at a distance of 7 cm from the nearer end of the string. Show that the force is approximately equal to a weight of mass 50 g, and find the tension of the string.
   (I.S.)

5. ABCD is a square. Forces of 2, $3\sqrt{2}$, and 9 N act at A in the directions AB, AC, and AD respectively. Find the additional force at A that will balance these forces.

6. Forces of 3, $\sqrt{3}$, 5, $2\sqrt{3}$, 6 N respectively act at a vertex of a regular hexagon towards the other five vertices. Find the

additional force that must be applied at the vertex to maintain these forces in equilibrium.

7. Five forces acting at a point are in equilibrium. Four of them have magnitudes 1, 2, 3, 4 N respectively and make angles of 0°, 60°, 120°, 210° with a given straight line. Find the magnitude and direction of the other force, and verify by means of a drawing.

8. ABCDEF is a regular hexagon. Show that forces represented by **AB**, **AC**, 2**DA**, **AE**, and **AF** are in equilibrium.

9. ABCD is a square, and BEC is an equilateral triangle lying in the same plane with E outside ABCD. If the forces represented by p**AB**, q**AD**, **AC**, and **EA** are in equilibrium, find p and q.

10. If D, E, F are the mid-points of the sides BC, CA, AB of the triangle ABC, show that three forces acting at a point represented by **AD**, **BE**, **CF** are in equilibrium.

Friction

1.28. When one body slides on another it is found, by experience, that a force tending to resist motion comes into play. This force is called the force of friction.

If we place a block of wood A (Fig. 1.26), of known weight, on a table and attach to it a piece of string passing over a pulley and carrying a light scale pan at the other end we can find the laws which govern the action of the force of friction.

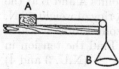

FIG. 1.26

On placing a small weight on the pan B no motion is produced in A. The friction between A and the table must then be equal to the weight of the pan B and the added weight. It is found that we can go on gradually increasing the weight on B until at a certain point A begins to move. This shows that only a limited amount of friction can be called into play.

As the force tending to move A increases from zero, so the force of friction increases from zero at the same rate until a certain maximum or limiting value is reached, and then motion takes place. The total weight of B is then equal to the force of friction. If we now place a weight on A so as to increase the reaction between A and the table, and repeat the experiment, we find that more weight has to be added to B before motion takes place, i.e. the maximum friction has increased. If we repeat the experiment with different weights on A we get a series of values for the maximum friction, which obviously depends on the reaction between A and the table. On dividing the total weight of B by the total weight of A in each case, we find that the number obtained is very nearly constant.

Similar experiments can be performed using blocks of different materials. The results of such experiments are embodied in the following:

Laws of Friction

1. *The direction of the friction is opposite to the direction in which the body tends to move.*

2. *The magnitude of the friction is, up to a certain point, exactly equal to the force tending to produce motion.*

3. *Only a certain amount of friction can be called into play. The maximum amount is called limiting friction.*

4. *The magnitude of the limiting friction (for given surfaces) bears a constant ratio μ to the normal reaction between the surfaces. This ratio μ depends on the nature of the surfaces, and is called the Coefficient of (Statical) Friction.*

5. *The amount of friction is independent of the areas and shape of the surfaces in contact provided the normal reaction is unaltered.*

6. *When motion takes place the friction still opposes the motion. It is independent of the velocity and proportional to the normal pressure, but is slightly less than the limiting friction.*

It must be clearly understood that these laws are experimental and, with the exception of the first three, are subject to limitations. Thus, for very great pressures the surfaces where they are in contact may be crushed and Law 4 will no longer hold.

Law 4 tells us that if F is the limiting friction (i.e. the force of friction when motion is just about to take place), R the normal reaction, and μ the coefficient of statical friction, then

$$F/R = \mu \text{ or } F = \mu R.$$

Great care must be taken not to assume that the friction *always* equals μR. It only has this value when motion is about to take place; otherwise it may have any value *from zero up to μR*.

1.29. Angle of friction

If the normal reaction R and the force of friction F be compounded into a single force (Fig. 1.27) this force is called the

Fig. 1.27

Resultant or Total Reaction, and it makes an angle $\tan^{-1} F/R$ with the normal.

As F increases from zero the angle made by this resultant with the normal increases until the friction F reaches its maximum value μR. In this case the tangent of the angle between the resultant and the normal is $\mu R/R$ or μ. When the friction is limiting the angle made by the resultant reaction with the normal is called the *Angle of Friction*, and is denoted by λ. We see that $\tan \lambda = \mu$.

The resultant reaction can make any angle with the normal up to this value, but cannot be inclined at a greater angle. Whatever direction along the surface the body tends to move, the limiting position of the corresponding resultant reaction will lie on a cone whose semivertical angle is λ or $\tan^{-1}\mu$.

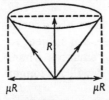

Fig. 1.28

This cone is called the *Cone of Friction* (Fig. 1.28). The resultant reaction must always lie within or on the surface of this cone, and in the latter case the equilibrium is limiting.

1.30. Equilibrium of a particle on a rough inclined plane

Suppose a particle of weight W be placed on a rough plane whose inclination to the horizontal is gradually increased (Fig. 1.29). At any inclination α the component of the weight down

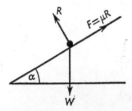

FIG. 1.29

the plane is $W \sin \alpha$. The normal reaction between the particle and the plane is $W \cos \alpha$. The limiting or minimum friction is $\mu W \cos \alpha$. Hence when $W \sin \alpha = \mu W \cos \alpha$, or $\tan \alpha = \mu$, motion is just about to take place.

The particle will therefore begin to slide down under its own weight when the angle of inclination is such that $\tan \alpha = \mu$, i.e. when the inclination of the plane is equal to the angle of friction.

1.31. Particle on rough horizontal plane acted on by an external force

If the force is horizontal (as in Fig. 1.30) and equal to P, then for motion to take place P must be greater than μR. But $R = W$, since there is no vertical motion, and hence P must be greater than μW.

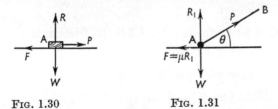

FIG. 1.30   FIG. 1.31

If the force P is inclined upwards, along AB, as in Fig. 1.31, at an angle θ, it will have an upward vertical component, and this reduces the pressure between the particle and the plane.

The normal reaction R_1 is now $W - P \sin \theta$, and the corresponding limiting friction is $\mu(W - P \sin \theta)$. When the motion is just about to take place we must therefore have

$$P \cos \theta = \mu(W - P \sin \theta).$$

$$\therefore \qquad P(\cos \theta + \mu \sin \theta) = \mu W$$

$$\therefore \qquad P\left(\cos \theta + \frac{\sin \lambda}{\cos \lambda} \sin \theta\right) = \frac{\sin \lambda}{\cos \lambda} W$$

where $\lambda = $ angle of friction;

$$\therefore \qquad P \frac{\cos \theta \cos \lambda + \sin \theta \sin \lambda}{\cos \lambda} = \frac{\sin \lambda}{\cos \lambda} W$$

$$\therefore \qquad P \cos (\theta - \lambda) = W \sin \lambda$$

$$\therefore \qquad P = W \frac{\sin \lambda}{\cos (\theta - \lambda)}.$$

The value of P will be a minimum when $\cos (\theta - \lambda)$ is a maximum, i.e. when $\theta = \lambda$, and then $P = W \sin \lambda$.

If P is inclined *downwards* along CA, as in Fig. 1.32, it has

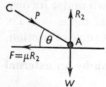

FIG. 1.32

a downward vertical component which increases the normal reaction and hence increases the friction. To move the particle with the least possible force the force should therefore be applied in an upward direction at an angle to the horizontal equal to the angle of friction.

When P is applied downwards (Fig. 1.32) the friction is

$$\mu(W + P \sin \theta)$$

and for the motion to take place, we must have

$$P \cos \theta > \mu(W + P \sin \theta)$$

$$\therefore \qquad P\left(\cos \theta - \frac{\sin \lambda \sin \theta}{\cos \lambda}\right) > \frac{W \sin \lambda}{\cos \lambda}$$

$$\therefore \qquad P > \frac{W \sin \lambda}{\cos (\theta + \lambda)}.$$

Now if $(\theta+\lambda)$ is nearly $90°$ the denominator of the right-hand side is very small. For motion to take place P must be very large. If $(\theta+\lambda)$ becomes equal to $90°$ the particle will not move, however large P may be.

Also if $(\theta+\lambda)$ is greater than $90°$ its cosine is negative, and it is impossible for P to move the particle. The negative value means that P must act in the opposite direction, i.e. along AC.

EXAMPLE

A particle of mass 30 kg resting on a rough horizontal plane is just on the point of motion when acted on by horizontal forces of 24 N and 32 N at right angles to each other. Find the coefficient of friction between the particle and the plane, and the direction in which the friction acts.

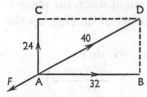

FIG. 1.33

In problems of this kind we must find the *resultant* force tending to move the particle; the particle tends to move in the direction of this resultant and the friction acts in the opposite direction.

Let AB, AC (Fig. 1.33) be the directions of the forces, A the particle. The resultant of the forces is

$$\sqrt{(24^2+32^2)} = 40 \text{ N},$$

and it acts along AD, making an angle $\cos^{-1}\frac{4}{5}$ with the 32 N. The friction acts in the direction DA. Since limiting friction $= \mu R$, where R is the normal reaction on the particle,

$$40 = 30\times9\cdot8\mu, \text{ or } \mu = \frac{4}{29\cdot4} = 0\cdot14.$$

1.32. Particle on rough inclined plane acted on by an external force

I. *When the inclination of the plane is less than the angle of friction*

In this case the friction is enough to prevent the particle sliding down under its own weight.

To find the force P, applied in a vertical plane through the line of greatest slope, required to move the particle up or down the plane, we proceed as follows.

(a) If P acts upwards at an angle θ to the plane, as in Fig. 1.34, the normal reaction R_1 equals $W \cos \alpha - P \sin \theta$.

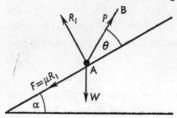

FIG. 1.34

Hence the limiting friction is $\mu(W \cos \alpha - P \sin \theta)$ and acts down the plane. The component of the weight down the plane is $W \sin \alpha$. When the particle is on the point of moving up the plane we have therefore

$$P \cos \theta = W \sin \alpha + \mu(W \cos \alpha - P \sin \theta),$$

$$\therefore \quad P\left(\cos \theta + \frac{\sin \lambda}{\cos \lambda}\sin \theta\right) = W\left(\sin \alpha + \frac{\sin \lambda \cos \alpha}{\cos \lambda}\right)$$

$$\therefore \quad P\frac{\cos (\theta - \lambda)}{\cos \lambda} = W\frac{\sin (\alpha + \lambda)}{\cos \lambda}$$

$$\therefore \quad P = W\frac{\sin (\alpha + \lambda)}{\cos (\theta - \lambda)}.$$

This is a minimum when $\theta = \lambda$, and then $P = W \sin (\alpha + \lambda)$. If $\theta = 0$, i.e. if P acts parallel to the plane,

$$P = W\frac{\sin (\alpha + \lambda)}{\cos \lambda}.$$

(b) If P acts downwards at an angle θ to the plane, along AC (Fig. 1.35), the limiting friction is again $\mu(W \cos \alpha - P \sin \theta)$,

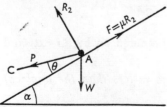

FIG. 1.35

but it now acts up the plane, as the particle is on the point of motion down the plane.

$$\therefore \qquad P \cos \theta + W \sin \alpha = \mu(W \cos \alpha - P \sin \theta)$$

$$\therefore \qquad P\left(\cos \theta + \frac{\sin \lambda \sin \theta}{\cos \lambda}\right) = W\left(\frac{\sin \lambda \cos \alpha}{\cos \lambda} - \sin \alpha\right)$$

$$\therefore \qquad P \frac{\cos (\theta - \lambda)}{\cos \lambda} = W \frac{\sin (\lambda - \alpha)}{\cos \lambda}$$

$$\therefore \qquad P = W \frac{\sin (\lambda - \alpha)}{\cos (\theta - \lambda)}.$$

P is again a minimum when $\theta = \lambda$, and its value is then

$$W \sin (\lambda - \alpha).$$

II. *When the inclination of the plane is greater than the angle of friction*

In this case the particle will slide down unless supported by external force. We have now to consider: (i) the force required to *move* the particle *up* the plane; (ii) the force required to *support* it.

(i) This is exactly the same as in I (i). The force acting parallel to the plane is, putting $\theta = 0$,

$$W \frac{\sin (\alpha + \lambda)}{\cos \lambda}.$$

If acting upwards at an angle θ,

$$P = W \frac{\sin (\alpha + \lambda)}{\cos (\theta - \lambda)}.$$

P is a minimum when $\theta = \lambda$, and then its value is $W \sin (\alpha + \lambda)$.

(ii) If P acts upwards at an angle θ (Fig. 1.36), then, as

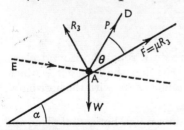

FIG. 1.36

above, the normal reaction R_3 is $W \cos \alpha - P \sin \theta$, and the limiting friction is

$$\mu(W \cos \alpha - P \sin \theta).$$

The friction now acts *up* the plane as the particle is on the point of moving *down*.

$$\therefore \qquad P \cos \theta + \mu(W \cos \alpha - P \sin \theta) = W \sin \alpha,$$

$$\therefore \qquad P\left(\cos \theta - \frac{\sin \lambda \sin \theta}{\cos \lambda}\right) = W\left(\sin \alpha - \frac{\sin \lambda \cos \alpha}{\cos \lambda}\right),$$

$$\therefore \qquad P \frac{\cos (\theta + \lambda)}{\cos \lambda} = W \frac{\sin (\alpha - \lambda)}{\cos \lambda},$$

$$\therefore \qquad P = W \frac{\sin (\alpha - \lambda)}{\cos (\theta + \lambda)}. \qquad (i)$$

P will be a minimum when $\theta = -\lambda$, i.e. when P acts along EA. This can also be shown as follows.

If P acts along EA it has a component perpendicular to the plane which increases the normal reaction; the limiting friction becomes $\mu(W \cos \alpha + P \sin \theta)$ and acts *up* the plane.

$$\therefore \qquad P \cos \theta + \mu(W \cos \alpha + P \sin \theta) = W \sin \alpha$$

$$\therefore \qquad P\left(\cos \theta + \frac{\sin \lambda \sin \theta}{\cos \lambda}\right) = W\left(\sin \alpha - \frac{\sin \lambda \cos \alpha}{\cos \lambda}\right)$$

$$\therefore \qquad P \frac{\cos (\theta - \lambda)}{\cos \lambda} = W \frac{\sin (\alpha - \lambda)}{\cos \lambda}$$

$$\therefore \qquad P = W \frac{\sin (\alpha - \lambda)}{\cos (\theta - \lambda)}. \qquad (ii)$$

This is a minimum when $\theta = \lambda$, and then $P = W \sin (\alpha - \lambda)$.

EXAMPLES 1.5

1. A body of mass 20 kg is resting on a rough horizontal plane, the coefficient of friction being 0·5; find the least force which, acting (i) horizontally, (ii) at an angle of 30° with the horizontal, would move the body.

2. A body of mass 40 kg is resting on a rough horizontal plane and can just be moved by a force of 98 N acting horizontally; find the coefficient of friction.

3. Find the least force required to move a mass of 20 kg along a rough horizontal plane when the coefficient of friction is 0·25.

4. A small wooden block of weight W is pushed along a rough floor by a force acting at the centre of its upper face and inclined at an angle θ to the vertical. Prove that the block will not move if θ is less than the angle of friction. Prove that, if θ is greater than the angle of friction λ the least force which will move the block is

$$W \sin \lambda \operatorname{cosec} (\theta - \lambda).$$

5. The length of an inclined plane is 5 m, and the height is 3 m; a force of 49 N acting parallel to the plane will just prevent a mass of 10 kg from sliding down. Find the coefficient of friction.

6. A body of mass 10 kg rests in limiting equilibrium on a rough plane whose slope is 30°; the plane is raised until its slope is 60°. Find the force parallel to the plane required to support the body.

7. A body of mass 20 kg is placed on a rough inclined plane whose slope is $\sin^{-1} \frac{3}{5}$; if the coefficient of friction between the plane and the body is 0·2, find the least force acting parallel to the plane required: (i) to prevent the body sliding down; (ii) to pull it up the plane.

8. A weight of mass 40 kg is on the point of sliding down a rough inclined plane when supported by a force of 196 N acting parallel to the plane, and is on the point of moving up the plane when acted on by a force of 294 N parallel to the plane. Find the coefficient of friction.

9. A weight of mass 610 kg is placed on a rough inclined plane of slope $\tan^{-1} \frac{11}{60}$ and coefficient of friction $\frac{1}{6}$, and is attached to a rope whose direction makes an angle $\tan^{-1} \frac{5}{12}$ with the upper surface of the plane. Find, to the nearest integer, the extreme values of the tension of the rope consistent with equilibrium.

10. Find the least force which will move a mass of 80 kg up a rough plane inclined to the horizontal at 30° when the coefficient of friction is 0·75.   (I.S.)

11. If the least force which will move a weight up a plane of inclination α is twice the least force which will just prevent the weight slipping down the plane, show that the coefficient of friction between the weight and the plane is $\frac{1}{3} \tan \alpha$.

12. The least force which will move a weight up an inclined plane is P. Show that the least force, acting parallel to the plane, which will move the weight upwards is

$$P\sqrt{(1+\mu^2)},$$

where μ is the coefficient of friction.   (I.S.)

13. A plane is inclined at 20° to the horizontal, and a mass of 100 kg is to be dragged up, the coefficient of friction being 0·25. Find the direction and magnitude of the least force required.

(H.S.C.)

14. Two inclined planes have a common vertex, and a string passing over a smooth pulley at the vertex is attached to two equal weights, one on each plane. If one plane is smooth and the other rough, find the relation between the angles of inclination of the two planes when the weight on the smooth plane is on the point of moving down.

15. Particles of mass 2 kg and 1 kg are placed on the equally rough slopes of a double inclined plane, whose angles of inclination are respectively 60° and 30°, and are connected by a light string passing over a small smooth pulley at the common vertex of the planes; if the heavier particle is on the point of slipping downwards, show that the coefficient of friction is $5\sqrt{3}-8$.

16. A weight of mass 20 kg is placed on a rough plane inclined at 22° to the horizontal. It is found that the least force which, acting downwards, along the slope of the plane, will cause it to move is 24 N. Find: (i) the coefficient of friction, and (ii) the least force which, acting along the slope of the plane, will just cause the weight to move upwards. (I.S.)

17. A weight W is dragged up a line of greatest slope of a rough plane of inclination α by a force P inclined at an angle β above the plane. If the coefficient of friction is μ, find the value of P which will just drag W. Find the work done by this force P in dragging W up a length l of the plane. If $W = 50$ N, $\alpha = 15°$, $\beta = 30°$, $\mu = 0·20$, $l = 30$ m, calculate the work done. (I.E.)

18. A body of weight W can be just supported on a rough inclined plane by a horizontal force P; it can also be just supported by a force Q acting up the plane. Find the cosine of the angle of friction in terms of P, Q, and W only. (I.S.)

19. A particle of weight W is placed on a rough inclined plane, the inclination of which exceeds the angle of friction. The least horizontal force required to prevent motion down the plane is W, the least horizontal force required to produce motion up the plane is $W\sqrt{3}$. Find the inclination of the plane and the angle of friction. Find also the magnitude and direction of the least force that can maintain the particle in equilibrium. (I.S.)

20. Two equal weights attached by a light string rest on the surface of a rough sphere of radius R, one of the weights being at the

highest point of the sphere. Find the greatest possible length of the string if the angle of friction is equal to α, and the friction of the string can be neglected.   (Ex.)

21. A block of wood, mass 2 kg, rests on a horizontal plank 1·8 m long. It is found that when the end of the plank is raised 0·6 m the block will just slide; find the coefficient of friction. If the vertical height of the end is increased to 0·9 m, find the least force perpendicular to the plank which will maintain equilibrium. (H.C.)

22. A plane is inclined to the horizontal at an angle of 30°; on it is a load of mass 20 kg. Find the force parallel to the plane which will prevent the load slipping down. If the plane is rough, the coefficient of friction being $\frac{1}{4}$, find the least force parallel to the plane which will just drag the load up the plane.   (H.C.)

23. The force P acting along a rough inclined plane is just sufficient to maintain a body on the plane, the angle of friction λ being less than α, the angle of the plane. Prove that the least force, acting along the plane, sufficient to drag the body up the plane is

$$P \frac{\sin (\alpha+\lambda)}{\sin (\alpha-\lambda)}. \hspace{2cm} \text{(I.E.)}$$

24. A body of weight W rests on a rough plane of inclination α. Show that the least magnitude of a force that will move it up the plane is $W \sin (\alpha+\lambda)$, where λ is the angle of friction. If the direction of the force be kept constant, show that if $\alpha > \lambda$ the magnitude of the force may be reduced to $W \sin (\alpha-\lambda) \sec 2\lambda$ before the body moves down the plane.   (C.W.B.)

CHAPTER TWO

STATICS OF A RIGID BODY—
PARALLEL FORCES—MOMENTS—
COUPLES

2.1. In the previous chapter we have considered the action of forces on a particle. We shall now begin to consider the action of forces on a rigid body. In these cases it is evident that we may have to find the resultant of two forces which are parallel and not in the same straight line. Since such forces do not meet in a point, we cannot obtain their resultant by direct application of the parallelogram of forces. The rules for obtaining the resultant of parallel forces are, however, obtained from the parallelogram law as explained in the next two paragraphs.

Two parallel forces are said to be *like* when they act in the same direction; when they act in opposite parallel directions they are said to be *unlike*.

2.2. To find the resultant of two like parallel forces acting on a rigid body

Let P and Q be the forces acting at points A and B (Fig. 2.1) of the body, and let them be represented by the lines AC and BD.

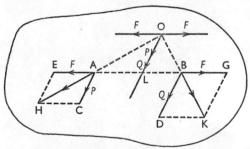

Fig. 2.1

At A and B introduce two equal and opposite forces, F, acting in the line AB, represented by AE and BG. The introduction of these forces will not disturb the action of P and Q, since, the body being rigid, the force F at A may be supposed transferred to B, where it would balance the other force F. Complete the parallelograms AEHC and BGKD, and produce the diagonals HA, KB to meet at O. Draw OL parallel to AC or BD to meet AB in L.

The forces P and F at A have a resultant represented by AH which may be supposed to act at O. Similarly, Q and F at B have a resultant represented by BK which may also be supposed to act at O.

These forces may now be resolved at O into their components P along OL, F parallel to AE, and Q along OL, F parallel to BG. The two forces F are in equilibrium.

Hence the original forces P and Q are equivalent to a force $(P+Q)$ acting along OL, i.e. parallel to the original directions of P and Q, and acting at L in AB.

To find the position of the point L.

The triangles OLA, ACH are similar by construction.

$$\therefore \quad \frac{OL}{LA} = \frac{AC}{CH} = \frac{P}{F}$$

$$\therefore \quad P \times LA = F \times OL. \tag{i}$$

The triangles OLB, BDK are similar.

$$\therefore \quad \frac{OL}{LB} = \frac{BD}{DK} = \frac{Q}{F}$$

$$\therefore \quad Q \times LB = F \times OL. \tag{ii}$$

Hence from (i) and (ii)

$$P \times LA = Q \times LB,$$

i.e. the point L divides AB *internally* in the inverse ratio of the forces.

We note if $P = Q$ that L bisects AB.

2.3. To find the resultant of two unlike parallel forces acting on a rigid body

Let P and Q be the forces (P being the greater) acting at points A and B (Fig. 2.2) of the body, and let them be represented by AC and BD.

At A and B introduce two equal and opposite forces, F, acting in the line AB and represented by AE and BG.

Since the body is rigid, these forces will be in equilibrium, and will not disturb the action of P and Q.

Complete the parallelograms AEHC, BGKD, and produce the diagonals AH, KB to meet at O. (The diagonals will always meet unless they are parallel, which is the case when P and Q are equal.)

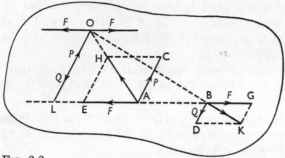

FIG. 2.2

Draw OL parallel to CA and BD to meet BA produced in L. The forces P and F at A have a resultant represented by AH which may be supposed to act at O. Similarly, Q and F at B have a resultant represented by BK which may also be supposed to act at O.

These forces may now be resolved at O into their components, P along LO, F parallel to AE, and Q along OL, F parallel to BG.

The two forces F are in equilibrium.

Hence the original forces P and Q are equivalent to a force $(P-Q)$ acting along LO, i.e. parallel to P in the same direction as P, and acting through L in BA produced.

To find the position of the point L.

The triangles OLA, HEA are similar.

$$\therefore \qquad \frac{OL}{LA} = \frac{HE}{EA} = \frac{P}{F}$$

$$\therefore \qquad P \times LA = F \times OL.$$

The triangles OLB, BDK are similar.

$$\therefore \qquad \frac{OL}{LB} = \frac{BD}{DK} = \frac{Q}{F}.$$

$$\therefore \qquad Q \times LB = F \times OL.$$

$$\therefore \qquad P \times LA = Q \times LB,$$

i.e. L divides the line AB *externally* in the inverse ratio of the forces.

2.4. *Case of failure*

In the figure of the preceding paragraph, if $P = Q$, the triangles AEH, BGK are equal in all respects. In this case $\angle HAE = \angle GBK$, so that the lines AH and KB being parallel will not meet at any point such as O, and the construction fails. Hence when the two forces are equal, unlike, and parallel there is no single force which is equivalent to them.

Such a pair of forces constitute what is called a *Couple*. They cannot be replaced by anything simpler.

The properties of couples will be considered later (2.14 and 3.16).

2.5. Centre of parallel forces

Let a system of like parallel forces W_1, W_2, W_3, ... act at the points A_1, A_2, A_3, ... (Fig. 2.3).

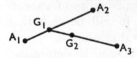

FIG. 2.3

The resultant of the forces W_1, W_2 acting at A_1 and A_2 is equal to $W_1 + W_2$, and always passes through a point G_1 in $A_1 A_2$ (such that $W_1 \times A_1 G_1 = W_2 \times G_1 A_2$), whatever the direction of the two forces. Similarly, the resultant of like parallel forces $W_1 + W_2$ at G_1 and W_3 at A_3 is equal to $W_1 + W_2 + W_3$, and always passes through a fixed point G_2 in $G_1 A_3$, such that

$$(W_1 + W_2)G_1 G_2 = W_3 \times G_2 A_3.$$

Hence, provided all the forces are like and parallel, their resultant is equal to the sum of the forces and always passes through a point whose position is fixed relative to A_1, A_2, etc. Its position does not depend on the common direction of the parallel forces.

This point is called the *Centre of the Parallel Forces*, and it is evident that the above argument holds whether the forces are all in one plane or not.

2.6.
A very important application of the theorem of the preceding paragraph occurs in connection with the weight of a

body. Every particle of matter is attracted towards the centre of the earth. This force of attraction is proportional to the mass of the particle, and is called its weight (see Dynamics, 3.11).

A body may be considered as made up of a very large number of particles, and if the size of the body is small compared with that of the earth the forces on all the particles of it will be very nearly parallel.

For bodies of ordinary size we shall consider them as being parallel.

Hence the points A_1, A_2, etc. in Fig. 2.3 may be taken to represent the particles of the body, and as the weights of these particles form a system of parallel forces, their resultant (which is equal to the weight of the body) always passes through some point G, whose position relative to the body is fixed and does not depend on the orientation of the body. This point is called the *Centre of Gravity* or *Centre of Mass* of the body.

The position of the centre of gravity of bodies of various shapes will be considered in Chapter 7. It will be seen that this point need not be a point in the body, but often is.

For some simple bodies the position is easy to determine, e.g. a uniform thin rod, a rectangle or parallelogram, a triangle. As these cases occur in a large number of problems, we shall consider them now.

2.7. Centre of gravity of a thin uniform rod

Since the rod is uniform, equal lengths of it, however small, have the same weight.

Let AB (Fig. 2.4) represent the rod, and let G be its mid-point.

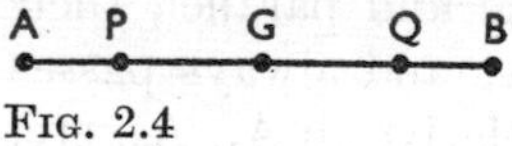

A   P   G   Q   B

Fig. 2.4

Take any point P between G and A, and a point Q between G and B, such that $PG = GQ$.

The centre of gravity of equal particles of the rod at P and Q is evidently at G, since the resultant of equal like parallel forces at P and Q passes through G.

Also for every particle such as P between G and A there is an equal particle at an equal distance from G, lying between G and B.

The centre of gravity of each of these pairs of particles is at G, and hence the centre of gravity of the whole rod is at G.

It will be noticed that the direction in which the parallel forces (the weights of the particles) act makes no difference; i.e. the resultant of the weights of all the particles passes through G in whatever position the rod is placed.

2.8. Centre of gravity of a thin plate or lamina in the shape of a parallelogram

Let ABCD (Fig. 2.5) be the parallelogram. Suppose it is divided into a very large number of very narrow strips, such as PQ, parallel to AD.

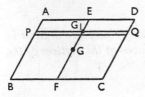

FIG. 2.5

Each of these strips may be considered as a thin uniform rod, and its centre of gravity will be at its mid-point G_1.

Hence the centre of gravity of the whole figure will lie on the line joining the mid-points of the strips, i.e. the line EF joining the mid-points of AD and BC. Similarly, by supposing the figure divided into strips parallel to AB, we see that the centre of gravity must lie on the line joining the mid-points of AB and DC.

The centre of gravity is therefore at G, where these lines intersect. G is also, of course, the point of intersection of the diagonals of the parallelogram.

2.9. Centre of gravity of a thin triangular plate or lamina

Let ABC (Fig. 2.6) be the triangular lamina.

Suppose it is divided into a very large number of narrow strips, such as B_1C_1 parallel to BC.

The centre of gravity of each strip is at its mid-point. Hence the centre of gravity of the whole triangle lies in the line joining the mid-points of the strips, i.e. in the median AD.

Similarly, by supposing the strips taken parallel to AC, we

see that the centre of gravity lies in the median BE. The centre of gravity is therefore at G, the point of intersection of the medians.

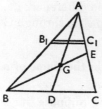

FIG. 2.6

We know from Geometry that this point is $\frac{1}{3}$ of the way up each median, i.e. $DG = \frac{1}{3}DA$.

2.10. *The centre of gravity of any uniform triangular lamina is the same as that of three equal particles placed at the vertices of the triangle.*

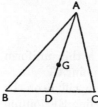

FIG. 2.7

Let ABC (Fig. 2.7) be the lamina. The resultant of like parallel forces, each equal to W acting at B and C, is a parallel force $2W$, acting through D, the mid-point of BC. The resultant of like parallel forces $2W$ at D and W at A is a force $3W$ acting at G in AD, where $AG = 2GD$. Hence G, which is the centre of gravity of three equal particles placed at A, B, and C, lies in AD and $GD = \frac{1}{3}AD$, i.e. it is the same point as the centre of gravity of the lamina.

EXAMPLES 2.1

1. Like parallel forces of 40 and 70 N act at points A and B, which are 22 cm apart. Find the magnitude of their resultant and the point in which it cuts AB. Find also the magnitude of the resultant and the point where it cuts AB produced if the forces are unlike.

2. Like parallel forces of 9 and 12 N act at points A and B which are 42 cm apart. Find the magnitude of their resultant and the point in which it cuts AB. Find also the magnitude and position of the resultant when the forces are unlike.

3. Unlike parallel forces of 12 and 8 N act at points A and B, which are 12 cm apart. Find the magnitude of the resultant and the point where it cuts AB produced.

4. The resultant of two like parallel forces, one of which is 8 N, is 20 N, and acts at a distance of 6 cm from the 8 N force. Find the magnitude of the other force and the distance of its line of action from the 8 N force.

5. Find the magnitude of two like parallel forces, acting at a distance of 4 m apart, which are equivalent to a force of 100 N acting at a distance of 1 m from one of the forces.

6. Four equal like parallel forces act at the corners of a square; show that their resultant passes through the centre of the square.

7. Three equal like parallel forces act at the vertices of a triangle; show that their resultant passes through the point of intersection of the medians of the triangle.

8. Like parallel forces of magnitudes P, P, $2P$ act at the vertices A, B, C respectively of a triangle ABC. Show that their resultant passes through the mid-point of the line joining C to the mid-point of AB.

9. P and Q are like parallel forces. If Q is moved parallel to itself, through a distance x, prove that the resultant of P and Q moves through a distance $Qx/(P+Q)$.

10. Three equal like parallel forces act at the mid-points of the sides of a triangle; show that their resultant passes through the point of intersection of the medians of the triangle.

11. Like parallel forces of magnitude 4, 2, 4 units act at the corners A, B, C respectively of a square ABCD. What is the magnitude of the parallel force that must be applied at D in order that the resultant of the four forces should pass through the centre of the square?

    If the like parallel force applied at D has a magnitude of 5 units, find the position of the line of action of the resultant.

12. Like parallel forces of magnitude 2, 5, 3 N act at the corners A, B, C, respectively, of a triangle ABC, where AB = 4 cm, BC = 3 cm, CA = 5 cm. Find the position of the line of action of the resultant.

2.11. Moment of a force

A system of forces acting on a rigid body may tend to rotate the body.

If a single force acts on a rigid body, of which one point is fixed, then, unless the line of action of the force passes through the fixed point the body will tend to turn about that point.* This introduces the idea of the *turning effect* or *moment* of a force which is usually defined as follows:

The moment of a force about a given point is the product of the force and the perpendicular drawn from the given point to the line of action of the force.

Fig. 2.8

Thus the moment of a force P, whose line of action is as shown in Fig. 2.8, about a point O, is $P \times ON$, where ON is the perpendicular drawn from O to the line of action of P.

It is clear that if the line of action of P passes through O its moment about that point is zero.

If O is a fixed point in a body, whose section by a plane containing O and the line of action of P is shown in the figure, the product $P \times ON$ is a measure of the tendency of P to turn the body about O. The turning power is increased if *either*: (1) P is increased; *or* (2) its distance from O is increased.

The moment of a force about a given point may be positive or negative, according to the direction in which the force tends to turn the body about the point.

In the figure the force P tends to turn the body in a direction opposite to that in which the hands of a clock move. In such cases the moment is said to be *positive*. If the force tends to turn the body in a clockwise direction its moment is said to be *negative*.

* Strictly, other forces come into play at the point where the body is fixed.

When a number of forces are acting on a body the *algebraic sum of their moments* is obtained by giving the value of the moment of each force its proper sign and adding them together.

The moment of a force has both magnitude and direction, and is therefore a vector quantity.

The unit of the moment of a force is the unit of force multiplied by the unit of length, e.g. dyne-centimetre or newton-metre.

2.12. Graphical representation of a moment

If the length AB marked off on the line of action of the force P (Fig. 2.9) represents the magnitude of P, then the moment of P about O is represented by $AB \times ON$.

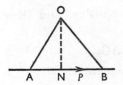

Fig. 2.9

But the area of the triangle AOB is $\frac{1}{2}AB \times ON$; hence twice the area of the triangle AOB represents the moment of P about O. We shall now use this graphical method of representation to prove the fundamental theorem on the moments of coplanar forces.

2.13. The algebraic sum of the moments of two forces about any point in their plane is equal to the moment of their resultant about that point. (Varignon's Theorem)

There are two cases to be considered.

(i) *Let the forces meet in a point.*

Let the forces be P and Q, acting at A, as shown in Figs. 2.10A and B, and let O be any point in their plane.

Draw OC parallel to direction of P to meet the line of action of Q in C. Take the length AC to represent the magnitude of Q, and on the same scale let AB represent P. Complete the parallelogram ABCD, and join OA, OB.

AD represents the resultant R of P and Q.

In either figure,

the moment of P about O is represented by 2 $\triangle$AOB,

,,   ,, Q ,,   O   ,,   ,,   2 $\triangle$AOC,

,,   ,, R ,,   O   ,,   ,,   2 $\triangle$AOD.

Also $\triangle$AOB $=$ $\triangle$ADB, same base and parallels.

$= \triangle$ADC.

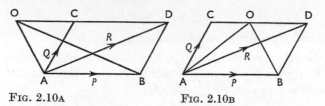

FIG. 2.10A   FIG. 2.10B

In Fig. 2.10A the moments are both positive, and their algebraic sum is represented by

$$2 \triangle AOB + 2 \triangle AOC = 2 \triangle ADC + 2 \triangle AOC$$
$$= 2 \triangle OAD = \text{moment of } R.$$

In Fig. 2.10B the moments are in opposite directions, that of P being positive and Q negative; their algebraic sum is represented by

$$2 \triangle AOB - 2 \triangle AOC = 2 \triangle ADC - 2 \triangle AOC$$
$$= 2 \triangle OAD$$
$$= \text{moment of } R.$$

(ii) *Let the forces be parallel.*

Let P and Q be two parallel forces acting as in Figs. 2.11A and 2.11B and O any point in their plane.

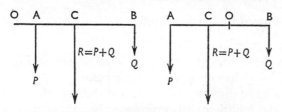

FIG. 2.11A   FIG. 2.11B

Draw OAB perpendicular to the forces to meet their lines of action in A and B.

The resultant R (equal to $P+Q$) is parallel to P and Q, and acts through a point C in AB, such that

$$P \times AC = Q \times CB.$$

In Fig. 11.11A the sum of the monents of P and Q about O is

$$P \times OA + Q \times OB$$
$$= P(OC-AC) + Q(OC+CB)$$
$$= (P+Q)OC - P \times AC + Q \times CB$$
$$= (P+Q)OC$$
$$= \text{moment of } R \text{ about O.}$$

In Fig. 11.11B the sum of the moments of P and Q about O is

$$P \times OA - Q \times OB$$
$$= P(OC+AC) - Q(BC-OC)$$
$$= (P+Q)OC + P \times AC - Q \times BC$$
$$= (P+Q)OC$$
$$= \text{moment of } R \text{ about O.}$$

2.14. *If the forces form a couple* there is no single resultant and the theorem does not apply.

In this case it is easy to show that the sum of the moments of the forces forming the couple is the same about any point in the plane of the forces.

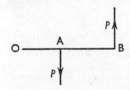

FIG. 2.12

Let P, P be the forces acting as shown in Fig. 2.12, and let O be any point in their plane.

Draw OAB perpendicular to the forces to meet their lines of action in A and B.

The sum of the moment about O is

$$P \times OB - P \times OA$$
$$= P(OB-OA)$$
$$= P \times AB$$

and this is independent of the position of O.

The product $P \times$ AB, where P is the magnitude of either of the forces of the couple, and AB is the perpendicular distance between the forces, is called the *Moment of the Couple*.

It should be noticed that the moment of a couple is equal to the moment of either of the forces of the couple about any point in the line of action of the other, and may be positive or negative, depending on the sense of rotation of the couple.

2.15. It is evident that the theorem of paragraph 2.13 may be extended to any number of forces which have a single resultant. For the theorem holds for any two of the forces not forming a couple; it also holds for the resultant of these two and any other of the forces not forming a couple with it, and so on successively until all the forces have been included. The resultant of all the forces but one cannot form a couple with the last remaining force, since we are assuming that the system has a single resultant.

We thus arrive at the general *Principle of Moments*, which may be stated as follows:

If any number of coplanar forces acting on a rigid body have a resultant, the algebraic sum of their moments about any point in their plane is equal to the moment of their resultant about that point.

2.16. If a system of coplanar forces is in equilibrium their resultant is zero, and its moment about any point must therefore be zero.

Hence, *when a system of coplanar forces is in equilibrium the algebraic sum of their moments about any point in their plane is zero.*

The converse of this is not necessarily true. For since the moment of a force about any point in its own line of action is zero, the sum of the moments of a system of coplanar forces about any point in the line of action of their resultant is zero. Hence, the fact that the sum of their moments about *one* point is zero does not necessarily mean that they are in equilibrium, for the point might be on the line of action of their resultant.

EXAMPLES 2.2

1. An equilateral triangle ABC is 8 cm high, it has forces of 2, 4, and 8 N acting along the sides AB, BC, CA respectively. Find the moment of this system of forces about each angular point. (C.E.)

2. Four forces, 3, 4, 5, and 6 N respectively, act in a clockwise direction along the sides of a square, each side of which is 20 cm. long. Find the value of the moments of these forces, about: (i) the centre of the square, and (ii) the point of intersection of the forces 3 and 6 N. State clearly the units of the answers.   (C.E.)

3. Forces of 4, 5, and 6 N act along the sides BC, CA, AB of an equilateral triangle ABC of side 2 m in the directions indicated by the order of the letters. Find the sum of their moments about the point of intersection of the medians of the triangle.

4. Four forces of 2, 4, 2, and 4 N act along the sides AB, CB, DC, DA respectively of a square ABCD of side 3 m. Find the sum of their moments: (i) about A, and (ii) about the centre of the square.

5. A, B, C, D are four points in order in a horizontal line at equal distances of 1 m apart. A force of 2 N acts at B perpendicular to AD and downwards; a force of 4 N acts at C upwards in a direction making an angle of 30° with CD; and a force of 1 N acts upwards at D perpendicular to AD. Find the sum of their moments: (i) about A; (ii) about C.

6. Forces of 1, 2, 3, 4, 5, and 6 N act along the sides of a regular hexagon AB, BC, CD, DE, EF, FA respectively. If the side of the hexagon is 2 cm, find the sum of the moments of the forces: (i) about the centre of the hexagon; (ii) about A.

2.17. The *principle of moments* is of extreme importance, and it will be used continually throughout the remainder of the book. In the present chapter we shall illustrate its use in dealing with some simple cases where a rigid rod, resting on supports, or pivoted about some point, is acted on by forces in addition to its own weight.

Instead of equating the algebraic sum of the moments to zero, it is often convenient to equate the sum of those acting in one direction round the point to the sum of those acting in the opposite direction.

It might be noted here that the Principle of Moments and the results established earlier in this chapter can be *verified experimentally* by applying known forces to a rigid body, such as a bar, and making appropriate measurements. As was suggested in 1.5, the Principle of Moments can be used as the basis of statics and the whole science developed from this starting-point.

EXAMPLE (i)

A uniform rod AB, *of length* 12 *m, and of mass* 50 *kg, rests on two supports, one at* A *and the other* 2 *m from* B. *Masses of* 4, 5, *and* 10 *kg are attached at points* 2 *m,* 4 *m, and* 8 *m respectively from* A. *Find the thrusts on the supports.*

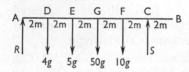

Fig. 2.13

Let C (Fig. 2.13) be the position of the other support, G the centre of gravity of the rod, and D, E, F, the points where the masses are attached. Let R, S N be the reactions at A and B.

The weight of the rod is 50g N and acts at G. The weights acting at D, E, F are 4g, 5g, 10g respectively.

Taking moments about A,

$$10S = 4g \times 2 + 5g \times 4 + 50g \times 6 + 10g \times 8$$
$$= (8 + 20 + 300 + 80)\, g$$
$$= 408\, g$$
$$\therefore \qquad S = 40{\cdot}8\, g = 399{\cdot}8.$$

Taking moments about C,

$$10R = 10\, g \times 2 + 50\, g \times 4 + 5\, g \times 6 + 4\, g \times 8$$
$$= (20 + 200 + 30 + 32)\, g$$
$$= 282g$$
$$R = 28{\cdot}2g = 276{\cdot}4.$$

Since we know that the sum of R and S must equal the sum of all the weights (including that of the rod), we might have obtained R by subtracting S from this total (69 g). In practice it is better, however, to find each reaction separately, and the fact that their sum must equal that of the weights provides a check on the working; otherwise a mistake in calculating the first will cause both results to be wrong.

EXAMPLE (ii)

A uniform beam is 12 m long and has a mass of 50 kg and masses of 30 kg and 40 kg are suspended from its ends; at what point must the beam be supported so that it may rest horizontally?

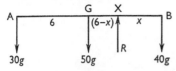

Fig. 2.14

Let AB (Fig. 2.14) be the beam, G its centre of gravity.

The required point X is the point about which the moments of the three weights balance. There must be a supporting force R acting on the beam at X. R must be vertical and equal to $120g$ N.

Let BX $= x$ m, then, taking moments about X,

$$40x = 50(6-x)+30(12-x)$$
$$= 300-50x+360-30x$$
$$120x = 660$$
$$\therefore \qquad x = 5\tfrac{1}{2}.$$

We can also obtain this position of X by taking moments of *all* the forces about one end of the rod.

Hence, taking moments about B,

$$Rx = 50g\times6+30g\times12$$
$$= (300+360)\,g$$
$$\therefore \qquad 120x = 660$$
$$\therefore \qquad x = 5\tfrac{1}{2}.$$

EXAMPLE (iii)

A uniform rod AB, 3·6 m long and of mass 25 kg, is pivoted at a point 0·9 m from A. A mass of 100 kg is suspended from A. What force applied at B, in a direction perpendicular to the rod, will keep it in equilibrium with A below B and AB inclined at 60° to the horizontal?

Let G (Fig. 2.15) be the mid-point of the rod, C the position of the pivot. Let R N equal the reaction acting on the rod at the pivot.

The rod will be in equilibrium when the sum of the moments about C of the weight acting vertically through G and the force P N acting at B perpendicular to AB, is equal to the moment of the 980 N force about C.

To obtain the moments of the force about C, we must find the *perpendicular* distances of their lines of action from C.

The perpendicular distance of the line of the 980 N is 0·9 cos 60° =0·45 m, and this is also the distance of the line of the 245 N force.

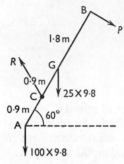

FIG. 2.15

The perpendicular distance of the line of P from C is 2·7 m.

$$\therefore \qquad 2 \cdot 7P + 245 \times 0 \cdot 45 = 980 \times 0 \cdot 45$$
$$\therefore \qquad 2 \cdot 7P = 735 \times 0 \cdot 45.$$
$$\therefore \qquad P = 122 \cdot 5.$$

EXAMPLES 2.3

1. A uniform rod, 1·8 m long and of mass 10 kg, rests horizontally on supports at its ends; if a weight of mass 3 kg is attached at a point 1·2 m from one end, find the pressures on the supports.

2. A uniform beam AB, 10 m long and of mass 40 kg, rests on two supports, one at A and the other 2 m from B. If a weight of mass 20 kg is attached to the beam at a point 6 m from A, find the pressures on the supports.

3. Two men carry a load of 100 kg which hangs from a light pole of length 2·4 m, each end of which rests on a shoulder of one of the men. The point from which the load is hung is 0·6 m nearer to one man than the other. What is the pressure on each shoulder?

4. A uniform beam 3 m long, with a weight of mass 25 kg hanging from one end balances in a horizontal position about a point 0·9 m from this end. Find the mass of the beam.

5. Masses of 1 kg, 2 kg, 3 kg, 4 kg, and 5 kg are suspended from a uniform rod AB 1·8 m long, of mass 3 kg and supported at its ends, at distances of 0·3 m, 0·6 m, 0·9 m, 1·2 m, and 1·5 m from A. Find the pressures on the supports.

6. Mass of 1 kg, 2 kg, 3 kg, 4 kg are suspended from a uniform rod of length 1·5 m and mass 3 kg, at distances of 0·3 m, 0·6 m, 0·9 m, 1·2 m from one end. Find the position of the point about which the rod will balance.

7. Find the magnitude and line of action of the resultant of parallel forces 3, 6, 8 in one direction and 12 in the opposite direction, acting at points A, B, C, D respectively in a straight line, where AB = 1 m, BC = 3 m, CD = 5 m.

8. A beam AB, 3 m long and of mass 6 kg, is supported at A and at another point. A load of 1 kg is suspended at B, loads of 5 kg and 4 kg at points 1 m and 2 m from B. If the pressure on the support at A is 40 N, where is the other support?

9. A heavy uniform beam, of length 3·6 m and mass 30 kg is suspended in a horizontal position by two vertical strings, each of which can just sustain a tension of 196 N. Within what distance from the centre can a weight of mass 7·5 kg be suspended without breaking either string?

10. A uniform bar AB, of mass 3 kg and length 75 cm, rests on two supports C and D, distant 10 and 60 cm from A. At points E and F, distant 20 and 50 cm from A respectively are suspended masses of 7 and 2 kg. Find the reactions at C and D, and the moment about the mid-point of the rod of all the forces acting on the bar to the right of that point.

11. A uniform rod, of length 1·8 m and mass 1 kg, rests in a horizontal position with its ends on two supports, each of which will bear the weight of a mass of 6 kg and no more. Find on what part of the rod a weight of mass 8·5 kg can be placed without breaking either support. (I.S.)

12. Prove that if a passenger of weight W advances a distance a along the top of a motor bus, a weight Wa/b is transferred from the back springs to the front springs, where b is the distance between the axles. (I.E.)

13. A uniform bar, 0·6 m long and of mass 17 kg, is suspended by two vertical strings. One is attached at a point 7·5 cm from one end, and can just support the weight of a mass of 9 kg without breaking; the other is attached 10 cm from the other end, and can just support a mass of 10 kg. A weight of mass 1·7 kg is now attached to the rod; find the limits of the positions in which it can be attached without breaking either string. (I.S.)

14. Coplanar forces of 10, 3, 7 units act vertically upwards at distances (measured positive to the right) of 5, −9, 2 m respectively from a fixed point O in their plane. A force of 20 units acts

vertically downwards through O. Find the resultant. If the force through O be increased to 30 units, find the resultant.

15. A rod AB, of length $(a+b)$ and weight W, has its centre of gravity at distance a from A. It rests on two parallel knife-edges at distance c apart in the same horizontal plane, so that equal portions of the rod project beyond each knife-edge. Prove that the pressures on the knife-edges are respectively

$$(a-b+c)W/2c \text{ and } (b-a+c)W/2c. \qquad \text{(I.E.)}$$

16. A heavy uniform rod, of mass 10 kg and length 1·2 m, is supported in a symmetrical position by two props 0·9 m apart. A mass of 2 kg is now suspended from one end of the rod. Calculate the pressures on the two props. (H.S.D.)

17. A light horizontal rod, 30 cm long, is supported by two vertical props, each 7·5 cm from an end of the rod, and is loaded with 16 kg at each end. What weights hung from the ends will produce in one prop a pressure double and in the other prop a pressure half of that produced by the 16 kg load? (H.C.)

18. Four metres of a plank, 12 m long and of mass 100 kg, project over the side of a quay. What mass must be placed on the end of the plank so that a man of mass 75 kg may be able to walk to the other end without the plank tipping over? (H.C.)

19. A horizontal beam ABCD rests on two supports at B and C, where $AB = BC = CD$. It is found that the beam will just tilt when a mass of p kg is hung from A or when a mass of q kg is hung from D. Find the mass of the beam, and prove that its centre of gravity divides AD in the ratio $2p+q : p+2q$. (H.C.)

20. A uniform beam AB is 1·8 m long and has a mass of 24 kg. It rests on two vertical supports at C and D, CD being a distance of 0·9 m and the pressure on the support at C and is double that on the support at D. Find the lengths of AC and DB, assuming that A is nearer to C than D.

21. A uniform beam rests in a horizontal position supported at a point distant 0·6 m from one end and carrying a mass of 10 kg suspended from this end. The pressure on the support is 300 N. Determine the mass and length of the beam.

22. A stiff heavy rod ABGCD, whose mass is 12 kg and whose centre of gravity is at G, is suspended by vertical strings attached to B and C, each of which could just support the weight of the rod. $AB = 5$ cm; $BG = 7·5$ cm; $GC = 10$ cm; $CD = 7·5$ cm. Find the tensions in the strings when masses of M_1 kg, M_2 kg are suspended from A and D respectively, and find the values of M_1 and M_2 when both strings are on the point of breaking. (I.S.)

23. A uniform beam, 3·6 m long and of mass 25 kg, rests on two supports at equal distances from the ends. Find the maximum value of this distance so that a man of mass 77 kg may stand anywhere on the beam without tilting it.

24. A uniform plane lamina in the form of a regular hexagon ABCDEF is free to rotate in its own plane, which is vertical, about its centre O. Masses of 1, 2, 3, 4, 5, and 6 kg are attached to the vertices A, B, C, D, E, F respectively. Find the inclination of AD to the vertical for which the system is in equilibrium.

(N.U. 3 and 4)

25. A light horizontal rod, 6 m long, is loaded with three 10-kg masses at points respectively 0·9, 2·1, and 4·5 m from one end, and it is subject to an upward thrust equal to 49 N at the mid-point. Find the resultant of these parallel forces acting on the rod, and if the rod is supported at its two ends deduce the pressures on the supports. (C.W.B.)

2.18. The lever

A lever consists essentially of a rigid bar which can turn about a fixed point called the *fulcrum*.

As mentioned earlier (1.5), the principle of the lever was known to Archimedes (287–212 B.C.), and until the sixteenth century, when the Parallelogram of Forces was discovered, it was the fundamental principle of Statics.

This principle is simply the principle of moments, i.e. when the lever is in equilibrium the algebraic sum of the moments about the fulcrum of the forces acting on it is zero.

We shall now consider some practical forms of lever.

2.19. The balance

This consists essentially of a rigid beam as in Fig. 2.16, where the extremities of the dotted line AB represent the points where the scale pans are attached.

The fulcrum is usually a knife-edge made of agate fixed through the beam and resting on an agate plate at O.

The scale pans hang from agate plates which rest on agate knife-edges at the points A and B.

The beam is constructed so that its centre of gravity G is below the line AB, and the fulcrum O is placed very close to AB in such a manner that, when the beam is horizontal, G and H (the mid-point of AB) are vertically below O. This ensures

that when the beam is horizontal the weights of the scale pans
and their contents act at equal distances from the fulcrum.

This is expressed by saying that the arms of the balance are
equal, a most important condition.

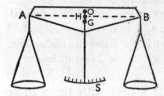

FIG. 2.16

A pointer, rigidly attached to the beam and at right angles
to AB, moves over a fixed scale S, and shows whether the beam
is horizontal.

The weights of the scale pans and their attachments must be
equal. We shall now show that, when these conditions are
satisfied, the beam can rest only in a horizontal position when
equal weights are placed in the pans, and that it can rest
inclined at a definite angle to the horizontal when the weights
are unequal.

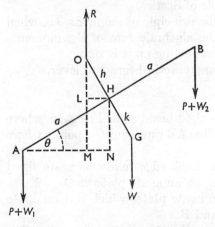

FIG. 2.17

The points A, B, O, H, and G are represented in Fig. 2.17,
where the beam AB is inclined at an angle θ to the horizontal.

Let P = the weight of each scale pan

W = the weight of the beam

W_1, W_2 = the weights placed in the pans

a = the length of each arm (i.e. AH = BH = a)

h = OH

k = HG.

The beam is acted on by forces

$P+W_1$, $P+W_2$ vertically downwards at A and B,

W vertically downwards at G,

and the vertical upward reaction R at O.

In the figure the angle LOH = θ, AN = $a \cos \theta$,

LH = $h \sin \theta$ and the distance of G from the vertical through H is $k \sin \theta$.

The distance of $(P+W_1)$ from O is AM = $a \cos \theta - h \sin \theta$.

,,   ,, $(P+W_2)$,,   O   = $a \cos \theta + h \sin \theta$.

,,   ,, W ,,   O   = $h \sin \theta + k \sin \theta$.

Taking moments about O, we get

$$(P+W_1)(a \cos \theta - h \sin \theta)$$
$$= W(h+k) \sin \theta + (P+W_2)(a \cos \theta + h \sin \theta)$$

$$\therefore \sin \theta [W(h+k) + (P+W_2)h + (P+W_1)h]$$
$$= \cos \theta [(P+W_1)a - (P+W_2)a]$$

$$\therefore \qquad \tan \theta = \frac{(W_1 - W_2)a}{(W_1 + W_2 + 2P)h + W(h+k)}$$

This result shows that if the weights in the pans are equal, i.e. $W_1 = W_2$, θ is zero, and the beam can rest only in a horizontal position.

If the weights are slightly different the beam will rest at a definite angle to the horizontal.

It should be noticed that if h and k were both zero, i.e. if the centre of gravity of the beam and the centre of suspension coincided in the line AB, the beam could rest in any position when equal weights were placed in the pans and could rest only in a vertical position if the weights were different.

2.20. The requisites of a good balance are that it must be

(i) *true*,   (ii) *sensitive*,   (iii) *stable*,   (iv) *rigid*.

(i) The balance will be true if the arms are equal, the weights of the scale pans are equal, and if the centre of gravity of the beam, the mid-point of the beam, and the fulcrum are in a straight line perpendicular to the beam.

(ii) In a sensitive balance the beam must, for a very small difference between the weights in the scale pans, be inclined at an appreciable angle to the horizontal. Using the result obtained in the last paragraph, the ratio $\dfrac{\tan \theta}{W_1 - W_2}$ may be taken as a measure of the sensitivity, which is therefore equal to

$$\frac{a}{(W_1 + W_2 + 2P)h + W(h+k)}.$$

This expression shows that the sensitiveness diminishes as the weights increase.

If, however, $h = 0$, i.e. if all the knife-edges are in the same plane, the sensitivity becomes a/Wk and is independent of the weights in the pans.

This condition is usually aimed at, but the slight bending of the beam prevents its being attained exactly.

For given weights the sensitivity increases with a, i.e. with the length of the arms; it also increases as W decreases, so that it is an advantage to have a long light beam. Too light a beam might mean loss of rigidity, and in better balances the beam is usually of an open girder type.

Diminishing k also increases the sensitivity.

(iii) Stability affects the time taken for the balance to come to rest at its position of equilibrium, and is really a dynamical question. Actually the condition is best satisfied when the moment of the forces about O is greatest, i.e. if W_1 is the weight in each pan, when $[2(P+W_1)h + W(h+k)] \sin \theta$ is greatest.

This is the case for a given value of θ when h and k are greatest.

Since the balance is most sensitive when h and k are small, and most stable when they are large, it is clear that great sensitivity and quickness in weighing cannot be attained together.

2.21. If a balance is not true and the cause is only that the scale pans differ in weight the error can be adjusted by putting paper or sand in the lighter pan or by means of a small screw working in the end of the beam near the points where the pans are attached.

If the arms of the balance are unequal no adjustment of this kind will make it true.

Whatever the cause of inaccuracy, however, the correct weight of a body can be obtained as follows:

Place the body in one pan and balance it by placing sand in the other pan.

Remove the body and put weights in the pan until they balance the sand.

These weights must be equal to the weight of the body.

This is sometimes called *Borda's method*.

2.22. *Double weighing*

Let the lengths of the arms be a_1 and a_2, and the weights of the pans P_1 and P_2. Suppose that the weight of the beam acts at a

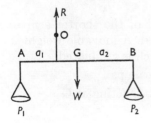

FIG. 2.18

distance x from the fulcrum O (Fig. 2.18) but that the beam remains horizontal when the balance is unloaded.

Then
$$P_1 a_1 = Wx + P_2 a_2. \tag{i}$$

Now suppose a weight W_1 on pan P_1 requires a weight W_2 on pan P_2 to make the beam horizontal; we get

$$(P_1 + W_1)a_1 = Wx + (P_2 + W_2)a_2 \tag{ii}$$

Now place W_1 in pan P_2 and find the weight W_3 which must be put in pan P_1 to balance it; we get

$$(P_1 + W_3)a_1 = Wx + (P_2 + W_1)a_2 \qquad \text{(iii)}$$

From (i) and (ii), $W_1 a_1 = W_2 a_2.$

From (i) and (iii), $W_3 a_1 = W_1 a_2.$

From these equations $W_1{}^2 = W_2 W_3$

or $W_1 = \sqrt{(W_2 W_3)}$

i.e. the true weight of the body is the geometric mean of the weights required to balance it when it is placed in each pan in turn.

Also, $\dfrac{a_1}{a_2} = \dfrac{W_2}{W_1} = \dfrac{W_1}{W_3}$

giving the ratio of the arms.

2.23. The common steelyard

This consists of a heavy rod AB (Fig. 2.19) supported at a fixed fulcrum O, nearer to one end than the other.

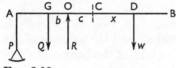

FIG. 2.19

A scale pan is attached to the end of the shorter arm to carry the body whose weight is required. A movable weight w can slide along the longer arm OB, which is graduated

Let $P = $ the weight of the scale pan

             $Q = $ the weight of the rod, acting at G

             $a = $ OA

             $b = $ OG.

Let C be the point at which w must be placed to keep the rod horizontal when there is no weight in the scale pan, and $OC = c$.

Then $wc = Pa + Qb.$ \qquad (i)

If a weight W be placed in the scale pan the weight w will balance it, when moved to D, a distance x from C, given by

$$w(c + x) = (P + W)a + Qb. \qquad \text{(ii)}$$

From (i) and (ii) $wx = Wa.$

Hence if from C we mark off distances a, $2a$, $3a$, etc., a weight in the pan which balances w at the first mark is equal to w; if it balances w at the second mark its weight is $2w$ and so on.

If w has a mass of 1 kg, the graduations correspond to kilogrammes in the pan, and they can be subdivided to show grammes.

At a distance $10a$ from C the sliding weight is equivalent to a mass of 10 kg in the pan, i.e. the steelyard can weigh anything with a mass between 0 and 10 kg.

In many cases a carrier is attached to the end of the longer arm, and slotted weights can be placed on the carrier. In this case C will again be the point at which the sliding weight w must be placed to keep the rod horizontal, when there is no weight in the pan.

The graduations are made in the same way as before, but if with w at C, a weight X on the carrier balances a weight W on the pan, then

$$X \times \text{OB} = W \times \text{OA}.$$

In this way a small weight may be made equivalent to a much larger weight in the scale pan.

2.24. The principle of the lever is made use of in many practical appliances. The usual object of these is to exert a large force at one point by applying a smaller force at another point.

Thus with a bar AB pivoted at C, where $\text{BC} = 20 \times \text{AC}$ (Fig. 2.20), a weight of 20 N at A can be supported by a force

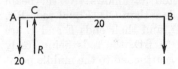

FIG. 2.20

of 1 N at B, and a slightly greater force will lift the weight at A. Scissors and cutting pliers are examples of double levers of this type.

If the rod AB is pivoted at A, a force of 1 N at B directed upwards, can support a weight of 21 N at C. A wheelbarrow or a beam with one end on the ground used for lifting a heavy girder are examples of this arrangement. A pair of nut-crackers is one example of a double lever of this type used for producing a large crushing force at C.

If AB is pivoted at A and a force F applied to C it can overcome a force of $\frac{1}{21} F$ oppositely directed at B. The end B, however, moves through twenty-one times the distance moved by C, and this arrangement is used to magnify movement, e.g. in the treadle of a lathe the foot is applied to C, and the end B is caused to move through a considerable distance by a small movement of the foot.

EXAMPLES 2.4

1. Explain how the position of the zero marks and the distance between the graduations for the sliding weight are determined in a common steelyard. If the sliding weight has a mass of 100 g, and if 50 g at the end of the longer arm of the steelyard balances 4kg at the end of the shorter arm when the sliding weight is at zero mark, find what fraction of the length of the longer arm is the distance between the graduations for the sliding weight to measure consecutive kg at the end of the shorter arm.   (I.A.)

2. Prove that if the fulcrum of a balance be not immediately above the centre of gravity but the arms are equal a correct result is obtained by weighing the body alternately in the two pans and taking the arithmetic mean of the results. Show further that if, in addition, the arms be unequal and differ by an amount h, the total span being l, then, to obtain a correct result the arithmetic mean must be diminished by $\frac{1}{2}(W_1 \sim W_2)(h/l)$, where W_1, W_2 are the apparent weights in the two cases.

3. A weighing machine is constructed as follows: A stiff beam ABCD is pivoted at C, BC being less than CD. Equal rods BF, DE are suspended from B and D, and their ends F and E are joined by a rod of length equal to BD, the rods being freely jointed at B, D, E, F; a scale pan is attached to the middle point of FE. The counterpoise is a weight P which can slide along AC. If M is the position of P when the machine is in equilibrium without any load, show how to graduate the machine.   (I.A.)

4. A tradesman has a pair of scales, which do not quite balance, and makes them balance by attaching a small weight to one of the pans. Show that if he tries to serve a customer with any weight of a commodity by weighing part of it against half the weight on one pan, and the rest against half the weight in the other pan, he will always cheat himself.   (H.S.C.)

5. A uniform rod 0·9 m long is smoothly pivoted on a fixed horizontal peg at its centre C; strings attached to its ends A and B pass respectively through fixed smooth rings A' and B' and carry weights of masses 7 kg each; A' is vertically above C and B' vertically below C, each being 45 cm from C. A weight of mass x kg is hung on the rod at A. Prove that if the rod is to rest in other than a vertical position x must not be less than a certain value, and determine for what value of x the rod will be in equilibrium in a horizontal position. (I.A.)

6. A balance having light arms of unequal length and scale pans of unequal weight does not balance when unloaded. A body known to weigh a units appears to weigh x units, and another body known to weigh b units appears to weigh y units. Show that the true weight of a body which appears to weigh z units is

$$\frac{bx-ay+(a-b)z}{x-y}.$$ (H.S.D.)

7. A heavy non-uniform bar of length 3 m can be balanced about its mid-point when a mass of 2 kg is suspended from one end; if a mass of 6 kg is suspended from the same end the bar will balance about a point 0·3 m from the centre. Find the weight of the bar and the position of its centre of gravity. Show how the bar might be graduated as a steelyard with a movable fulcrum, and prove that the distances of the graduations measured from the end at which the masses are weighed form a series in harmonical progression. (I.E.)

8. Two uniform rods AB, BC of the same material and thickness are rigidly jointed together at B so that the angle ABC is 120°. The bent lever so formed is then pivoted at B so that it is free to turn in a vertical plane, and in the position of equilibrium BC is horizontal. Show that if a weight is attached to C so that in equilibrium AB becomes horizontal this weight must be $\frac{3}{2}$ of the weight of BC. (I.S.)

9. ABC is a horizontal lever pivoted at its mid-point B and carrying a scale pan of weight W_0 at C; AD is a light bar pivoted at A to the lever and at D, vertically above A, to a horizontal bar FDE, which is freely movable about its end F, which is fixed. The weight of this bar is W_1, and its centre of gravity is at a distance d from F and FD $= c$. Show how to graduate this bar with a movable weight w for varying weights W placed in the scale pan at C. If cm graduations correspond to weights of mass 200 g and the mass of $w = 100$ g, find the value of c. In this case find the relation between W_0 and W_1 when $d = 2$ cm, and the zero mark is 2 cm from F. (I.E.)

10. A heavy thin rod AB of length l can be made to balance across a small smooth peg C when a weight $2W$ is suspended from A. Alternatively, it can be made to balance across the peg with a weight $3W$ suspended from B. If the distance AC in the first case is the same as the distance BC in the second, show that the distance of the centre of gravity of the rod from A lies between $\frac{2}{5}l$ and $\frac{1}{2}l$.

If the two equal distances above are each $\frac{1}{4}l$ and if the weights $2W$ and $3W$ are suspended from A and B respectively, find the distance from A to the peg when the rod balances.   (L.A.)

CHAPTER THREE

FORCES IN A PLANE ACTING ON A RIGID BODY

3.1. Rigid body subject to three forces

We shall consider first problems where there are only three forces acting on a body.

If a rigid body is in equilibrium under the action of three forces in a plane the lines of action of these forces must all be parallel or all meet in a common point.

Let P, Q, R be the forces. If they are not all parallel two of them, say P and Q, must meet in some point O.

The resultant of P and Q must then be some force passing through O.

But since the three forces are in equilibrium, this resultant must balance R.

Hence R must be equal and opposite to the resultant of P and Q, and in the same straight line with it, and must therefore pass through O.

If the forces are all parallel the resultant of any two of them must be equal and opposite to, and in the same straight line as, the third force.

3.2. *From the preceding theorem we see that, unless the three forces are parallel, we can use the methods which apply to forces acting on a particle,* i.e. we can use Lami's Theorem, or the triangle of forces graphically, or we can resolve in two directions at right angles. In some cases it may be quicker to take moments about some convenient point. In all cases it is important to draw a figure with the three forces appropriately shown, either all parallel or meeting in a point.

3.3. The following points, some of which have been mentioned earlier, must be carefully remembered, as they are of fundamental importance:

(i) The weight of a body acts vertically downwards through its centre of gravity.

(ii) When a body is leaning against a *smooth* surface the reaction on the body is normal to the surface.

(iii) When a rod is resting on a *smooth* peg the reaction of the peg on the rod is perpendicular to the rod.

(iv) The tension in a light string is the same throughout the string, and this tension is unaffected by the string passing over *smooth* pegs or pulleys. If the pulley is rough the tension is different on the two sides of the pulley.

(v) The resultant of two equal forces bisects the angle between them. Thus, when a string passes over a smooth peg the thrust on the peg bisects the angle between the portions of the string on each side of the peg.

(vi) When a rigid body is *freely* suspended from a fixed point O the centre of gravity G of the body must lie in the vertical through O.

The resultant reaction at O must balance the weight, and for this to be possible the two forces must be in the same straight line. This result follows whether O is a point in the body itself, as in Fig. 3.1, or whether the body is attached to O by two strings, as in Fig. 3.2. If other forces act on the body G is not necessarily vertically below O.

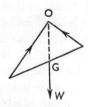

FIG. 3.1   FIG. 3.2

The above considerations, together with the fact that when there are only three non-parallel forces they must meet in a point, enable us to draw an accurate figure showing the position of the body. This is illustrated in the examples given below.

EXAMPLE (i)

A uniform beam AB, 6 m long, has a mass of 40 kg. The end A, about which the beam can turn freely, is attached to a vertical wall, and the beam is kept in a horizontal position by a rope attached to a point of the beam 1·25 m from A and to a point of the wall vertically above A. If the tension of the rope is not to exceed the weight of a mass of 120 kg, show that the height above A of the point of attachment of the string to the wall must not be less than $1\frac{2}{3}$ m. (I.S.)

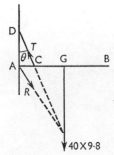

Fig. 3.3

Let G (Fig. 3.3) be the centre of the beam, C and D the points of attachment of the rope to the beam and to the wall.

Let $AD = x$ m and let the angle $ADC = \theta$.

Then
$$\cot \theta = \frac{x}{1\cdot25} = \tfrac{4}{5}x$$

$\therefore$
$$\operatorname{cosec}^2 \theta = 1 + \frac{16x^2}{25} \text{ or } \sin \theta = \frac{5}{\sqrt{(25+16x^2)}}.$$

The reaction R N acting on the beam at A must pass through the point of intersection of the tension T acting at C and the weight.

Taking moments for the beam about A, if T N be the tension in the rope, we get
$$T \times x \sin \theta = 3 \times 40 \times 9\cdot8$$

$\therefore$
$$T = \frac{120 \times 9\cdot8}{x \sin \theta}.$$

Now $T \not> 120 \times 9\cdot8$, and hence
$$\frac{120}{x \sin \theta} \not> 120$$

$\therefore \qquad x \sin \theta \not< 1$

$\therefore \qquad 5x \not< \sqrt{(25+16x^2)}$

$\qquad 25x^2 \not< 25 + 16x^2$

$\therefore \qquad x \not< \tfrac{5}{3}$ or $1\tfrac{2}{3}$.

EXAMPLE (ii)

A heavy uniform rod AB, *of weight* W, *is hinged at* A *to a fixed point.
It is pulled aside by a horizontal force* P *so that it rests inclined at an
angle of* 30° *to the vertical. Find the magnitude of the force* P *and the
reaction at the hinge.*

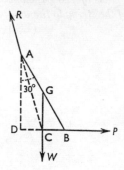

FIG. 3.4

Let G (Fig. 3.4) be the mid-point of the rod. The weight W
acts vertically downwards through G.

Let the verticals through G and A cut the line of action of P in
C and D, then the reaction R at A must pass through C, the point
of intersection of W and P.

Method (*a*). Taking moments about A for the rod, we get

$$P \times \text{AD} = W \times \text{CD}.$$

Now $\qquad \text{AD} = \text{AB} \cos 30° \quad \text{and} \quad \text{CD} = \tfrac{1}{2}\text{AB} \sin 30°.$

$$\therefore \qquad P = W \frac{\tfrac{1}{4}\text{AB}}{\frac{\sqrt{3}}{2}\text{AB}} = \frac{\sqrt{3}}{6}W.$$

Also, if X and Y are the horizontal and vertical components of R

$$X = P = \frac{\sqrt{3}}{6}W$$

and $\qquad\qquad\qquad Y = W$

$$\therefore \qquad R = \sqrt{(X^2 + Y^2)} = W\sqrt{(\tfrac{3}{36} + 1)} = W\sqrt{\tfrac{13}{12}}.$$

If θ is the inclination of R to the horizontal,

$$\tan \theta = \frac{Y}{X} = \frac{6}{\sqrt{3}} = 2\sqrt{3}.$$

Method (*b*). The triangle ADC has its sides parallel to the forces *W*, *P*, *R* taken in order, and can therefore be used as a triangle of forces.

$$\therefore \qquad \frac{P}{DC} = \frac{R}{AC} = \frac{W}{AD}.$$

As above
$$AD = \frac{\sqrt{3}}{2}\, AB \text{ and } CD = \tfrac{1}{4}AB.$$

Also
$$AC^2 = AD^2 + CD^2 = (\tfrac{3}{4} + \tfrac{1}{16})AB$$

$$\therefore \qquad AC = \frac{\sqrt{13}}{4}\, AB$$

$$\therefore \qquad P = W\,\frac{\tfrac{1}{4}AB}{\frac{\sqrt{3}}{2}AB} = \frac{\sqrt{3}}{6}\, W$$

and
$$R = W\,\frac{\frac{\sqrt{13}}{4}AB}{\frac{\sqrt{3}}{2}AB} = W\sqrt{1\tfrac{3}{12}}.$$

Also
$$\tan ACD = \frac{AD}{DC} = \frac{\frac{\sqrt{3}}{2}AB}{\tfrac{1}{4}AB} = 2\sqrt{3}.$$

EXAMPLE (iii)

A uniform heavy rod AB *has the end* A *in contact with a smooth vertical wall, and one end of a string is fastened to the rod at a point* C, *such that* AC = $\tfrac{1}{4}$AB, *and the other end of the string is fastened to the wall vertically above* A. *Find the length of the string, if the rod rests in a position inclined to the vertical.*

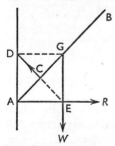

FIG. 3.5

Draw the rod AB inclined to the wall as in Fig. 3.5, and let G be its mid-point.

Since the wall is smooth, the reaction at A is normal to the wall, i.e. horizontal.

Let this normal meet the line of the weight in E; then, since this reaction, the weight and the tension of the string are the only forces acting on the rod, they must meet in a point.

Hence the direction of the string must pass through E. Join EC and produce it to meet the wall in D, then D must be the point of attachment to the wall and CD is the length of the string.

Now, since GEA, EAD are right angles, and C is the mid-point of AG, it is clear that the figure AEGD is a rectangle, and that AG and ED are its diagonals.

$$\therefore \qquad\qquad\qquad CD = AC,$$

i.e. the length of the string is 0·25 of the length of the rod.

The tension in the string and the reaction at A can be found by resolving horizontally and vertically, or by using the triangle ADE to serve as the triangle of forces. They will involve the inclination of the rod to the vertical, which may have any value less than 90°.

Note. If the rod rests with B uppermost, as in Fig. 3.5, it is clear that the point where the string is attached to the rod must lie between A and G. For the line joining E to this point must meet the wall in the point where the string is attached to the wall, and this is obviously impossible if C is at or above G.

If the string is attached between G and B or at B the end B must be lower than A, as in the following example.

EXAMPLE (iv)

A uniform rod AB, of length a, hangs with one end A against a smooth vertical wall, being supported by a string, of length l, attached to the other end of the rod and to a point of the wall vertically above A. Show that, if the rod rests inclined to the wall at an angle θ,

$$\cos^2 \theta = \frac{l^2 - a^2}{3a^2}.$$

Since the wall is smooth, the reaction R at A (Fig. 3.6) is horizontal.

Now the line of the string has to pass through the point where the line of the weight meets R, and this is obviously impossible unless B is below A.

Hence draw AB downwards and let G be its middle point. Let the vertical through G meet R in D, join BD and produce it to meet the wall in C, then BC represents the string and BC = l.

Let AC = h, then DG = $\frac{1}{2}h$, since G is the mid-point of AB and GD is parallel to AC.

In the triangle AGD,

$$\cos \theta = \frac{GD}{AG} = \frac{\frac{1}{2}h}{\frac{1}{2}a}. \tag{i}$$

In the triangle ACB,

$$\cos CAB = \frac{h^2 + a^2 - l^2}{2ah}$$

but

$$\cos CAB = -\cos \theta$$

$\therefore$

$$-\cos \theta = \frac{h^2 + a^2 - l^2}{2ah}.$$

Substituting $\qquad h = a \cos \theta$ from (i), we get

$$-\cos \theta = \frac{a^2 \cos^2 \theta + a^2 - l^2}{2a^2 \cos \theta}$$

$\therefore$

$$-2a^2 \cos^2 \theta = a^2 \cos^2 \theta + a^2 - l^2$$

$\therefore$

$$\cos^2 \theta = \frac{l^2 - a^2}{3a^2}.$$

Again, the tension in the string and the reaction R at A can be found by resolving horizontally and vertically, or by using the triangle CAD as the triangle of forces.

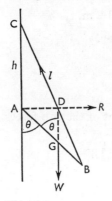

Fig. 3.6

EXAMPLE (v)

A rectangular board, 1·8 m by 1·2 m, is suspended with its longer side horizontal by means of a light string 4·8 m long passing through smooth rings at the upper corners and over a smooth peg. Find the tension of the string and the thrust on each ring if the weight of the board is W.

Let ABCD (Fig. 3.7) represent the board, and O the position of the peg, the rings being at A and B.

Since the string is continuous and only passes over smooth surfaces, the tension T is the same throughout.

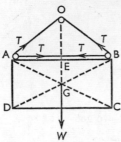

FIG. 3.7

The resultant of the two tensions at O bisects the angle AOB, and it must also balance the weight W acting vertically through G, the centre of gravity of the board.

Hence G must be vertically below O.

If OG cut AB in E, AE = 0·9 m.

Also   AO+OB = 3 m, and OA = OB,

∴   AO = 1·5 m

∴   OE = $\sqrt{(2\cdot25-0\cdot81)} = 1\cdot2$ m.

If the angle AOE = θ, then resolving vertically,

$$2T \cos \theta = W$$

and $\cos \theta = \frac{4}{5}$

∴ $T = \frac{5}{8}W.$

The thrust R on the ring at A is the resultant of the two forces T acting at an angle OAE, where $\cos \text{OAE} = \frac{3}{5}$.

∴ $R^2 = T^2+T^2+2T^2\times\frac{3}{5} = \frac{16}{5}T^2 = \frac{16}{5}\times\frac{25}{64}W^2$

∴ $R = \frac{\sqrt{5}}{2}W.$

The direction of R bisects the angle OAE.

EXAMPLE (vi)

A heavy uniform rod, 26 cm long and of mass 10 kg, is suspended from a fixed point by strings fastened to its ends, their lengths being 24 cm and 10 cm. Find the angle at which the rod is inclined to the vertical and the tensions in the strings.

Let AB (Fig. 3.8) represent the rod, and O the point of suspension. The weight of the rod is 98 N.

Since there are only three forces acting on the rod, the line of the weight must pass through O, i.e. the mid-point G of the rod must be vertically below O.

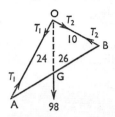

Fig. 3.8

Let $\angle OAB = \theta$.

Since $24^2 + 10^2 = 26^2$

the angle AOB is a right angle, and

$$GO = GA = GB$$

∴ $\angle AOG = \theta$.

Let T_1 and T_2 N be the tensions in OA and OB. Resolving along OA and OB, we have immediately

$$T_1 = 98 \cos \theta = \tfrac{12}{13} \times 98 = 90 \cdot 5$$
$$T_2 = 98 \sin \theta = \tfrac{5}{13} \times 98 = 37 \cdot 7.$$

EXAMPLE (vii)

A rod whose centre of gravity divides it into two portions, a and b, rests inside a smooth sphere in a position inclined to the horizontal. Show that, if θ be its inclination to the horizontal and 2α the angle it subtends at the centre of the sphere,

$$\tan \theta = \frac{b-a}{b+a} \tan \alpha.$$

Let AB (Fig. 3.9) be the rod, C the centre of the sphere. Since the sphere is smooth, the reactions at A and B must be normal to the sphere, and therefore pass through C. Hence the line of the weight must pass through C, i.e. the centre of gravity G must be vertically below C.

The result then follows from the geometry of the figure.

Let $AG = a$, $GB = b$. Since the angle $ACB = 2\alpha$, the angles A and B are each equal to $90° - \alpha$.

Draw CD perpendicular to AB. Then angle $GCD = \theta$. Also $AD = \frac{1}{2}(a+b)$ and $GD = \frac{1}{2}(a+b)-a = \frac{1}{2}(b-a)$.

Hence
$$\tan \theta = \frac{GD}{CD}$$

$$= \frac{\frac{1}{2}(b-a)}{\frac{1}{2}(b+a)\cot \alpha} = \frac{b-a}{b+a}\tan \alpha.$$

If the reactions are required they can now be found by taking moments, or by resolving in two perpendicular directions.

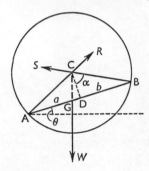

FIG. 3.9

Let R, S be the reactions at A and B respectively.
Taking moments for the rod about A, we get

$$S(a+b)\sin(90°-\alpha) = Wa\cos \theta$$

$$\therefore \qquad S = W\frac{a\cos \theta}{(a+b)\cos \alpha}.$$

Taking moments about B, we get

$$R(a+b)\sin(90°-\alpha) = Wb\cos \theta$$

$$\therefore \qquad R = W\frac{b\cos \theta}{(a+b)\cos \alpha}.$$

EXAMPLE (viii)

A rod, which is not uniform, rests in a vertical plane with its lower end A on a smooth plane inclined at an angle α to the horizontal, and the upper end B against a smooth vertical wall. If G is the centre of gravity of the rod, and β its inclination to the horizontal, show that

$$\frac{AG}{GB} = \frac{\sin \alpha \sin \beta}{\cos(\alpha+\beta)}. \qquad \text{(H.S.D.)}$$

Let the reactions at A and B, which are normal to the plane and wall respectively, meet at C (Fig. 3.10). Then G must be vertically below C.

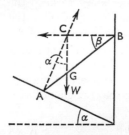

FIG. 3.10

Now $\angle CBG = \beta$, $\angle ACG = \alpha$, and $\angle CAG = 90° - \beta - \alpha$.

Hence
$$\frac{AG}{\sin \alpha} = \frac{CG}{\sin CAG} = \frac{CG}{\cos (\alpha + \beta)}$$

and
$$\frac{GB}{\sin 90°} = \frac{CG}{\sin \beta}$$

$$\therefore \qquad \frac{AG}{GB} = \frac{\sin \alpha \sin \beta}{\cos (\alpha + \beta)}.$$

Note. In most of these problems the result follows from the geometry of the figure, if it is drawn so that the three forces meet at a point.

EXAMPLES 3.1

1. A uniform rod can turn freely about a hinge at one of its ends, and is pulled aside from the vertical by a horizontal force acting at the other end of the rod and equal to half the weight of the rod. Find the inclination to the vertical at which the rod will rest.

2. If, in Question 1, the horizontal force is three-quarters the weight of the rod, find the inclination to the vertical, and also the reaction at the hinge.

3. A ladder AB rests against a smooth vertical wall at A and is supported by a socket in which its lower end B is placed; the vertical through G (the centre of gravity of the ladder and the load) meets the horizontal through A in K, and the same horizontal cuts the vertical through B in L. Prove that the triangle BKL will serve as a triangle of forces for the weight and the reactions at A and B. (I.S.)

4. A uniform beam AB of weight W can turn in a vertical plane about a hinge at A, and to the other end B is tied a rope which passes over a smooth pulley C vertically above A so that AC = AB. Find the tension of the rope necessary to keep the beam at an angle of 60° with the horizontal. Find also the direction and magnitude of the reaction at the hinge.   (I.A.)

5. AB is a uniform bar of weight W, movable about a smooth horizontal axis fixed at A; to B is attached a light cord which passes over a pulley C fixed vertically over A, and supports a mass of weight P at its free end. Show by applying the triangle of forces that, in the position of equilibrium,

$$CB = 2\frac{P}{W} \times AC.$$ (I.A.)

6. A uniform heavy rod AB freely hinged to a fixed point A in a smooth wall is kept in a horizontal position parallel to the wall by a light cord, attached to its free end B and fastened to a point P in the line LM in which the wall meets the ceiling. Prove that for various positions of P on LM (the length of the cord being adjusted so that the rod is always horizontal) the tension in the cord is proportional to its length BP.   (H.S.D.)

7. A pole rests with its lower end P in a socket, and is supported by a rope joining a point Q of the pole to a point R vertically above the socket. Prove that if the vertical through the centre of gravity of the pole cuts QR in S, the triangle PRS forms a triangle of forces for the weight, the tension in the rope, and the reaction of the socket.   (H.C.)

8. A uniform rod 3 m long is suspended by a light string of length 5 m passing over a smooth peg, and rests horizontally, the string being attached to the ends of the rod. If the rod has a mass of 7 kg, find the tension of the string.   (H.C.)

9. A uniform rod AB of mass 10 kg is smoothly hinged at A and rests in a vertical plane with the end B against a smooth vertical wall. If the rod makes an angle of 40° with the wall, find the pressure on the wall and the magnitude and direction of the reaction at A.   (H.C.)

10. A uniform lamina of weight W in the form of an isosceles triangle ABC right-angled at B is freely hinged to a fixed point at A and rests with AC vertical and C above A, equilibrium being maintained by a horizontal string attached to C. Find the tension in the string and the magnitude and direction of the reaction at A.   (H.C.)

11. A uniform equilateral triangular lamina ABC of weight W has the vertex A hinged to a fixed point, about which it can turn freely in a vertical plane, and rests with AB vertical, B being above A, and the vertex C in contact with a smooth vertical wall. Find the reaction between the lamina and the wall, and the magnitude and direction of the reaction at A. (H.C.)

12. A uniform rod ACB of weight W is supported with its end A against a smooth vertical wall, with the end B uppermost, by means of a string attached to C and to a point D in the wall on the same level as B. If the inclination of CD to the wall is 30°, find the tension of the string and the reaction at the wall, and prove that $AC = \frac{1}{3}AB$.

13. A uniform rod AB is in equilibrium at an angle α with the horizontal with its upper end A resting against a smooth peg and its lower end B attached to a light cord, which is fastened to a point C on the same level as A. Prove that the angle β at which the cord is inclined to the horizontal is given by the equation

$$\tan \beta = 2 \tan \alpha + \cot \alpha,$$

and that $$AC = \frac{AB \sec \alpha}{1 + 2 \tan^2 \alpha}.$$ (H.S.C.)

14. A uniform heavy rod, whose length is equal to the diameter of a smooth hemispherical bowl which is fixed with its axis vertical, rests with one end in the bowl and the other end outside the bowl. Prove that the inclination of the rod to the horizontal is about 32° 32′. (I.S.)

15. A sphere of mass 5 kg and radius 63 cm is hung by a string 24 cm long from a point in a smooth vertical wall. Find the tension in the string. (H.S.D.)

16. A uniform flagstaff, 12 m long and of mass 120 kg, has its lower end attached to the ground by a swivel; it is being raised by a rope attached to its highest point. If the inclination of the rope to the horizontal is 20° when that of the flagstaff is 50°, find graphically or otherwise, the tension of the rope and the magnitude and direction of the reaction of the swivel. (I.A.)

17. A uniform bar AB, of weight $2W$ and length l, is free to turn about a smooth hinge at its upper end A, and a horizontal force is applied to the end B so that the bar is in equilibrium with B at a distance a from the vertical through A. Prove that the reaction at the hinge is equal to

$$W \left[\frac{4l^2 - 3a^2}{l^2 - a^2} \right]^{\frac{1}{2}}.$$ (I.S.)

18. A rectangular block hangs suspended from a support by two wires of equal length attached to two points symmetrically situated on the upper face, the upper ends being attached to the same point of the support. Show that the tension in the wires is increased if their lengths are shortened. If the block is cubical, of edge 0·9 m and mass 2000 kg, and the points of attachment of the wires are 0·6 m apart, find the shortest possible length of the wires, given that the breaking strain of each is 15 000 N. (I.E.)

19. A uniform circular plate of weight W, whose centre is C and plane vertical, is freely movable in its own plane about a horizontal axis fixed at a point A of the circumference. The line AC is to be kept at a given inclination α to the vertical by causing the plate to rest against a fixed smooth peg at a point B on the circumference. Find the position of B such that the pressure on the peg is least, and find this least pressure. (I.A.)

20. A uniform circular hoop, of radius a and weight W, free to move in a vertical plane, passes through a smooth fixed ring at P, and a point Q of the hoop is attached by an inextensible string of length l to a fixed point O vertically above P. Prove that in the equilibrium position the tension of the string is $W(l+a)/PO$, provided that $PO > l$. (I.A.)

21. A picture of mass 5 kg is hung from a nail by a cord 1·5 m long fastened to two rings 0·9 m apart. Find the tension in the cord. (I.A.)

22. A uniform ladder of mass 20 kg rests at an angle 25° with the vertical, with one end against a smooth vertical wall and the other end on a rough horizontal floor, the vertical plane through the ladder being perpendicular to the wall. Find the magnitudes of the reactions at the floor and the wall. (H.S.D.)

23. A sphere, of radius a and weight W, rests on a smooth inclined plane supported by a string of length l with one end attached to a point on the surface of the sphere and the other end fastened to a point on the plane. If the angle of inclination of the plane to the horizontal be α, prove that the tension of the string is

$$\frac{W(a+l)\sin\alpha}{\sqrt{(l^2+2al)}}.$$ (H.S.D.)

24. A uniform rod AB can turn freely in a vertical plane about the end A, which is fixed, and the rod is held at an inclination θ to the vertical by means of a string attached to B. Find the direction of the string that the tension in it may be as small as possible, and determine the reaction of the hinge in this case.

25. An equilateral triangular lamina is suspended by threads fastened to two of its angular points. The direction of the threads pass through the centre of gravity of the lamina, and one of them is horizontal and has a tension of 5 N. Find the tension of the other thread and the weight of the lamina.

26. A uniform rod, 3 m long and of mass 20 kg, is placed on two smooth planes inclined at 30° and 60° to the horizontal. Find the pressures on each plane and the inclination of the rod to the horizontal when in equilibrium.

27. A rod of weight W, whose centre of gravity divides its length in the ratio $2:1$, lies in equilibrium inside a smooth hollow sphere. If the rod subtends an angle of 2α at the centre of the sphere and makes an angle θ with the horizontal, prove that $\tan\theta = \frac{1}{3}\tan\alpha$. Find the reactions at the ends of the rod in terms of W and α.

(C.A.)

28. Two small rings of weights $3W$ and $5W$ are capable of sliding on a smooth circular wire of radius a fixed in a vertical plane. The rings are connected together by a light inextensible string of length $8a/3$ which passes over a smooth peg fixed at a height $a/3$ vertically above the highest point of the wire. The rings rest on opposite sides of the vertical through the peg. Find the reaction of the wire on each ring and show that the tension in the string is $15W/4$.

    If the wire is uniform and of weight W, find the horizontal and vertical components of the external force required to keep the wire in position, indicating the directions clearly. (L.A.)

3.4. Rigid body subject to more than three forces

We have now to consider the general case where there are more than three coplanar forces acting on a rigid body. The forces need not meet in a point.

From the principle of the transmissibility of force we assume that the point of application of any force may be taken as **any** point on its line of action. For example, if four forces act on a rigid body at the points A, B, C, and D, as shown in Fig. 3.11, they may be assumed to act at A_1, B_1, C_1, and D_1, points on the lines of action of the four forces. It follows that what is important is not the shape or size of the body but the magnitudes of the forces and their relative positions.

The conditions of equilibrium therefore depend only on the forces and not on the body. For this reason we frequently speak of the conditions of equilibrium of a given system of forces without specifying the body on which they act.

Questions are often set asking for the resultant of certain forces acting round the sides of a square or polygon. The sides of the figure merely fix the lines of action of the forces, which

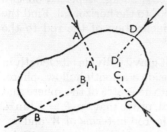

Fig. 3.11

need not be acting on a body of that shape; their resultant will be the same whatever the shape of the body on which they act, provided it is rigid.

We shall first deduce the conditions of equilibrium for any number of coplanar forces from the theorem in the next paragraph. A more general method of treatment is given in 3.21.

3.5. Any system of coplanar forces acting on a rigid body, and not in equilibrium, can be reduced either to a single force or a single couple

We can always reduce any three of the forces, say P, Q, and R, to two. For we can always compound P with either Q or R, unless P forms a couple with each of them.

In this case Q and R are equal, parallel, and *like* forces (for each is in the opposite direction to P), and therefore Q and R can be compounded.

By taking another force of the system with the two forces obtained from P, Q, and R, we can again reduce the three forces to two, and by repeating this process we shall obviously reduce the system to two forces which, if not in equilibrium, must either form a couple or have a single resultant.

3.6. If the system reduces to a single force this is usually denoted by R; if it reduces to a couple the moment of this will be denoted by G.

It is evident that the reduction of any system of forces does not cause any change in the sum of the components in any direction, for on compounding any two forces the component of the resultant is equal to the sum of the components of the two forces. Hence, on reducing a system of forces, the sum of the components of the last two forces in any direction is equal to the sum of the components of the original forces in that direction, and is also equal to the component of the single resultant R to which the system finally reduces. If the system reduces to a couple the sum of the components of the original forces in any direction must be zero.

In the same way the sum of the moments about any point is unaffected by the process of reduction; the moment of R or of the couple about any point is equal to the sum of the moments of the separate forces about that point. (*Cf.* 2.13.)

It is clear that R is given in magnitude and direction by the vector sum of the system of forces. If the vector sum is zero, $R = 0$; the system will then, in general, reduce to a couple of moment G. If in addition $G = 0$ the system will be in equilibrium.

3.7. Conditions of equilibrium

Now, if R is zero its component in any direction must be zero; but to ensure, conversely, that R *is* zero we must know that its components in *two* different directions are zero. If we know that the component in one direction only is zero this direction *might* be perpendicular to the direction of R, but if the component in a second direction is also zero, then R must be zero.

Also, if the components of R in any two directions are zero the sums of the components of the separate forces in these directions must also be zero.

Since the moment of a couple is the same about all points in its plane, we see that, if G is to be zero, the sum of the moments of the separate forces of the system about *any* point in their plane must be zero.

Conversely, if the sum of the moments of the forces about *any* point in their plane is zero the system cannot reduce to a couple.

It must be noted, however, that the sum of the moments

about one or two points being zero does not prove that the system is in equilibrium, for these two points might happen to be on the line of action of the single resultant. If, however, the sum of the moments about three points, not in the same straight line, is zero, then the system must be in equilibrium, as it cannot reduce to either a force or a couple.

The necessary conditions for a system of forces to be in equilibrium are (i) that the sum of the components in any given direction should be zero, and (ii) the sum of the moments about any given point should be zero, but they are not separately *sufficient* to ensure that the forces *are* in equilibrium.

3.8. The simplest set of conditions which is *sufficient* to ensure that a system of forces is in equilibrium is as follows:

(i) the sums of the components of the forces in any *two* directions must be zero; *and*

(ii) the algebraic sum of the moments of the forces about any point in their plane must be zero.

An alternative form of (i) is that the vector sum of all the forces must be zero.

Condition (i) ensures that the system does not reduce to a single force, and condition (ii) ensures that it does not reduce to a couple.

It is clear that (i) and (ii) are really equivalent to *three* conditions.

Another set of conditions already mentioned above is that the sums of the moments of the forces about three points not in the same straight line must be zero. This, again, is equivalent to three conditions.

3.9. In solving problems in Statics we are dealing with a system of forces in equilibrium, and we can therefore use one of the sets of conditions mentioned above.

As a rule we obtain three equations connecting the unknown forces and angles as follows:

(*a*) Equate to zero the algebraic sum of the components of all the forces in some convenient direction.

(*b*) Equate to zero the algebraic sum of the components of

all the forces in some other direction (usually perpendicular to the direction in (*a*)).

(*c*) Equate to zero the algebraic sum of the moments of the forces about any point in their plane.

Note. The directions chosen in (*a*) and (*b*) are usually, but not necessarily, the horizontal and vertical. The point about which we take moments is usually chosen so as to exclude as many forces as possible, i.e. the point through which most of the forces pass.

3.10. The method of obtaining the equations mentioned in the last paragraph is illustrated in the following examples.

It will be noticed that there are, in some cases, geometrical relations between lengths or angles involved which give additional equations to those obtained by resolving and taking moments. In some problems the difficulties are not really in applying mechanical principles but in the geometrical and trigonometrical knowledge required to obtain the result asked for.

EXAMPLE (i)

A heavy uniform beam, hinged to a vertical wall, has a mass of 150 kg and is 4·5 m long. A tie attached to the other end keeps the beam horizontal, and is fixed to the wall 3·6 m above it. A mass of 200 kg is hung from this end. Find the tension of the tie and the thrust on the beam.

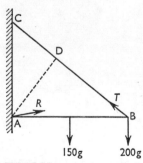

FIG. 3.12

Let AB (Fig. 3.12) represent the beam, AC the wall, and BC the tie. Since

$$AB = 4·5 \text{ m}, \quad AC = 3·6 \text{ m}$$
$$BC = 0·9\sqrt{41} \text{ m}, \text{ and } \sin ABC = 4/\sqrt{41}.$$

If AD is the perpendicular from A to BC,
$$AD = AB \sin ABC = 18/\sqrt{41} \text{ m.}$$

The forces acting on the beam are shown, R being the reaction at the hinge A, and T the force at B equal to the tension in the tie.

Taking moments about A so as to eliminate R, we get

$$\frac{18}{\sqrt{41}} T = 150 \text{ g} \times 2 \cdot 25 + 200 \text{ g} \times 4 \cdot 5$$

$$\therefore \qquad T = \frac{275 \times 9 \cdot 8 \times 4 \cdot 5}{18} \sqrt{41} = 673 \cdot 75 \sqrt{41} \text{ N.}$$

The only horizontal forces acting on the beam are the horizontal components of the tension in the tie and of the reaction at A. These must be equal and opposite, and the thrust on the beam must be equal to either of them.

Hence the thrust is

$$T \cos ABC = 673 \cdot 75 \sqrt{41} \times \frac{5}{\sqrt{41}}$$
$$= 3368 \cdot 75 \text{ N.}$$

EXAMPLE (ii)

One end of a uniform rod of weight W is attached to a hinge, and the other end is supported by a string attached to the other end of the rod and to a point on the same level as the hinge, the rod and string being inclined at the same angle to the horizontal. Find the tension in the string and the action at the hinge.

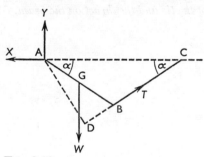

FIG. 3.13

Let AB (Fig. 3.13) be the rod, G its mid-point, and BC the string, AC being horizontal.

Let $\angle BAC = \angle BCA = \alpha$ and $AB = l$.

Then $AC = 2l \cos \alpha$, and if AD be the perpendicular from A to BC we get $AD = 2l \cos \alpha \sin \alpha$.

Hence, if T be the tension in the string, taking moments about A,

$$T \times 2l \cos \alpha \sin \alpha = W \frac{l}{2} \cos \alpha \text{ or } T = \frac{W}{4 \sin \alpha}.$$

If X and Y be the horizontal components of the reaction at A, then resolving horizontally,

$$X = T \cos \alpha = \frac{W \cos \alpha}{4 \sin \alpha}$$

and resolving vertically,

$$Y = W - T \sin \alpha = \tfrac{3}{4} W.$$

If R is the resultant reaction at A,

$$R = \sqrt{(X^2 + Y^2)} = \frac{W}{4} \sqrt{(9 + \cot^2 \alpha)}.$$

If θ is the inclination of R to the horizontal,

$$\tan \theta = \frac{Y}{X} = \tfrac{3}{4} \times \frac{4 \sin \alpha}{\cos \alpha} = 3 \tan \alpha.$$

EXAMPLE (iii)

A pole with one end resting on the ground is kept vertical by two ropes attached to fixed points on the ground at equal distances d from the foot of the pole, and in the same plane with it. The ropes are fastened to the pole at heights a, b, and the tensions are adjusted so that the vertical reaction on the ground is twice the weight of the pole. Prove that the total reaction on the ground makes an angle θ with the vertical, where

$$\tan \theta = \tfrac{1}{4} d \left(\frac{1}{a} \sim \frac{1}{b} \right).$$

Let OAB (Fig. 3.14) be the pole, O the lower end, and A and B the points of attachment of the ropes AC, BD, so that $OC = OD = d$.

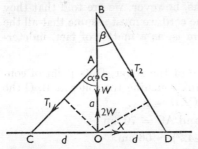

FIG. 3.14

Let $AO = a$, $OB = b$, $\angle OBD = \beta$, $\angle OAC = \alpha$, and let the tensions in AC and BD be T_1 and T_2.

Taking moments about O,

$$T_1 d \cos \alpha = T_2 d \cos \beta. \tag{i}$$

Now since the vertical reaction on the ground is $2W$, where W is the weight of the pole, the sum of the vertical components of the tensions must be equal to W.

$$\therefore \qquad\qquad T_1 \cos \alpha + T_2 \cos \beta = W. \tag{ii}$$

Hence, using (i),

$$T_1 \cos \alpha = T_2 \cos \beta = W/2.$$

The horizontal component, X, of the reaction on the ground must equal the difference between the horizontal components of the tensions, for resolving horizontally we get:

$$X = T_1 \sin \alpha - T_2 \sin \beta = \frac{W}{2} (\tan \alpha - \tan \beta)$$

$$= \frac{W}{2}\left(\frac{d}{a} - \frac{d}{b}\right).$$

The vertical reaction is $2W$, so that

$$\tan \theta = \frac{X}{2W} = \frac{d}{4}\left(\frac{1}{a} - \frac{1}{b}\right).$$

EXAMPLE (iv)

A roller of 42 cm radius lies on the ground; a uniform plank 60 cm long rests flat on the roller with one end on the ground and the other projecting over the roller, the length of the plank being at right angles to the axis of the roller. If the roller and the plank are prevented from slipping by a cord 70 cm long attached to the axle of the roller and to that end of the plank which lies on the ground, prove that the tension of the cord is $\frac{9}{50}$ of the weight of the plank.

Note. No mention is made of friction between the bodies and the ground or between themselves. As, however, we are told that they are prevented from slipping by the cord we must assume that all the contacts are smooth. The problem is, as a matter of fact, indeterminate otherwise.

Let C (Fig. 3.15) be the centre of the roller, D its point of contact with the ground, AB the plank touching the roller at B, G the mid-point of the plank, and let $\angle CAD = \alpha$.

Since   CD = 42 cm, and AC = 70 cm

                AD = $\sqrt{(70^2 - 42^2)}$ = 56 cm,

$\therefore$ AB = 56 cm.

Figure 3.15A shows the forces acting on the plank and Fig. 3.15B the forces on the roller.

The reaction R acting on the plank at its contact with the roller is along the radius CB, and the weight W of the plank acts vertically through G.

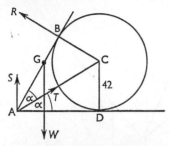

FIG. 3.15A   FIG. 3.15B

Taking moments about A for the plank, to eliminate the reaction S at the ground and the tension T, we get

$$56R = W \times 30 \cos 2\alpha.$$

Now $\sin \alpha = \frac{3}{5}$, $\cos \alpha = \frac{4}{5}$ and $\cos 2\alpha = \cos^2 \alpha - \sin^2 \alpha = \frac{7}{25}$

$$\therefore \qquad R = W \times \frac{30 \times 7}{56 \times 25} = \frac{3}{20}W.$$

The forces tending to move the roller horizontally are the horizontal components of R and of the tension T in the cord. Since these must be equal and opposite, we get

$$T \cos \alpha = R \sin 2\alpha$$

$$\therefore \qquad T = 2R \sin \alpha = 2 \times \frac{3W}{20} \times \frac{3}{5} = \frac{9W}{50}.$$

EXAMPLE (v)

A uniform rod of weight W and length $2a$ is free to turn about one end A. It is supported by means of a light string of length l, one end of which is attached to a point vertically above and at a distance h from A, while the other end is attached to a smooth ring of weight w through which the rod passes. Show that in the equilibrium position the string is inclined at an angle θ to the vertical, where

$$Wal(h \cos \theta - l) = w(l^2 - 2lh \cos \theta + h^2)^{\frac{3}{2}}.$$

Let C (Fig. 3.16) be the upper end of the string, D the position of the ring, G the mid-point of the rod, and let the angle CAD $= \alpha$, the angle CDG $= \phi$, and AD $= x$.

Draw AE (Fig. 3.16A) perpendicular to CD, then AE $= h \sin \theta$. Let T be the tension in the string.

Figure 3.16A shows the forces acting on the ring, and Fig.

3.16B the forces on the rod. S is the reaction acting on the rod at the hinge. Equal and opposite forces R act on the ring and rod perpendicular to the rod.

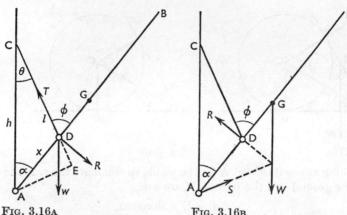

FIG. 3.16A   FIG. 3.16B

For the ring, resolving along and perpendicular to AB, we get

$$T \cos \phi = w \cos \alpha \qquad \text{(i)}$$

and

$$T \sin \phi = w \sin \alpha + R. \qquad \text{(ii)}$$

For the rod only, taking moments about A, we get

$$Rx = Wa \sin \alpha. \qquad \text{(iii)}$$

Substituting in (ii), for T from (i) and R from (iii), we get

$$w \cos \alpha \tan \phi = w \sin a + W \frac{a}{x} \sin \alpha$$

$\therefore \qquad wx(\cos \alpha \tan \phi - \sin \alpha) = Wa \sin \alpha$

$\therefore \qquad wx \sin (\phi - \alpha) = Wa \sin \alpha \cos \phi$

$\therefore \qquad wx \sin \theta = Wa \sin \alpha \cos \phi.$

But from the figure

$$\frac{x}{\sin \theta} = \frac{l}{\sin \alpha}.$$

Hence, $wx^2 = Wal \cos \phi.$

Again, from the figure

$$DE = x \cos \phi = h \cos \theta - l$$

and hence $Wal(h \cos \theta - l) = wx^3$

which gives the result, since

$$x^2 = h^2 - 2hl \cos \theta + l^2.$$

Note. In addition to equation (iii) we could have obtained two other equations for the rod by resolving along and perpendicular to the rod. These would have enabled us to find S in magnitude and direction.

EXAMPLE (vi)

A ladder rests at an angle α to the horizontal, with its ends resting on a smooth floor and against a smooth vertical wall, the lower end being joined by a string to the junction of the wall and the floor. Find the tension of the string, and the reaction at the wall and the ground. Find also the tension of the string when a man, whose weight is equal to that of the ladder, has ascended the ladder three-quarters of its length.

Let AB (Fig. 3.17) be the ladder, C the junction of the wall and the floor, and G the mid-point of the ladder.

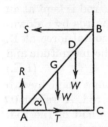

FIG. 3.17

Since the surfaces are smooth, the reactions at A and B are normal to the ground and wall; let these be R and S.

Let W be the weight of the ladder, acting vertically through G, and let T be the tension in the string.

Resolving vertically, $R = W.$ (i)

Resolving horizontally, $S = T.$ (ii)

Taking moments about B,

$$T \times \text{AB} \sin \alpha = R \times \text{AB} \cos \alpha - W \frac{\text{AB}}{2} \cos \alpha. \qquad \text{(iii)}$$

Substituting for R from (i),

$$T \sin \alpha = W \cos \alpha - \tfrac{1}{2} W \cos \alpha$$

∴ $$T = \tfrac{1}{2} W \cot \alpha.$$

From (ii) $S = T = \tfrac{1}{2} W \cot \alpha.$

When the man is at D, where $\text{AD} = \tfrac{3}{4}\text{AB}$, equations (i) and (iii) become

$$R = W + W = 2W$$

and $\quad T \times \text{AB} \sin \alpha = R \times \text{AB} \cos \alpha - W \dfrac{\text{AB}}{2} \cos \alpha - W \dfrac{\text{AB}}{4} \cos \alpha$

$\therefore \qquad\qquad T \sin \alpha = 2W \cos \alpha - \tfrac{3}{4}W \cos \alpha$

$\therefore \qquad\qquad\qquad T = \tfrac{5}{4}W \cot \alpha.$

EXAMPLES 3.2 ? uniform

1. A heavy bar AB, of mass 20 kg and length 2·4 m, is hinged at A to a point in a vertical wall, and is maintained in a horizontal position by means of a chain attached to B and to a point in the wall 1·5 m above A. If the bar carries a load of 10 kg at a point 1·8 m from A, calculate the tension in the chain, and the magnitude and direction of the action at A between the bar and the wall.   (I.E.)

2. A uniform rod ACB, 1·8 m long and of mass 4 kg, is free to turn about a hinge at C, which is 0·6 m from A. The rod is kept at an inclination of 45° to the vertical with A downwards by a downward vertical force of 49 N at A, and a horizontal force at B. Find the magnitude of the force at B, and the magnitude and direction of the reaction at the hinge C.   (I.E.)

3. A heavy uniform rod of weight W is hung from a point by two equal strings, one attached to each end of the rod. A weight w is hung half-way between the centre and one end of the rod. Prove that the ratio of the tensions in the strings is

$$\frac{2W+3w}{2W+w}.$$ (H.S.C.)

4. A door 2·25 m in height is hung from two hinges placed 25 cm from the top and bottom. The door has a mass of 18 kg, and its centre of gravity is 67·5 cm distant from the line of the hinges. Find the total force on each hinge, it being assumed that half the weight of the door is supported by each hinge.   (I.S.)

5. A gate hangs from two hinges A, B, its centre of gravity being G, and the vertical through G meets the horizontal through B in K. Prove that if the whole weight is supported at A the triangle ABK will serve as a triangle of forces to determine the reactions at A, B; also modify this construction when the weight is equally divided between A and B.   (I.S.)

6. A uniform beam rests with its ends on two smooth inclined planes which make angles of 30° and 60° with the horizontal respectively. A weight equal to twice that of the beam can slide along its length. Find the position of the sliding weight when the beam rests in a horizontal position.   (I.A.)

7. A smooth uniform ladder rests with its extremities against a vertical wall and a horizontal plane, and is held by a rope, one end of which is attached to a rung of the ladder one-quarter of the way up, the other end being fixed to a point of the base of the wall, vertically below the top of the ladder. Show that if the base and top of the ladder be distant a and b respectively from the base of the wall the ratio of the reactions P and Q between the ladder and the ground and wall respectively is given by

$$Q/P = 3a/5b.$$ (I.A.)

8. A ladder, 3 m long and of mass 17·5 kg, rests with the end A against a smooth vertical wall, and the other end B on the ground, which is smooth, at a distance of 1·8 m from the wall; it is maintained in this position by a horizontal cord attached at B. Find the tension of the cord if the centre of gravity of the ladder is 1·2 m from B. Find also the magnitude and direction of the force which, applied at A, will keep the ladder in position without the help of the cord. (I.S.)

9. A uniform plank AB, 1·8 m long and of given weight, is supported horizontally by two pegs at C and D respectively, where $AC = BD = 20$ cm. Another exactly equal plank A′B′ is placed over the first so that A′ projects 27·5 cm beyond A. Find the pressures R_1 and R_2 at C and D. A vertical force P is now applied at A′, but so that equilibrium is still maintained; if $R_1′$ and $R_2′$ are the new pressures at C and D, show that

$$\frac{R_1′ - R_1}{R_2 - R_2′} = \frac{75}{19}.$$ (H.S.D.)

10. The centre of gravity of a hemispherical bowl is on the radius round which the bowl is symmetrical, and divides this radius in the ratio $m : n$. If the bowl is placed with its curved surface in contact with a plane, rough enough to prevent sliding, inclined at an angle θ to the horizontal, find the inclination of the rim of the bowl to the horizontal. It is found that the bowl rests in equilibrium, with the plane of its rim vertical, when θ is approximately 25°. Calculate the ratio $m : n$. (I.S.)

11. A man wishes to pull a smooth lawn roller of diameter 50 cm and mass 100 kg over a kerbstone 10 cm high. Find the direction in which he should pull, in any position of the roller, so as to raise the roller with the least effort; and show that the greatest force he need exert is 784 N. (H.C.)

12. It is required to pull a garden roller of weight W and radius r, whose centre of gravity lies on its axis, up a step of height $\frac{1}{2}r$. The force is applied to a handle which acts directly on the axis

of the roller. What is the best direction in which to pull the
handle? Compare the force required to move the roller, when
applied in this direction, with that required when the handle is
pulled horizontally. Is there any tendency for the roller to slip
on the step or ground?   (H.S.D.)

13. A uniform bar AB, 0·9 m long and of mass 2 kg, has a cord 1·5 m
long attached to its ends. The cord passes through a smooth ring
O fixed in a smooth vertical wall, and the rod is placed in a
vertical plane perpendicular to the wall with the end A against
the wall and vertically below O. Prove that the rod will be in
equilibrium if OA is 0·6 m, and show that the tension of the
string is 14·7 N.   (H.C.)

14. A telegraph pole carries six wires, of which three go south, two
go north-east, and one west. If all the wires are in one horizontal
plane and are stretched to a tension of 900 N, find the pull
they exert on the pole. The wires are each 12 m above the ground
and a stay fixed to the pole at a point 9 m up is made fast to the
ground 4·5 m from the foot of the pole. Find the tension in this
stay if there is no tendency for the pole to overturn.   (N.U.S.)

15. A triangular lamina ABC in which the angles at B and C are 45°
is of mass 3 kg and is hung from the mid-point of the side BC.
Masses of 2 and 8 kg respectively are hung from A and B and a
mass M is hung from C. Find the value of M if the side BC
makes an angle of 60° with the vertical.   (Q.E.)

3.11. Jointed rods

We shall now consider some problems where two or more
rigid bodies are involved, and, in particular, cases where a
number of heavy rods are smoothly jointed together.

We shall not consider the stresses in the material of the rods.

In the ordinary problems on the equilibrium of jointed heavy
rods we only consider the forces acting at the joints (and, of
course, the weights of the rods), i.e. we consider the equilibrium
of the rods under the action of their own weight and the forces
exerted on their *ends* by the hinges.

3.12. Consider a heavy rod AB (Fig. 3.18) freely jointed at A
and B.

Its weight acts vertically through its centre of gravity G,
and it is obvious that, to keep it in equilibrium by means of
forces applied at A and B, these forces must either be vertical

or they must meet, as shown, on the line of action of the weight, i.e. the vertical through G.

In neither case can they act along the rod (unless it is vertical). If, however, the rod is *light*, then forces applied at the ends which keep it in equilibrium must be equal and opposite and act *along the rod*.

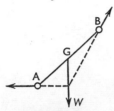

FIG. 3.18

It is the confusion between the cases of the heavy rod and the light rod which causes most of the trouble experienced by students in problems on heavy rods. In the case of a heavy rod the one direction in which the forces at the ends *cannot* act is along the rod. We usually consider the horizontal and vertical components of the force exerted by a hinge on the rod, and in order to show these clearly in the diagram, it is better in drawing it not to make the rods meet, but to leave a space between them.

3.13. EXAMPLE (i)

Two equal uniform rods AC and CB are smoothly jointed at C, and have their other ends attached to supports at two points, A and B, at the same level. If each rod has a mass of 40 kg and is inclined at 60° to the horizontal, find the action on the hinge C.

Let D, E (Fig. 3.19) be the mid-points of the rods; then the weights of 40g N act vertically at D and E.

Let X, Y be the horizontal and vertical components of the action of the hinge at C on the rod AC. Its action on BC will consist of components equal to those on AC but in opposite directions. These forces and the forces acting on the rods at the supports at A and B are shown in the figure.

Let l be the length of each rod.

Taking moments about A for AC,

$$Xl \sin 60° + Yl \cos 60° = 40 \, g \frac{l}{2} \cos 60°$$

∴ $X \tan 60° + Y = 20 \, g.$ (i)

Taking moments about B for BC,

$$Xl \sin 60° - Yl \cos 60° = 40 \, g \, \frac{l}{2} \cos 60°$$

$$X \tan 60° - Y = 20 \, g. \qquad\qquad (\text{ii})$$

From equations (i) and (ii) it is obvious that $Y = 0$.

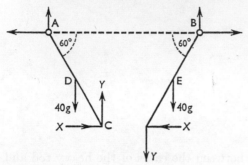

FIG. 3.19

Hence the reaction at C consists only of a horizontal force X N whose value is given by

$$X \tan 60° = 20 \, g$$

$$\therefore \qquad X = \frac{196}{\sqrt{3}} = \frac{196}{3}\sqrt{3}.$$

Note. Since the whole system is symmetrical about the vertical through C, we might have argued without writing down any equations that Y must be zero. It is clear from Fig. 3.19 that the forces on the two rods are identical only if Y is zero; otherwise the symmetry is destroyed.

EXAMPLE (ii)

Two uniform bars AB, AC *of equal length and weight* W *and* W' *hang in a vertical plane from two hinges* B *and* C *at the same level, the bars being smoothly jointed at* A. *Prove that the horizontal component of the reaction at* A *is* $\frac{1}{4}(W+W')a/h$, *where* 2a *is the distance* BC, *and* h *is the depth of* A *below* BC. *Find also the vertical component of the reaction.* (I.A.)

Let D, E (Fig. 3.20) be the mid-points of the rods, and X, Y the horizontal and vertical components of the action of the hinge at A on the rods.

These must be in opposite directions on the two rods, but it does not matter whether we show them as in the figure or in reversed directions. If we take them in the opposite directions to those in which they really act we shall simply obtain *negative* values for them.

Let the angle ABC $= \alpha$, then the angle ACB $= \alpha$ also. Let l be the length of each rod.

Taking moments about B for AB, we get:

$$X \, l \sin \alpha + Yl \cos \alpha = W \frac{l}{2} \cos \alpha$$

$$\therefore \qquad X \tan \alpha + Y = \tfrac{1}{2} W. \qquad\qquad \text{(i)}$$

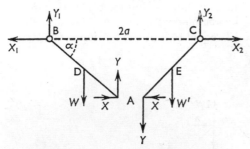

FIG. 3.20

Taking moments about C for AC, we get:

$$Xl \sin \alpha - Yl \cos \alpha = W' \frac{l}{2} \cos \alpha$$

$$\therefore \qquad X \tan \alpha - Y = \tfrac{1}{2} W'. \qquad\qquad \text{(ii)}$$

Adding (i) and (ii),

$$2X \tan \alpha = \tfrac{1}{2}(W + W')$$

$$\therefore \qquad X = \tfrac{1}{4}(W + W') \cot \alpha$$

$$\therefore \qquad X = \frac{(W + W')a}{4h}, \text{ since } \cot \alpha = \frac{a}{h}.$$

Subtracting (ii) from (i)

$$2Y = \tfrac{1}{2}(W - W') \text{ or } Y = \tfrac{1}{4}(W - W').$$

Note. The other four equations obtained by resolving horizontally and vertically for the rod AB and for the rod BC would enable us to find X_1, Y_1 and X_2, Y_2, the components of the reactions on the rods at the hinges B and C.

EXAMPLE (iii)

A square figure ABCD is formed of four equal heavy uniform rods jointed together, and the system is suspended from the joint A, and kept in the form of a square by a string connecting the joints at A and C. Find the tension in the string, and the magnitude and direction of the action at either of the joints B or D.

Draw the diagram as in Fig. 3.21, leaving gaps at A, B, C, and D. All the forces acting on AD and CD are shown; identical sets of forces act on AB and BC from symmetry.

Let X, Y be the horizontal and vertical components of the action of the hinge at D on AD. The reaction on CD will consist of components equal to those on AD but in opposite directions.

The forces acting on AD are the forces X, Y at D, its weight W and certain forces at A. Since the supporting force R acting on the whole system at A is $4W$, by symmetry $2W$ acts on AD and $2W$ acts on AB. Similarly, a force $\frac{1}{2}T$, equal to half the tension in the string,

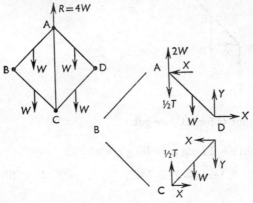

Fig. 3.21

acts downwards on AD at A. In addition, there is the reaction at the hinge A; the vertical component of this is zero from symmetry, and the horizontal component must equal X in the direction shown, since the only other horizontal force acting on AD is X at D.

Resolving vertically,

$$2W + Y = W + \tfrac{1}{2}T \qquad \text{(i)}$$

and taking moments about A, if l is the length of the rod, we get

$$Xl \cos 45° + Yl \sin 45° = W(\tfrac{1}{2}l \sin 45°)$$

$$\therefore \qquad X + Y = \tfrac{1}{2}W. \qquad \text{(ii)}$$

The forces acting on CD are the forces X, Y at D as shown, its weight W, the tension $\frac{1}{2}T$ upwards as explained above, and the reaction at the hinge C. The vertical component of this reaction is zero from symmetry, and its horizontal component must equal X in the direction shown, since the only other horizontal force acting on CD is X at D.

Resolving vertically

$$\tfrac{1}{2}T = W + Y \qquad \text{(iii)}$$

and taking moments about C we get

$$Xl \cos 45° - Yl \sin 45° = W(\tfrac{1}{2}l \cos 45°)$$
$$\therefore \qquad X - Y = \tfrac{1}{2}W. \qquad\qquad\qquad\text{(iv)}$$

From (ii) and (iv) $Y = 0$ and $X = \tfrac{1}{2}W$. (We could not foresee here that $Y = 0$.)

Substituting in (i) or (iii), $T = 2W$.

EXAMPLE (iv)

Four equal uniform rods are freely jointed to form a rhombus ABCD. *The rhombus is suspended from the joint* A, *and is maintained in the form of a square by means of a rod of negligible weight joining the midpoints of* AC *and* CD. *Find the vertical and horizontal components of the reactions at the joints* B *and* D.

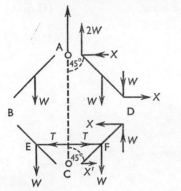

FIG. 3.22

Since EF has no weight, the forces exerted on it by the rods at E and F must be along its length and equal and opposite (Fig. 3.22). Also since the figure would tend to collapse so that B and D moved inwards, it is evident that there is a thrust T in EF.

Since the supporting force on the whole system at A is $4W$, by symmetry $2W$ acts on AD and $2W$ acts on AB.

The vertical component of the reaction on AD at the joint A is zero from symmetry; the horizontal component is denoted by X as shown. At the other joint D the horizontal component of the reaction on AD must also be X in the direction shown, and the vertical component must be W downwards (resolving vertically).

Taking moments about A for the rod AD, if l is the length of AD we get

$$Xl \cos 45° = Wl \sin 45° + W(\tfrac{1}{2}l \sin 45°)$$
$$\therefore \qquad X = \tfrac{3}{2}W.$$

The vertical and horizontal components of the reaction at D are therefore W and $\frac{3}{2}W$, and by symmetry they will be the same at B.

If we were asked for the thrust in EF we could find it by taking moments about C for CD; this gives

$$T\frac{l}{2}\cos 45^\circ + W\frac{l}{2}\sin 45^\circ = \frac{3}{2}Wl\cos 45^\circ + Wl\sin 45^\circ$$

$$\therefore \qquad\qquad \tfrac{1}{2}T + \tfrac{1}{2}W = \tfrac{3}{2}W + W$$

$$\therefore \qquad\qquad\qquad T = 4W.$$

EXAMPLE (v)

Two equal heavy beams AB, AC are smoothly jointed at A, and B is joined by a string to the mid-point of AC; the beams rest with B and C on a smooth horizontal plane; if the angle BAC = 60°, find the tension in the string in terms of the weight of a beam.   (H.S.D.)

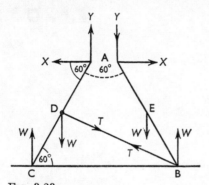

Fig. 3.23

Let D (Fig. 3.23) be the mid-point of AC, E that of AB.

Since AB = AC and the angle BAC = 60°, the triangle ABC is equilateral. Hence BD is perpendicular to AC.

Since the beams are of equal weight W and the lines of action of the weights are equidistant from B and C, the vertical reactions at B and C are each equal to W.

(This can also be shown by taking moments about C and B for *both beams*.)

Let T be the tension in the string, and l the length of each beam.

Taking moments about A for AC,

$$T \times \frac{l}{2} + W \times \frac{l}{4} = W \times \frac{l}{2}$$

$$\therefore \qquad\qquad \frac{T}{2} = \frac{W}{4} \text{ or } T = \frac{W}{2}.$$

If we required the reaction at A it could be obtained as follows:

Let X, Y be the horizontal and vertical components of the action of the hinge at A on AC.

Resolving horizontally for AC,

$$X = T \cos 30° = \frac{\sqrt{3}}{4} W.$$

Resolving vertically for AC,

$$Y + W = W + T \sin 30°$$

$$\therefore \qquad Y = \tfrac{1}{2}T = \tfrac{1}{4}W.$$

The resultant reaction R at A is given by

$$R = \sqrt{(X^2 + Y^2)} = W\sqrt{(\tfrac{3}{16} + \tfrac{1}{16})} = \tfrac{1}{2}W.$$

It is inclined to the horizontal at an angle

$$\tan^{-1}\frac{Y}{X} = \tan^{-1}\frac{1}{\sqrt{3}}$$

i.e. at an angle of 30°.

EXAMPLE (vi)

Six equal weightless rods hinged together at their ends form a regular hexagon $ABCDEF$. The rod AB is held horizontally and a weight W is hung from the mid-point of DE, the hexagonal shape being maintained by a light rod CF. *Find the stress in the rod* CF.

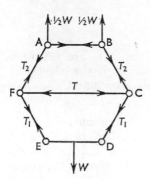

FIG. 3.24

Let ABCDEF (Fig. 3.24) represent the framework.

Since the rods are light and (except ED) acted on by forces at their ends only, the stresses in all the rods except ED must be along the length of the rods.

Let T_1 be the tension in CD and EF, and T_2 that in BC and FA. The rod ED is in equilibrium subject to its weight W and forces T_1 at E along EF and T_1 at D along DC. The vertical components of the tensions T_1 must support the weight W, and hence

$$2T_1 \cos 30° = W,$$

$$\therefore \qquad T_1 = \frac{W}{\sqrt{3}}.$$

The vertical stress at A and at B is evidently $\frac{1}{2}W$. Hence resolving vertically for the hinge B,

$$T_2 \cos 30° = \tfrac{1}{2}W$$

$$\therefore \qquad T_2 = \frac{W}{\sqrt{3}}.$$

If T is the thrust in CF, resolving vertically and horizontally for the hinge C, we get

$$T_1 \sin 60° = T_2 \sin 60° \text{ or } T_1 = T_2$$

and $\qquad T = T_1 \cos 60° + T_2 \cos 60°$

$$= \frac{W}{2\sqrt{3}} + \frac{W}{2\sqrt{3}} = \frac{W}{\sqrt{3}}$$

$$\therefore \qquad T = W/\sqrt{3}.$$

EXAMPLES 3.3

1. Three equal uniform rods, each of weight W, are smoothly jointed so as to form an equilateral triangle. If the triangle be supported at the mid-point of one of its sides, find the actions at the joints.

2. A square ABCD is formed by four equal uniform rods, freely jointed together, and the system is supported at the lower joint C, and kept in shape by a light rod joining C and A. Find the thrust in the rod, and the magnitude and direction of the action at either of the joints B or D.

3. A rhombus ABCD is formed by four equal uniform rods freely jointed together, and the system is suspended from the joint A, and kept in shape, with the angle CAD = 30°, by a string connecting A and C. Find the tension in the string and the magnitude and direction of the reaction at B or D.

4. AB and BC are two uniform exactly similar rods, each of weight W, freely hinged together at B, and carrying small rings of negligible weight which enable the ends A and C to move without friction on a fixed horizontal wire. The rods are placed so as to include a right angle, with the joint B below the wire, and are prevented from closing up by means of a rigid stay of negligible weight joining the mid-points of the rods. Find the stress in this stay and the reactions at A, B, and C.   (H.C.)

5. AB and BC are two rods of equal length freely jointed at B, the weight of AB is W and that of BC is $2W$. They are placed in a vertical plane inclined to one another at 90° with the ends A and C on a horizontal plane. What horizontal forces must be applied at A and C to maintain equilibrium?   (I.E.)

6. Two rods AB and BC, a m and b m long, of the same material and cross-section, are freely jointed at B and hang with their ends A and C attached to two points at the same level, and at such a distance apart that ABC is a right angle. If the material of the rods has a mass of M kg per m, find the reactions at the joint B, and at the points of attachment A and C.   (I.E.)

7. Two uniform ladders, each of length a and weight W, are hinged at their upper ends, and stand on a smooth horizontal plane. A weight W is hung from a rung of one of the ladders at a distance b from its lower end, and the ladders are prevented from slipping by means of a rope of length $2c$ attached to their lower ends. Find the pressure of each ladder on the ground, and the tension in the rope.

8. Two uniform rods AB, BC of the same material and thickness, but of different lengths, are freely jointed at B, and the ends A and C are fixed in the same vertical line. Show that the stress at the joint B acts along BD, the bisector of the angle ABC, and that its magnitude is

$$\tfrac{1}{2}W \frac{BD}{AC}$$

where W is the weight of the two rods.

9. A step-ladder of weight $2W$ consists of two equal parts, jointed at the top, and held together by a rope half-way between the top and bottom, so that when the rope is tight the angle between the two halves of the ladder is $2 \tan^{-1} \tfrac{6}{13}$. A man of weight $5W$ mounts the ladder and then stops two-thirds of the way up. Neglecting the friction between the ladder and the ground, find the tension in the rope and the reaction at the hinge.   (I.S.)

10. Two uniform boards rest against smooth parallel walls, their lower ends being in contact on a smooth horizontal floor. Prove that if the weights of the boards are w, w', and their inclinations to the vertical θ, θ', the condition for equilibrium is

$$w \tan \theta = w' \tan \theta'. \qquad \text{(I.A.)}$$

11. Two heavy plane rectangular areas, of the same lengths, but of different widths, are hinged together along their equal sides and placed on a smooth horizontal plane with the hinge horizontal and uppermost. They are kept from sliding by a string attached to the mid-points of their lower edges. If their weights are W_1, W_2, and their inclinations to the plane are θ and ϕ respectively, prove that the action between them at the hinge makes an angle with the horizontal whose tangent is

$$\frac{W_1 \tan \phi - W_2 \tan \theta}{W_1 + W_2}. \qquad \text{(H.S.D.)}$$

12. AB, BC, CD, DE, EA are five equal uniform rods, each of weight W, freely jointed at their extremities and suspended from the joint A in the form of a regular pentagon, this configuration being maintained by light strings joining A to C and to D. Find the reactions at B and E, and show that the tension in either string is $2W \cos 18°$. (H.S.D.)

13. Three equal rods AB, BC, CD and a rod AD of double their length are freely hinged together at A, B, C, D, and the framework is suspended from the mid-point of BC. If w is the weight of each of the equal rods and $2w$ that of the longest rod, find the magnitudes of the forces on the hinges. Show that the line of action of the forces at A and B meet at a depth $BC/\sqrt{3}$ below BC. (H.S.C.)

14. A frame is formed of three uniform rods, BC, CA, AB of lengths 1·5, 2, 2·5 m and of mass 1 kg per m, smoothly joined by weightless pins at their ends. It is suspended from such a point D in the longest side AB that it rests in equilibrium with AB horizontal. Find the distance of D from the mid-point of AB and the stresses at the joints. (Ex.)

15. AC and BC are two light rods freely jointed at C, and freely jointed to a wall at A and B so that AC is horizontal and the angle ACB $= \alpha°$, and the point B is vertically below A. A weight W is suspended from C. Find the tension in AC and the thrust in BC. Prove that the tension in AC, if also a weight W' be suspended from the mid-point of BC, is $\frac{1}{2}(W' + 2W) \cot \alpha$. (H.S.D.)

16. A disc of radius a and weight W rests, as shown in Fig. 3.25, between two light rods which are smoothly hinged together at O with their ends A, B resting on a smooth table, and are maintained in equilibrium by a string AB; OA = OB = c, $\angle$AOB = 2α; the whole figure is in a vertical plane. Find the tension in the string.   (H.S.C.)

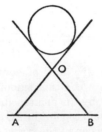

Fig. 3.25

17. BA, AC are heavy uniform rods, of mass 6 kg and 3 kg respectively, smoothly hinged to each other at A and to a light rod BC. The hinges at B and C are supported by vertical strings so that BC is horizontal with A below BC; the perpendicular from A on BC is of length 0·9 m, and its foot divides BC so that BD is 0·6 m and DC is 1·2 m. Prove that the reaction at A is horizontal; find the reaction and also the thrust in BC.   (H.S.D.)

18. Two uniform rods AB, BC, alike in all respects and each of weight W, are rigidly jointed at B so that ABC is a right angle, and the end A is hinged freely to a fixed point from which the two rods hang in equilibrium. Show that AB makes an angle $\tan^{-1}\frac{1}{3}$ with the vertical. If the rods are freely jointed at B, but A and C are connected by a light inextensible string of such length that ABC is a right angle, show that the tension in the string is $3W/2\sqrt{5}$.   (H.C.)

19. AB and BC are two uniform rods of weights W and W' respectively. They are freely hinged together at B and the end A is freely pivoted to a fixed point A, while the end C is constrained to move on a fixed horizontal wire, passing through A, by means of a small, smooth ring of negligible mass. Show that the horizontal force which must be applied at C to keep the rods in the position in which the angles CAB and ACB are θ and ϕ and B is below AC, is

$$\tfrac{1}{2}(W + W')\cos\phi\cos\theta\operatorname{cosec}(\theta+\phi).$$ (H.C.)

20. A straight rod ABC of negligible weight is horizontal, and is pivoted at a smooth hinge A. A load of mass 100 kg hangs from C.

The rod is supported at its mid-point B by a uniform rigid rod BD of negligible weight, D being a smooth hinge fixed vertically below A, with $AD = AB$. Find the horizontal and vertical components of the stresses at D and B.

21. Two uniform rods AB, BC, of equal lengths but of different weights, are freely jointed at B, and jointed at A and C to two fixed points in the same horizontal line, at such a distance apart that ABC is a right angle. Show that the tangent of the angle which the direction of the reaction at B makes with the rod BA is the ratio of the weight of AB to that of BC. (I.S.)

22. Two uniform rods AB, AC, of weight W_1, W_2, and of equal length, are smoothly hinged at A, and rest with B, C on a smooth horizontal plane, being kept in equilibrium by an inextensible string joining BC. A weight w is suspended from a point in AC at a distance of $\frac{3}{4}AC$ from A. Prove that the tension of the string is

$$\tfrac{1}{4}(W_1 + W_2 + \tfrac{1}{2}w)\tan \tfrac{1}{2}A. \qquad \text{(I.S.)}$$

23. Each half of a step-ladder is 1·65 m long, and the two parts are connected by a cord 70 cm long attached to points in them distant 40 cm from their free extremities. The half with the steps has a mass of 8 kg and the other half a mass of 2 kg. Find the tension in the cord when a man of mass 77 kg is standing on the ladder 45 cm from the top, it being assumed that the cord is fully stretched and that the reactions between the ladder and ground are vertical. (I.S.)

24. Two uniform bars AB, BC are smoothly jointed at B and the end C is pivoted to a point on a wall by a smooth pivot. The bar AB is 1·5 m long and has a mass of 10 kg, the bar BC is 1·2 m long and has a mass of 5 kg. The two bars are kept in a horizontal line by a prop placed under AB. Find the position of the prop and the reaction at C. (N.U.3)

25. Two equal uniform rods AB and AC, each of length 1·2 m and mass 1·5 kg, are smoothly jointed at A. They are placed symmetrically on a smooth fixed sphere of radius 0·3 m, so that A is vertically above the centre. Show that the angle between the rods when in equilibrium is a right angle, and find the magnitude of the reaction at the joint. (N.U. 3 and 4)

26. A uniform rod, of length $2a$ and mass m, is pivoted at its lower end to a fixed point O. A light string fastened at one end to a fixed point at a distance $2a$ vertically above O passes over a small smooth groove in the free upper end of the rod and supports a mass M at its other end. Find the inclination of the upper part of the string to the vertical in the position of equilibrium. (N.U.3)

27. A circular cylinder is maintained in equilibrium, with its axis horizontal and touching along its length an inclined plane, by means of a rod AB of weight equal to that of the cylinder hinged to the plane below the cylinder at A and being a tangent to the cylinder at its upper end B. The vertical plane through the rod intersects the inclined plane in a line of greatest slope, and contains the centre of gravity of the cylinder. If all the surfaces are smooth prove that

$$\tan \alpha = \frac{\sin 2\theta}{5 - \cos 2\theta}$$

where α is the inclination of the plane to the horizontal and θ is the angle which the rod AB makes with the plane.   (H.S.D.)

28. Two uniform rods AB and CD, each of weight W and length a, are smoothly jointed together at O, where $OB = OD = b$. The rods rest in a vertical plane with the ends A and C on a smooth table, and the ends B and D connected by a light string. Prove that the reaction at the joint is $(aW/2b) \tan \alpha$, where α is the inclination of either rod to the vertical.   (C.W.B.)

29. Two equal rods OA, OB, each 5 m long, and of mass 5 kg, are connected at O, and their other ends are placed on a smooth horizontal plane, A, B, O, being in the same vertical plane. If a string 6 m long connects A and B, find: (i) the reaction at O; (ii) the reactions at A and B; (iii) the tension in the string.

30. A ladder of length l placed vertically against the wall of a house of height h is let down slowly to the ground by a rope from the top of the house, the foot of the ladder always pivoting against the foot of the wall, and the ladder moving in a vertical plane perpendicular to the wall. The rope is tied to the top of the ladder. Prove that the initial tension of the rope just as the top of the ladder leaves the wall is $W(h-l)/2h$.   (H.S.C.)

3.14. Resultant of coplanar forces

In Chapter 1 we showed how to find the resultant of any number of forces acting on a particle. We shall now consider the more general problem that arises when the forces act on a rigid body.

Usually we shall require to find the resultant of a number of forces whose magnitudes and lines of action are given.

Often we shall not be told on what the forces are acting, for as already mentioned, their resultant is independent of the shape and size of the body, provided it is rigid.

The magnitude of the resultant is, as a rule, best obtained by resolving the forces in two directions at right angles, adding the components in these directions, and compounding the two components so obtained into a single force.

The ratio of these two components gives the direction of the resultant, i.e. the tangent of the angle made by it with one of the directions in which the forces were resolved.

To fix the position of the line of action we can either find one point on it, and give this and the direction, or in some cases it is easier to find the points in which the resultant cuts two given lines; the latter are often conveniently obtained by taking moments. No general method should be laid down; the shortest method of obtaining the required result differs considerably in different cases.

Various methods are used in the following examples.

EXAMPLE (i)

Find the magnitude, direction, and line of action of the resultant of three forces 1, 2, 3 units acting in order round the three sides of an equilateral triangle of side 2a.

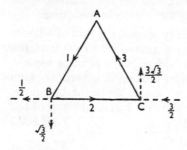

FIG. 3.26

Let the forces 1, 2, 3 act along the sides AB, BC, CA of the equilateral triangle ABC, as shown in Fig. 3.26.

Resolve the forces along and perpendicular to BC.

The force 1 is equivalent to $\frac{1}{2}$ acting at B in the direction CB, and $\sqrt{3}/2$ acting at B perpendicular to BC and downwards.

The force 3 is equivalent to $\frac{3}{2}$ acting at C in the direction CB, and $3\sqrt{3}/2$ acting at C perpendicular to BC and upwards.

The components along BC balance the force 2, and we are left with unlike parallel forces $\sqrt{3}/2$ at B and $3\sqrt{3}/2$ at C.

Their resultant is $\sqrt{3}$ acting upwards perpendicular to BC, at a point D in BC produced such that

$$\frac{3\sqrt{3}}{2}\text{CD} = \frac{\sqrt{3}}{2}\text{BD}$$

or $\qquad\qquad\qquad 3\text{CD} = \text{BD}$

or $\qquad\qquad\qquad \text{CD} = \tfrac{1}{2}\text{BC} = a.$

EXAMPLE (ii)

ABCD *is a given quadrilateral; forces are represented in magnitude, lines of action, and senses by the sides* AB, BC, DC *(cyclical order interrupted), and* DA. *What are the magnitude and line of action of the resultant?* (I.S.)

The resultant of the forces AB, BC (Fig. 3.27) is equal and parallel to AC and acts at B.

The resultant of this force and the force DA is equal and parallel to DC, and acts at E, where a parallel to CA through B meets DA produced.

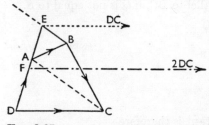

Fig. 3.27

We now have two parallel forces, each equal to DC, one acting along DC and one parallel to DC at E.

Their resultant is a force 2DC acting at F the mid-point of DE.

Alternatively, we may proceed as follows:

The magnitude and direction of the resultant of the four forces is given by their vector sum, viz.

$$\textbf{AB}+\textbf{BC}+\textbf{DC}+\textbf{DA} = \textbf{DC}+(\textbf{DA}+\textbf{AB}+\textbf{BC})$$
$$= \textbf{DC}+\textbf{DC}$$
$$= 2\textbf{DC}.$$

The resultant is therefore a force 2**DC** and its line of action may be found by taking moments about any point, say D.

Suppose the resultant 2**DC** cuts DA at F. Its moment about D is clockwise and is represented by $4\triangle$FDC. But the moment of the

four given forces about D is also clockwise and is represented by
$2\triangle ABD + 2\triangle BCD$.

$$\therefore \qquad 4\triangle FDC = 2(\triangle ABD + \triangle BCD)$$
$$= 2(\text{quad. } ABCD)$$
$$= 2\triangle EDC, \text{ since } BE \text{ is parallel to } CA.$$
$$\therefore \qquad 2FD = ED,$$

that is, F is the mid-point of DE.

EXAMPLE (iii)

*Four forces P, Q, R, and S act along the sides of a rectangle ABCD in
the directions AB, BC, CD, DA respectively. If AB = a and AD = b,
find the magnitude of the resultant of the system of forces and the distance
from A of the points in which the line of action of the resultant cuts the
sides AB, AD.*

Draw the rectangle ABCD and insert the forces as in Fig. 3.28.
The forces P and R are equivalent to a force $P-R$ parallel to AB,
provided P is not equal to R, and similarly the forces Q and S are
equivalent to a force $Q-S$ parallel to BC, if Q is not equal to S.

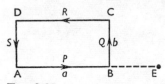

Fig. 3.28

The magnitude of the resultant is therefore

$$\sqrt{[(P-R)^2 + (Q-S)^2]}$$

If the resultant cuts AB produced at a point E distant x from A,
then since E is on the line of action of the resultant, the sum of the
moments of the forces about it is zero,

$$\therefore \qquad Q(x-a) = Rb + Sx$$
$$\therefore \qquad x(Q-S) = Rb + Qa$$
$$\therefore \qquad x = \frac{Rb + Qa}{Q-S}.$$

Similarly, if the resultant cuts DA produced at a point F distant
y from A, taking moments about F, we get

$$Py = Qa + R(b+y)$$
$$\therefore \qquad y(P-R) = Qa + Rb.$$
$$\therefore \qquad y = \frac{Qa + Rb}{P-R}.$$

Note.

If P is greater than R (as assumed above), F will be below A.

If R is greater than P the above value of y will be negative and F will be above A.

If S is greater than Q, E will be to the left of A instead of to the right, as shown in Fig. 3.28.

If $R = P$ and $S = Q$, the system reduces to a couple of moment $Pb+Qa$.

EXAMPLE (iv)

ABCDEF *is a regular hexagon; forces* P, 2P, 3P, 5P, 6P *act along* AB, BC, DC, EF, AF *respectively; show that a force can be determined to act along* ED *so that the six forces are equivalent to a couple, and find the moment of the couple.* (I.S.)

Draw the hexagon and insert the forces as in Fig. 3.29.

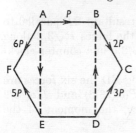

FIG. 3.29

Method (i)

A force along ED will form a couple with the given forces provided these forces have a resultant parallel to DE.

Now, sum of the components of the forces perpendicular to $ED = 5P \cos 30° - 6P \cos 30° - 2P \cos 30° + 3P \cos 30° = 0$.

Also, sum of the compounds of the forces parallel to $ED = -5P \sin 30° - 6P \sin 30° + 2P \sin 30° + 3P \sin 30° = -2P$.

Hence the resultant of the five forces is a force of magnitude $2P$ parallel to DE. Its position can be found by taking moments about some point.

Suppose this resultant is at a perpendicular distance x above the centre of the hexagon, and that the length of each side is a.

Taking moments about the centre, we get:

$$2P \times x = (-5P+6P-P-2P+3P)a\sqrt{3}/2$$

∴ $$x = a\sqrt{3}/4.$$

Hence the perpendicular distance of the resultant from

$$ED = a\sqrt{3}/4 + a\sqrt{3}/2 = 3a\sqrt{3}/4.$$

Hence with a force $2P$ along ED the five forces will form a couple of moment $2P(3a\sqrt{3}/4) = 3\sqrt{3}Pa/2$ in a counter-clockwise direction.

Method (ii)

The force $5P$ is equivalent to a force $5P \cos 30° = 5P\sqrt{3}/2$ along EA and $5P \sin 30° = 5P/2$ along DE.

The force $6P$ is equivalent to a force $6P \cos 30° = 3P\sqrt{3}$ along AE and $6P \sin 30° = 3P$ along BA.

The force $2P$ is equivalent to a force $2P \cos 30° = P\sqrt{3}$ along BD and $2P \sin 30° = P$ along AB.

The force $3P$ is equivalent to a force $3P \cos 30° = 3P\sqrt{3}/2$ along DB and $3P \sin 30° = 3P/2$ along ED.

In all, the five forces are equivalent to $5P/2-3P/2 = P$ along DE, $3P-P-P = P$ along BA, $P\sqrt{3}/2$ along AE and $P\sqrt{3}/2$ along DB.

The forces P along DE and BA have a resultant $2P$ parallel to DE through the centre of the hexagon, and the forces $P\sqrt{3}/2$ along AE and DB form a couple of moment $(P\sqrt{3}/2)a$ in a counter-clockwise direction.

If therefore we introduce a force $2P$ along ED the six forces are equivalent to two couples of moments $2P(a\sqrt{3}/2)$ and $Pa\sqrt{3}/2$ respectively, both counter-clockwise. Hence, the moment of the resultant couple is $3Pa\sqrt{3}/2$.

EXAMPLES 3.4

1. A, B, C are three points on a line ABCD, such that $AB = BC = a$. Forces of 3, 6, and 4 N respectively act at A, B, C in directions making angles of 60°, 120°, and 270° with AD. Show that they reduce to a single force, and find where its line of action cuts AD.   (I.S.)

2. A, B, C, D are four points in a straight line at equal intervals of 0·6 m. At A, B, D forces 2, 3, and 4 N respectively act perpendicular to AD upwards; at C a force of 9 N acts perpendicular to AD downwards. Show that the system is equivalent to a couple, and find where the 3-N force should have been applied to produce equilibrium.   (I.S.)

3. Forces 3, 3, and 5 N act respectively along the sides BA, AC, BC of an equilateral triangle of altitude 1·2 m. Find the distance from A of the line of action of the resultant.   (I.A.)

4. ABCD is a square, E and F are the mid-points of BC and CD respectively. Find the magnitude and direction of the resultant

of the forces, 2, 5, 10, 1 units along AB, AE, FA, AD respectively.

5. ABC is an equilateral triangle, forces of 4, 2, and 1 unit act along the sides AB, AC, BC respectively, in the directions indicated by the letters. Prove that their resultant is a force of $3\sqrt{3}$ units in a direction at right angles to BC, and find the point in which the line of action meets BC. (I.A.)

6. Find the resultant of the following forces acting along the sides of a square ABCD; 21 N along CD, 15 N along DA, 3 N along BA, 9 N along CB; and show that its line of action bisects two of the sides of the square. (I.S.)

7. Find the magnitude of the resultant of the following forces acting along the sides of a square ABCD; 11 N along DA, 7 N along CB, 19 N along CD, 5 N along BA; and prove that its line of action bisects AD and trisects CD. (I.A.)

8. AD is an altitude of triangle ABC in which BC = 6, CA = 7, AB = 5. A force of 12 N along DA is equilibrated by parallel forces at B and C. If the directions of all the forces are rotated about A, B, C respectively through the same angle so as to be perpendicular to AB, prove that the resultant of the forces is a couple and find its moment. (I.S.)

9. ABCD is a trapezium in which the parallel sides AD and BC are in the ratio 2 : 3, and AB = AD = DC; find the magnitude and position of the resultant of forces, $3P$ from B towards C, P from B towards A, $2P$ from D towards A, and $2\frac{1}{2}P$ from D towards C. (I.E.)

10. A unit force acts along the side AB of a square ABCD. Find the magnitudes and directions of the forces which must act along the remaining three sides in order to maintain equilibrium. Find the resultant if: (i) the force along BC is reversed; (ii) the forces along BC and AD are both reversed. (H.S.D.)

11. ABCD is a rectangular board in which AB = DC = 0·9 m, BC = AD = 1·8 m. One rope pulls it along AD with a force of 15 units, another rope pulls it along BC with a force of 25 units, a third rope pulls it along CD with a force of 16 units. Find what force along AB will make the resultant pass through G, the mid-point of the rectangle ABCD, and calculate the resultant completely. (I.S.)

12. ABCD is a square, of side a, traced on a lamina; E, F are points on BA and BC produced through A and C respectively, so that BE = $3a$, BF = $3a$. A system of forces acting on the lamina consists of P along AB, $2P$ along BC, $3P$ along CD, $4P$ along DA,

and $2\sqrt{2}P$ along EF. Prove that the resultant of the system is a couple of moment Pa.   (H.S.D.)

13. Forces 1, 2, 3, 4 N respectively act along the sides AB, BC, CD, DA of a square of side 1 m. Find the distance from the centre of the square of the line of action of the resultant. What additional force along the diagonal BD will make the whole system have a resultant passing through A?   (H.S.D.)

14. A rectangle ABCD can turn freely in a horizontal plane about its centre, which is fixed, and forces of 1, 2, and 3 N act along AB, BC, and DC respectively. If AB = 0·3 m, BC = 0·6 m, determine the force which must be applied along AD to keep the rectangle at rest.

15. A square lamina ABCD on a smooth horizontal table can turn about a smooth peg fixed at the point of trisection of the diagonal BD, which is nearer to B. It is acted on by a force of 5 units along AD and a force of 3 units along CB. Find the force which, acting at A in the direction AB, will keep it in equilibrium and the resultant pressure on the peg.   (H.S.C.)

16. Forces $3P$, $2P$, P, $2P$ act along the sides AB, CB, CD, AD of a square ABCD in the directions indicated by the order of the letters. Find the magnitude and the line of action of the resultant.   (H.C.)

17. ABCD is a square of side $2a$; P is the mid-point of AD and Q that of DC; and the following forces act: 20 from A to B, 20 from C to D, 40 from A to Q, 30 from P to B. Show that the resultant is a force of 50, and that the length of the perpendicular from A on its line of action is about $0·26a$. (Represent the resultant in a figure.)   (I.S.)

18. Four forces in equilibrium act along the sides of a quadrilateral ABCD. Show that the resultant of the forces acting along AB and BC acts along BD, and find the ratios of all the forces, when each of the angles ABC and BCD is 50°, and each of the angles BDC and BAC is a right angle.   (I.S.)

19. ABCD is a rectangle in which BC = nAB. Prove that the system of forces, P, nP, P, nP acting along AB, CB, CD, AD would maintain equilibrium. Find, when BC = 3AB, the magnitude and position of the resultant of forces 1, 3, 1, 5 N, acting along the sides AB, BC, CD, and DA of the rectangle.   (I.E.)

20. Forces $3P$, $4P$, P, $5P$ act along the sides AB, CB, CD, DA of a square lamina ABCD of side 12 cm in the order indicated. Prove that the resultant of the forces meets AD produced at a point

distant 6 cm from D; and find the distance from B at which the resultant meets AB. (N.U.3.)

21. Four forces P, P, Q, Q act along the sides AB, BC, CD, DA of a rhombus; find the sum of their moments about the centre O of the rhombus. Prove that their resultant is at a distance $\frac{1}{2}$ BD $(P+Q)/(P-Q)$ from O. Discuss the case when $P = Q$.
(C.W.B.)

22. A force P is represented in magnitude, direction and position by **AB,** where A and B are the points $(-2, 1)$ and $(2, 4)$ respectively. Another force Q compounded with P gives rise to a resultant R, of magnitude 3 units, acting along the line $8x+6y = 15$, Show that Q is one of two possible forces each of magnitude $\sqrt{34}$ units.

Find the equations of the lines of action of these two possible forces.

23. Forces of magnitude 4, 5, 6 and 7 N act along the sides AB, BC, CD, and DA respectively of a 30 cm square; and a force of 8 N acts from B to the mid-point of DC. Find the magnitude of the resultant and the distance of its line of action from A. (Q.E.)

24. Forces of magnitude F, $2F$, $3F$, $4F$, $5F$, $6F$ act along the sides of a regular hexagon, taken in order. Show that they are equivalent to a single force $6F$ acting parallel to one of the given forces, the distance of the line of action of that force and of the resultant from the centre of the hexagon being in the ratio $2:7$. (I.E.)

3.15. *The resultant of two forces, acting at a point* O *in directions* OA *and* OB *and represented in magnitude by* $l \times$ OA *and* $m \times$ OB *is represented in magnitude and direction by* $(l+m)$OC, *where* C *is a point in* AB *such that* $l \times$ CA $= m \times$ CB.

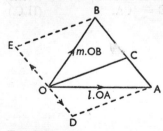

FIG. 3.30

For let C (Fig. 3.30) divide AB so that
$$l \times CA = m \times CB.$$
Complete the parallelograms OCAD and OCBE.

The force $l \times$ OA is equivalent to forces represented by $l \times$ OC and $l \times$ OD.

The force $m \times$ OB is equivalent to forces represented by $m \times$ OC and $m \times$ OE.

Hence the forces $l \times$ OA and $m \times$ OB are equivalent to a force represented by $(l+m)$OC, and forces represented by $l \times$ OD and $m \times$ OE.

But OD = CA and OE = CB and $l \times$ CA $= m \times$ CB.

Therefore these two latter forces are equal and opposite and are therefore in equilibrium.

Hence the resultant of $l \times$ OA and $m \times$ OB is represented by $(l+m)$OC.

In vector notation the resultant of the two forces is given by their vector-sum. We write

$$l\mathbf{OA}+m\mathbf{OB} = l(\mathbf{OC}+\mathbf{CA})+m(\mathbf{OC}+\mathbf{CB})$$
$$= (l+m)\mathbf{OC}+l\mathbf{CA}+m\mathbf{CB}$$
$$= (l+m)\mathbf{OC}$$

provided $\quad\quad l \times$ CA $= m \times$ CB.

If $l = m = 1$, we get $\mathbf{OA}+\mathbf{OB} = 2\mathbf{OC}$ provided CA = CB, that is, the resultant of forces OA, OB is a force 2OC, where C is the mid-point of AB. This is also obvious from the fact that in this case OC is half the diagonal of the parallelogram, of which OA and OB are adjacent sides.

EXAMPLE (i)

Forces represented by 2BC, CA, BA *act along the sides of a triangle* ABC. *Show that their resultant is* represented by 6DE, *where* D *bisects* BC *and* E *is a point on* CA *such that* CE $= \frac{1}{3}$CA.   (H.C.)

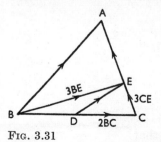

FIG. 3.31

In Fig. 3.31 the resultant of forces 2BC and BA is a force 3BE, where E is a point in AC such that 2CE = EA, i.e. CE $= \frac{1}{3}$CA.

Also CA = 3CE, and the resultant of forces 3BE along BE and 3CE along CA is a force 6DE along DE, where D is the mid-point of BC.

Alternatively, we may proceed as follows:

The vector sum of the three forces is

$$2\mathbf{BC}+\mathbf{CA}+\mathbf{BA} = \mathbf{CA}+(2\mathbf{BC}+\mathbf{BA})$$
$$= \mathbf{CA}+3\mathbf{BE} \text{ if } 2\mathbf{CE} = \mathbf{EA},$$
$$= 3\mathbf{CE}+3\mathbf{BE}$$
$$= 6\mathbf{DE} \text{ if } \mathbf{BD} = \mathbf{DC}.$$

Since the sum of the moments of the three given forces about D is zero, the resultant must pass through D and is therefore represented in magnitude, direction, and line of action by **6DE**.

EXAMPLE (ii)

ABCDEF is a regular hexagon and O is any point. Prove that the resultant of forces represented by OA, OB, OC, OD, OE, OF *is a force* 6OP, *where P is the centre of the circumcircle of the hexagon.*

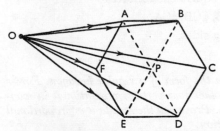

FIG. 3.32

In Fig. 3.32 P is the mid-point of AD, BE, and CF. Let O be any point and join OP.

The resultant of forces represented by OA and OD is a force 2OP.

The resultant of forces represented by OB and OE is a force 2OP.

The resultant of forces represented by OC and OF is a force 2OP.

Therefore the resultant of forces represented by OA, OB, OC, OD, OE, OF is a force represented by 6OP.

EXAMPLE (iii)

M *is the point of trisection of the side* AC *of a triangle* ABC *which is nearer to* A, *and* N *is the point of trisection of the side* AD *which is*

nearer to B. *Resolve a force, represented in magnitude and direction by*
MN, *into three forces acting each along a side of the triangle.*   (I.S.)

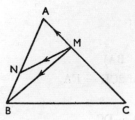

FIG. 3.33

Join MB (Fig. 2.33).

Since N divides AB in the ratio 2 : 1, the force MN is equivalent
to forces $\frac{2}{3}$MB and $\frac{1}{3}$MA acting at M.

The latter is along CA and equal to $\frac{1}{9}$CA.

Since M divides AC in the ratio 1 : 2, a force MB is equivalent to
forces $\frac{2}{3}$AB and $\frac{1}{3}$CB acting at B.

Hence a force $\frac{2}{3}$MB is equivalent to forces $\frac{4}{9}$AB and $\frac{2}{9}$CB acting
at B.

The force MN is therefore equivalent to forces

$$\tfrac{4}{9}AB, \ \tfrac{2}{9}CB, \ \tfrac{1}{9}CA,$$

acting along the corresponding sides.

EXAMPLE (iv)

ABCDEF *is a plane lamina in the form of a regular hexagon. Forces
act from* A *and* B *towards the other four vertices proportional in magnitude to the distance from them. Prove that the resultant is proportional
to* 6AE, *and find the line of action.*

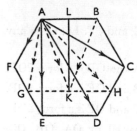

FIG. 3.34

The resultant of forces proportional to AF and AE (Fig. 3.34) is
a force proportional to 2AG, where G is the mid-point of EF.

Similarly, the resultant of forces proportional to AC and AD is
a force proportional to 2AH, where H is the mid-point of CD.

The resultant of forces proportional to 2AG and 2AH is a force proportional to 4AK, where K is the mid-point of GH.

Similarly, the resultant of the forces along BC, BD, BE, BF is a force represented by 4BK.

Finally, the resultant of forces 4AK and 4BK is represented by 8LK, where L is the mid-point of AB, and it is clear that LK = $\frac{3}{4}$AE.

Hence the resultant of the eight forces is a force proportional to 6AE and acts along LK.

Vectorially we write

$$\mathbf{AF+AE+AD+AC} = 2\mathbf{AG}+2\mathbf{AH}$$
$$= 4\mathbf{AK}$$

and the resultant passes through A.

Also

$$\mathbf{BF+BE+BD+BC} = 2\mathbf{BG}+2\mathbf{BH}$$
$$= 4\mathbf{BK}$$

and the resultant passes through B.

Hence the eight forces are equivalent to a force 4AK along AK and a force 4BK along BK, which in turn are equivalent to a force 8LK acting along LK.

EXAMPLES 3.5

1. The sides BC and DA of a quadrilateral ABCD are bisected in F and H respectively. Show that if two forces parallel and equal to AB and DC act at a point their resultant is parallel to HF and equal to 2HF.

2. ABC is a triangle and G the point of intersection of its medians; if O is any point in the plane of the triangle, prove that the resultant of forces represented by OA, OB, OC is represented by 3OG.

3. If O is the circum-centre, and H the ortho-centre of a triangle ABC, prove that the resultant of forces represented by HA, HB, HC will be represented in magnitude and direction by 2HO.

4. ABCD is a quadrilateral, of which A and C are opposite vertices. Two forces acting at A are represented in magnitudes and directions by AB and AD; and two forces acting at C are represented in magnitudes and directions by CB and CD. Show that the resultant force is represented in magnitude and direction by four times the line joining the mid-points of the diagonals of the quadrilateral.

5. ABCD is a quadrilateral; forces are completely represented by the lines AB, BC, AD, DC. Prove that their resultant is represented in magnitude and direction by 2AC, and that its line of action bisects BD.

6. O is any point in the plane of a triangle ABC, and D, E, F are the mid-points of the sides. Show that the system of forces represented by OA, OB, OC is equivalent to the system represented by OD, OE, OF.

7. ABCD is a quadrilateral and O is any point in its plane, E, F, G, H are the mid-points of AB, BC, CD, DA respectively. Prove that the resultant of forces represented by OA, OB, OC, OD is represented by 4OK where K bisects EG.

8. A point P on the circumference of a circle is joined to fixed points A and B on the circle. Forces 2PA, 3PB act along PA, PB respectively, and their resultant is represented in direction and magnitude by PQ. Find the locus of Q as P moves round the circle.   (H.S.C.)

9. Three forces AB, 2BC, 2AC act, in the directions indicated by the letters, along the sides of a triangle ABC. Find their resultant, and deduce that the line joining the mid-point of AB to the point of trisection of BC nearer to C cuts AC produced in E where AC = CE.   (I.S.)

10. A and B are fixed points; P moves in such a way that the resultant of forces represented by PA and PB is always double the former force. Find the locus of P.

11. If the resultant of forces represented by lines drawn from a point P to the corners of a quadrilateral be of constant magnitude, prove that the locus of P is a circle, and find its centre and radius.

12. ABCD is a parallelogram and E is a point in AD. Find a point F in BC such that the resultant of forces represented by AE and AF may act in the direction AC.

13. Forces equal to **AB**, **AC**, and **CB** act in the sides of a triangle ABC. Prove that their resultant acts in ED and is equal to **4ED**, where D, E are the mid-points of the sides BC, CA respectively.

14. O is any point on a median of a triangle ABC. Forces act from O towards A, B, and C proportional to the distances of O from these points. Prove that their resultant is represented in magnitude and direction by 3OG, where G is the centroid of the triangle.

15. Forces completely represented by AB, CB, CD, AD act in the sides of a quadrilateral ABCD. Prove that their resultant is

completely represented by 4HK, where H and K are the midpoints of AC and BD respectively. (C.W.B.)

16. ABC is an equilateral triangle and G is its centroid. Forces 1, 1, 4, 2, 2, 1 N act along BC, CA, AB, AG, BG, CG respectively. Determine the magnitude and direction of their resultant and the distance from A at which it cuts AB. (I.S.)

17. Prove that the resultant of any number of concurrent forces $l \times$ OA, $m \times$ OB, $n \times$ OC ... is $(l+m+n+ \ldots)$OG, where G is the centre of gravity of masses proportional to $l, m, n, \ldots$ placed at A, B, C, ... respectively. (H.S.C.)

18. ABCD is a trapezium with AB parallel to DC. Show that forces represented in magnitude, direction, and line of action by **AD**, **DC**, **CB**, **BA**, **AC** and **BD** have a resultant represented in magnitude and direction by 2**EF**, where E and F are the mid-points of AB and CD respectively. Show that the line of action cuts BA produced at a distance $\frac{1}{2}$CD from A.

19. Forces are represented in magnitude, direction, and line of action by 3**BC**, 2**AC**, and 7**BA**. The line of action of the resultant cuts AB at F and AC at E. Show that 2AF = 3FB, 7AE = 3EC, and that the resultant is represented in magnitude and direction by (50/3)**FE**. (H.S.C.)

20. If AD, BE are medians of any triangle ABC, show that the five forces represented in magnitude, direction, and line of action by **AB**, 2**CB**, 3**CA**, **BE**, and **DA** have a resultant represented completely by 5**CH**, where H is the point dividing AB internally in the ratio 3 : 7.

21. Forces are represented in magnitude, direction, and line of action by **BC**, **AC**, and 3**BA** in a parallelogram ABCD. Show that their resultant is represented in magnitude and direction by 2**BD** and find its line of action.

22. A point P lies in the plane of the rectangle ABCD. Forces represented completely by **AB**, 2**DC**, l**PA**, l**BP**, m**CP**, m**PD** are in equilibrium. Find l in terms of m.

Prove that P can lie at any point of a line parallel to AB and find the ratio in which this line divides AB.

3.16. Composition of couples

We have seen (3.14) that the moment of a couple is the same about any point in its plane. We shall now prove the following theorem, which enables us to obtain the resultant of any number of couples in one plane.

Two couples acting in the same plane are equivalent to a single couple whose moment is the algebraic sum of the moments of the separate couples.

Case (i). When the lines of action of the forces are all parallel.

Let P, P, Q, Q be the forces of the couples acting as in Fig. 3.35, and draw a straight line OABCD perpendicular to their lines of action to meet them in A, B, C, D.

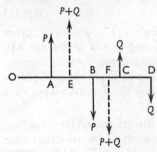

FIG. 3.35

The force P and Q at A and C are equivalent to a parallel force $(P+Q)$ at a point E in AC such that $P \times AE = Q \times EC$.

The forces P and Q at B and D are equivalent to a parallel force $(P+Q)$ at a point F in BD, such that $P \times BF = Q \times FD$, and this force is in the opposite direction to the first one.

Hence the two couples are equivalent to a single couple.

Also the moment of the resultant couple

= the sum of the moments about O of the two forces $P+Q$ at E and F.

But the moment of $P+Q$ at E about O

= the sum of the moments about O of P at A and Q at C.

Similarly, the moment of $P+Q$ at F about O

= the sum of the moments about O of P at B and Q at D.

Hence the moment of the resultant couple

= the sum of the moments of the four forces of the couples,

= the sum of the moments of the original couples.

Case (ii). When the lines of action of the forces are not all parallel.

Let P, P, Q, Q be the forces of the couples; and let one of the forces P meet one of the forces Q in O (Fig. 3.36), and the other two forces meet in O'.

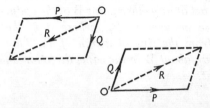

FIG. 3.36

The forces P, Q at O can be combined into a single force R, and so can the forces P and Q at O'. Also these single forces will be equal, parallel, and opposite, for they are both resultants of forces P and Q acting at the same angle but in opposite directions.

Hence the two couples are equivalent to a single couple.

Now, the moment of this resultant couple

= the moment about O of R at O'

= the sum of the moments about O of P and Q at O'

= the sum of the moments of the original couples.

The theorem having been proved for two couples, it follows that any number of couples in a plane are equivalent to a single couple whose moment is the algebraic sum of the moments of the separate couples.

3.17. From the theorem of the last paragraph we can deduce the following:

1. *Two couples acting in a plane, whose moments are equal and opposite, balance one another.*

For their resultant is a couple of zero moment, which means that the forces of the couple are each zero, or its arm is zero, and in the latter case it must consist of two equal and opposite forces in the same straight line which are obviously in equilibrium.

2. *Any two couples of equal moment and in the same plane are equivalent.*

This follows by reversing the directions of the forces of one of the balancing couples in (1).

3.18. *A force P acting at any point* A *of a rigid body may be transferred parallel to itself, to act at any other point* B *of the body, by introducing a couple whose moment is Pp, where p is the perpendicular distance of* B *from the line of action of P. This couple acts so as to turn the body about* B *in the same direction as P, acting at* A, *tends to move it.*

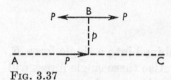

FIG. 3.37

Let AC (Fig. 3.37) be the line of action of P.

Apply at B two equal and opposite forces P acting along the line through B parallel to AC.

One of these, the one acting towards the right, is the original force P transferred to act at B.

The other forms with the original force a couple whose moment is Pp, where p is the perpendicular distance of B from AC.

3.19 EXAMPLE (i)

Prove that the combination of a couple with a force in the same plane is equivalent to changing the position of the line of action of the force.

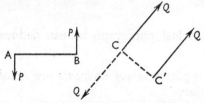

FIG. 3.38

Let the couple be formed by two forces P acting at A and B (Fig. 3.38), and let Q be the force acting at C.

We can replace the couple by any other couple of equal moment in the same plane.

We choose a couple of which the forces are of magnitude Q, one acting at C in the opposite direction to the force Q already acting there. The other force of the couple will act at C', where CC' is perpendicular to the direction of the original force Q, and CC' = (P/Q)AB.

We now have two forces at C balancing each other, and a single force Q at C'.

The result is therefore to move the line of action of Q parallel to itself through a distance M/Q, where M is the moment of the couple.

EXAMPLE (ii)

If three forces acting on a rigid body be represented in magnitude, direction, and line of action by the sides of a triangle taken in order they are equivalent to a couple whose moment is represented by twice the area of the triangle.

Let ABC (Fig. 3.39) be the triangle and P, Q, R the forces, so that P, Q, R are represented completely by the sides BC, CA, AB respectively.

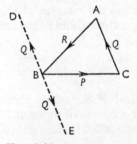

FIG. 3.39

Draw DBE parallel to AC, and introduce at B two equal and opposite forces, equal to Q, acting in the directions BD and BE.

The forces P, R and the force Q acting in the direction BD are in equilibrium by the triangle of forces, since they are acting at the point B.

We are thus left with two forces, each equal to Q, acting along the lines CA and BE.

These form a couple whose moment is $Q \times$ the perpendicular distance of B from CA.

Also, since CA represents Q, this moment is represented by CA $\times$ the perpendicular distance of B from CA, i.e. by twice the area of the triangle ABC.

Alternatively, we can say that the vector-sum of the three forces, **AB**+**BC**+**CA**, is zero. Hence the forces are equivalent to a couple. The moment of the couple is obtained by taking moments about any point, say B, and the result follows as above.

EXAMPLE (iii)

*Forces represented in magnitude and direction by l**AB**, m**BC**, l**CD**, m**DA** act along the corresponding sides of a quadrilateral ABCD. Show that they are equivalent to a couple if l = m or if ABCD is a parallelogram.*

If the four forces reduce to a couple their vector sum must be zero. The vector sum is

$$l\mathbf{AB}+m\mathbf{BC}+l\mathbf{CD}+m\mathbf{DA}$$
$$= l(\mathbf{AB}+\mathbf{BC}+\mathbf{CD}+\mathbf{DA})+(m-l)(\mathbf{BC}+\mathbf{DA})$$
$$= (m-l)(\mathbf{BC}+\mathbf{DA}).$$

This is zero if $l = m$ or if $\mathbf{BC}+\mathbf{DA} = 0$. If the latter condition is satisfied BC is equal and parallel to AD, so that ABCD is a parallelogram.

When either of these conditions is satisfied the moment of the four forces about any point, D say, is clearly not zero, and hence the forces reduce to a couple.

EXAMPLES 3.6

1. A uniform rod AB of length $2a$ and weight W can rotate about a smooth fixed horizontal axis at A. If a couple of moment N is applied to it and maintains it at an angle of 30° to the vertical, find N.

2. A uniform bar AB of length 3·6 m and mass 5 kg is clamped at the end A in a horizontal position, and a mass of 2·5 kg is suspended from the end B. Find the force and the couple that must act on the bar where it is clamped.

3. A uniform ladder of length l and weight W is held with its upper end resting against a smooth vertical wall, and with its lower end on a smooth horizontal surface. A man of weight W' stands on the ladder at a distance l' from its lower end. Show that if the ladder is kept from slipping by means of a couple, the moment of the couple is equal to

$$(\tfrac{1}{2}Wl+ W'l) \sin \theta$$

where θ is the inclination of the ladder to the vertical. (H.C.)

4. Prove that two couples whose forces are coplanar are in equilibrium if their moments are equal in magnitude and opposite in sign.

   A uniform square plate ABCD of weight W is kept in equilibrium with the corner A against a rough vertical wall by means of a horizontal force W acting at C, the point B being above A. Prove that the coefficient of friction cannot be less than unity, and find the inclination of AB to the vertical. (N.U.)

5. A uniform heavy bent rod ABCD, whose parts AB, BC, CD form three sides of a square, is smoothly hinged to a point fixed at A in a smooth wall and is supported in a vertical plane perpendicular to the wall, with AB, CD horizontal and CD below AB, by the pressure of the wall against D. Find the reactions of the hinge at A and of the wall at D.

   Show that the stresses in the rod at B and C each consist of a force and a couple; find the reactions of the parts AB, DC on the part BC, and verify the equilibrium of BC. (H.C.)

6. Forces $3P$, $4P$, and $5P$ respectively act along the sides of a right-angled triangle of sides 3 m, 4 m, 5 m, in the same circular sense viewed from any point of the triangle. Find the forces which, acting at the ends of the side 5 m long and at right angles to it, will maintain equilibrium with these forces.

7. ABCDEF is a regular hexagon. Show that forces represented completely by AB, CD, and EF are equivalent to a couple of moment equal to the area of the hexagon.

8. P, Q, R are the mid-points of the sides AB, BC, CA of a triangle ABC. Find the value of k if the forces **AB**, **BC**, **CA**, k**PQ**, k**QR**, k**RP** are in equilibrium. (L.A.)

9. ABCD is a square of side a m. Forces of 10, 5, 10, 15 N act along AB, BC, CD, DA respectively. Show that there are two forces such that if either is combined with these four forces the system will reduce to a couple of moment $5a$ N m. (L.A.)

10. ABCD is a rectangle with AB $= a$, BC $= b$, M is the mid-point of BC. Three forces are represented completely by k**AM**, k**MC**, k**CD**, where k is positive. Find the magnitude and direction of their resultant and the distance from A of its line of action.

    Find the magnitude of the couple that must be combined with the three forces in order that the resultant of the system shall pass through the mid-point of AB. (L.A.)

3.20. Resultant of coplanar forces

We shall now consider another method of reducing a system of coplanar forces.

From the theorem proved in the next paragraph we can deduce the theorem of 3.5, and also easily obtain the various sets of conditions necessary and sufficient for a system of forces to be in equilibrium.

3.21. *Any system of coplanar forces acting on a rigid body can, in general, be replaced by a single force acting at an arbitrary point in the plane of the forces together with a couple.*

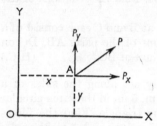

FIG. 3.40

Let the forces P_1, P_2, ... P_n act at the points A_1, A_2, ... A_n, and let O (Fig. 3.40) be any point in the plane of the forces. Take O as origin of coordinates, and let the coordinates of A_1, A_2, ... A_n, referred to rectangular axes through O be (x_1, y_1), (x_2, y_2), ... (x_n, y_n).

Consider any one of the forces P acting at the point A (x, y).

Resolve P into its components P_x, P_y, parallel to OX, OY.

We can transfer P_x parallel to itself to act at O by introducing a couple whose moment is yP_x, and we can transfer P_y parallel to itself to act at O by introducing a couple xP_y.

These couples are in opposite senses, and the algebraic sum of their moments is $xP_y - yP_x$.

Similarly for all the forces.

Let X be the algebraic sum of all the components of the forces transferred to act along the axis of x, and Y the sum of those components acting along the axis of y.

These can be compounded into a single resultant R acting at O, and the couples can be added (with their proper signs) to form a single couple of moment G, say, equal to the sum of the moments of all the couples.

If the resultant R makes an angle θ with the axis of x,
$$R^2 = X^2 + Y^2 \text{ and } \tan \theta = Y/X.$$

It should be noticed that the values of R and θ are independent of the position of the point O, since they do not contain the coordinates of any of the points A_1, A_2, ... A_n. In fact, R is given in magnitude and direction by the vector sum of the forces P_1, P_2, ... P_n.

The moment of the resultant couple is
$$G = \sum(xP_y - yP_x)$$

where P_x, P_y are the components of P parallel to the axis, and the $\sum$ denotes summation for all the forces.

It is evident that G is the sum of the moments of all the forces about the origin O, and its value will depend on the position of O.

3.22. *Conditions of equilibrium for a system of coplanar forces*

Let the forces be reduced to a single force R at any arbitrary point O and a couple G.

Then for equilibrium we must have $R = 0$ and $G = 0$.

If $R = 0$ we must have both $X = 0$ *and* $Y = 0$.

This gives us *three* conditions, which can be stated as follows:

The algebraic sums of the components of the forces in any two directions which are not parallel must be zero, and the algebraic sum of the moments of all the forces about any arbitrary point must be zero.

We obtained these conditions earlier (3.8).

3.23. *Change of base*

If we take another point O' with coordinates (x', y') as base instead of the origin the moment of the couple for this base may be obtained from that for the origin O by writing $x-x'$ and $y-y'$ for x and y in the value for G.

$$\therefore \qquad G' = \sum(x-x')P_y - \sum(y-y')P_x$$
$$= \sum xP_y - \sum yP_x - \sum x'P_y + \sum y'P_x$$
$$= G - x'\sum P_y + y'\sum P_x$$

since x', y' are constant.

$$\therefore \qquad G' = G - x'Y + y'X.$$

In this result G is the sum of the moments of the forces about the origin, x', y' are the coordinates of the base, X and Y the sums of the components of the forces parallel to the axes.

3.24. *Line of action of resultant*

If the system is not in equilibrium, and we reduce it to a force R at the point (x', y'), and a couple G', then

$$R^2 = X^2 + Y^2$$

and
$$G' = G - x'Y + y'X.$$

Now if R is zero, $X = 0$ and $Y = 0$, and the system reduces to a couple G, since this cannot vanish too.

If R is not zero we may, by properly choosing the base, make the couple G' vanish, so that the system reduces to the single force R. This is the case when the coordinates (x', y') of the base satisfy the equation

$$G - xY + yX = 0$$

i.e. the base must lie on this line.

Now this line makes an angle $\tan^{-1} Y/X$ with the axis of x, and is therefore parallel to R, and since R acts at the base (x', y'), this straight line is the line of action of R. The equation of the line of action of the resultant is therefore

$$G - xY + yX = 0.$$

3.25. Other forms of the conditions of equilibrium

(1) *A system of coplanar forces will be in equilibrium if the sum of the moments of all the forces about two different points (say O and C) is zero, and the sum of the components in some one direction, not perpendicular to OC, is also zero.*

Taking O as origin, C as base (x', y'), we have

$$G = 0$$

and
$$G' = G - x'Y + y'X = 0$$

and
$$X = 0.$$

These give $X = 0$, $Y = 0$, $G = 0$, provided x' is not zero, i.e. provided C is not on the y axis (so that X is perpendicular to OC).

(2) *A system of coplanar forces will be in equilibrium if the sum of the moments about three different points,* O, C, D, *not all in the same straight line, are each zero.*

Taking O as origin, C as (x', y'), and D as (x'', y''), these conditions give

$$G = 0$$

and
$$G - x'X + y'X = 0$$

and
$$G - x''Y + y''X = 0$$

∴
$$-x'Y + y'X = 0$$

and
$$-x''Y + y''X = 0.$$

Therefore $X = 0$ and $Y = 0$, unless $x'y'' - x''y' = 0$, i.e. unless O, C, D are in a straight line.

3.26. EXAMPLE (i)

Forces 2, 2, 3, 2 units act along the sides AB, CD, ED, EF respectively of a regular hexagon ABCDEF in the directions indicated by the order of the letters. Find the magnitude of the resultant and prove that it acts along AB. (I.S.)

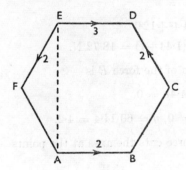

FIG. 3.41

Fig. 3.41 shows the forces acting along the sides of the hexagon. Take axes along and perpendicular to AB with A as origin. Let the side of the hexagon $= a$.

Sum of the components of the forces parallel to AB

$$= 2 - 2 \cos 60° + 3 - 2 \cos 60° = 3$$

and the sum of the components of the forces perpendicular to AB

$$= 2 \sin 60° - 2 \sin 60° = 0$$

which together show that the resultant has magnitude 3 and is parallel to AB.

But the sum of the moments of the forces about A

$$= 2 \times 2a \sin 60° - 3 \times 2a \sin 60° + 2 \times a \sin 60° = 0$$

and hence the resultant must pass through A.

Therefore the resultant has magnitude 3 units and acts along AB.

EXAMPLE (ii)

A force acting in the xy plane has moments −60 Nm, −156 Nm, and 84 Nm, about the origin, the point (8, 0) and the point (0, 10) respectively, the coordinates being in metres. Find the magnitude of the force and the points in which it cuts the coordinate axes. (I.E.)

Let X, Y N be the components of the force along the axes and G Nm its moment about the origin.

The moment about the point (x, y) is $G - xY + yX$ in Nm.

∴ $$G = -60$$
and $$-60 - 8Y = -156$$
and $$-60 + 10X = 84.$$
∴ $$Y = 12 \text{ and } X = 14·4$$
∴ $$R = \sqrt{(X^2 + Y^2)} = \sqrt{[14·4^2 + 12^2]}$$
$$= 12\sqrt{(1·44 + 1)} = 18·72 \text{ N.}$$

The equation of the line of action of the force R is

$$-60 - 12x + 14·4y = 0.$$

When $y = 0$, $x = -5$, and when $x = 0$, $y = 60/14·4 = 4·2$.

Hence the line of action of the force cuts the axes at the points $(-5, 0)$ and $(0, 4·2)$.

3.27. Other examples on coplanar forces

EXAMPLE (i)

Forces 1, 2, 3, 4 act in the sides AB, BC, CD, DA of a square ABCD. Reduce the system to a force through A and a force in BC.

Figure 3.42A shows the forces acting in the sides of the square ABCD. Fig. 3.42B shows the system replaced by a force R at A making an angle θ with AB, and a force F in BC. The two systems are equivalent if the sum of the components of each set of forces in any two perpendicular directions is the same, and if the sum of the moments of each set about any point is the same.

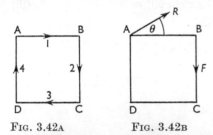

Fig. 3.42A   Fig. 3.42B

Resolving parallel and perpendicular to AB we get

$$R \cos \theta = 1 - 3 \tag{i}$$

$$R \sin \theta - F = 4 - 2 \tag{ii}$$

and taking moments about A

$$F \times a = 2a + 3a \tag{iii}$$

where a is the side of the square.

From equation (iii) $F = 5.$

Hence from equations (i) and (ii)

$$R \cos \theta = -2$$
$$R \sin \theta = 7.$$

Therefore $R = \sqrt{53}$ and $\tan \theta = -3 \cdot 5$ (θ lies between 90° and 180°). Hence the force R has magnitude $\sqrt{53}$ and makes an angle of $\tan^{-1} 3 \cdot 5$ with BA produced.

EXAMPLE (ii)

An equilateral triangular lamina ABC resting on a smooth horizontal plane is acted upon by a force of 5 N along BC, 3 N along AC, and 2 N along AD, where AD is perpendicular to BC. Find the force at B and the couple which will keep the lamina at rest.

Draw the triangle and insert the forces as in Fig. 3.43.

Let the additional force be R N at an angle θ to BC and the couple be of moment N Nm in a counterclockwise direction as shown.

Since the lamina is at rest when subject to all these forces, resolving parallel and perpendicular to BC we get:

$$R \cos \theta + 5 + 3 \cos 60° = 0 \tag{i}$$

$$R \sin \theta - 2 - 3 \sin 60° = 0 \tag{ii}$$

since the sum of the components of the two forces constituting the couple is zero.

Also taking moments about B,

$$N-2\times\tfrac{1}{2}a-3\times a \sin 60° = 0 \qquad\qquad \text{(iii)}$$

where $a =$ side of the triangle ABC.

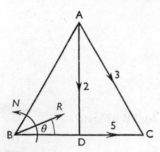

FIG. 3.43

From (iii) $N = a+3a\sqrt{3}/2 = 3\cdot598a.$

From (1) and (ii)

$$R \cos \theta = -5-\tfrac{3}{2} = -\tfrac{13}{2}$$

and $R \sin \theta = 2+3\sqrt{3}/2$

$\therefore$ $R^2 = \tfrac{169}{4}+4+\tfrac{27}{4}+6\sqrt{3} = 53+6\sqrt{3}$

$\therefore$ $R = 7\cdot962.$

Also, $\tan \theta = -\dfrac{4+3\sqrt{3}}{13} = -0\cdot7074.$

Since $\sin \theta$ is positive and $\cos \theta$ is negative, θ lies between 90° and 180° and, in fact, equals $180°-35°\ 16' = 144°\ 44'.$

EXAMPLE (iii)

Show that a given force may be resolved into three components, acting in three given lines which are not all parallel or concurrent.

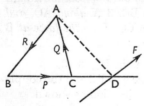

FIG. 3.44

Let the three lines form a triangle ABC (Fig. 3.44), and suppose the given force F cuts BC in D.

Then F can be resolved into two components acting along DA and BC respectively, and the force along DA can be resolved into two components along AB and CA respectively. The resolutions can be effected graphically, as in Fig. 3.45, where the dotted line is parallel to DA.

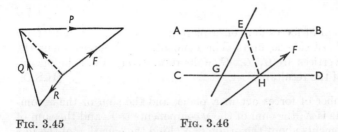

FIG. 3.45   FIG. 3.46

Similarly, if two of the lines AB, CD are parallel, as in Fig. 3.46, and EG is the third line.

Let F cut CD in H.

Then F can be resolved into two components along CD and HE respectively.

The force along HE can be then resolved into two components acting along GE and BA respectively.

EXAMPLES 3.7

1. Ox, Oy are rectangular axes, and P is a point whose coordinates are (3, 4). Find the intercepts made on Ox, Oy by the line of action of the resultant of a force of 7 units along OP, and a counterclockwise couple of moment 21 units.   (I.S.)

2. Forces of magnitudes 1, 2, 3, 4, 5, 6 units act in the same sense along the sides of a regular hexagon taken in order, and a force acts at the centre of the hexagon. If the several forces are equivalent to a couple, find the moment of the couple, and the magnitude and direction of the force at the centre.   (H.S.D.)

3. Forces of magnitude F, $2F$, $3F$, $4F$ act along the sides BA, BC, CD, DA of a quadrilateral ABCD in the directions indicated by the order of the letters, the quadrilateral being such that AB and BC are two sides of a square ABCE and D is the mid-point of CE. Find the magnitude and direction of the resultant, and find the distances from B at which its line of action meets AB and BC.   (I.E.)

4. Forces of magnitude F, $2F$, $3F$, $4F$, $5F$, $6F$ act along the sides of a regular hexagon, taken in order. Show that they are equivalent to a single force $6F$ acting parallel to one of the given forces, the distance of the line of action of that force and of the resultant from the centre of the hexagon being in the ratio $2:7$.   (I.E.)

5. A system of forces acts on a plate in the form of an equilateral triangle of side $2a$ units. The moments of the forces about the three vertices are G_1, G_2, G_3 units respectively. Find the magnitude of the resultant.   (H.S.C.)

6. A number of forces act in a plane, and the sum of the x components is X, the sum of the y components is Y, and the sum of the moments about the origin is N. Find the equation of the line of action of the resultant.   (H.S.D.)

7. ABCD is a square whose side is 2 m, P is the mid-point of AD and Q of CD. Forces of magnitude 10, 10, 30, 40 act along AB, CD, QB, CP in the directions indicated by the order of the letters. Find the magnitude of the resultant, and the distances from A of the points where its line of action meets AB and AD.

8. ABCDEF is a regular hexagon. Forces of 1, 3, 2, and 4 N act along AB, BE, ED, and DA respectively. Find the magnitude of the resultant. Take AB as x axis and AE as y axis, and find the equation of the line of action. Indicate by an arrow the direction of the resultant.   (I.E.)

9. A force in the plane of two rectangular axes has components X and Y in the directions of the axes of x and y respectively, and its line of action passes through the point (x', y'). Prove that it is equivalent to a force whose components are X and Y acting at the origin, together with a couple of moment $x'Y-y'X$. Parallel forces, P_1, P_2, P_3, ... act in the plane of these axes at points (x_1, y_1), (x_2, y_2), (x_3, y_3), ... respectively. Prove that the forces are in equilibrium whatever their common direction, if $\Sigma P = 0$, $\Sigma Px = 0$, $\Sigma Py = 0$.   (H.C.)

10. ABCD is a rectangle in which AB $=$ 5 m, BC $=$ 3 m. Forces of 2 N, 4 N, 3 N, 11 N act along AB, BC, DC, DA respectively, the sense in each case being indicated by the order of the letters. If this system is reduced to a force acting at the point of intersection of AC and BD, together with a couple, determine the magnitude and direction of the force, and the moment and sense of the couple.   (I.S.)

11. Forces act at the corners of a square of side 8 cm, as shown in Fig. 3.47. Find by calculation the magnitude and direction of the resultant force acting at the centre and the resultant couple.

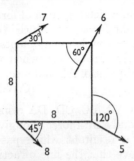

Fig. 3.47

12. A plane regular hexagon OABCDE has a side of 5 cm. Forces of 1, 2, 3, 4, 5, and 6 N act along the sides OA, AB, BC, CD, DE, and EO respectively, in the sense indicated by the order of the letters. Find the resultant force at O and the resultant moment about O.   (I.C.)

13. Prove that a system of coplanar forces is in equilibrium if the sum of the moments about each of three non-collinear points is zero. A couple of 10 Nm units acts on a square board ABCD of side 2 m. Replace the couple by forces acting along AB, BD, CA.   (N.U.3)

14. ABC is a triangle in which AB = AC = 4 cm, BC = 3 cm. E is the mid-point of AB, F a point on BC such that CF = 1 cm. Find three forces along the sides of the triangle ABC which will together be equivalent to a force of 4 N along EF, representing them in an accurate diagram on a scale of 1 N to 2 cm.   (Ex.)

15. ABC is a triangular lamina with AC = 6 cm, BC = 8 cm, and C a right angle. A force 10 N acts along AB. Prove that this can be completely balanced by two forces acting perpendicular to AC, BC respectively at their mid-points. Find the magnitudes of these forces.   (N.U.)

16. A system of forces acting on a rigid lamina in a plane is reduced, firstly, to a single force acting at a point A in the lamina, and secondly, to a single force acting at another point B and a couple of moment G. Prove that if the system were reduced to a single force acting at the mid-point of AB and a couple the moment of the couple would be $\frac{1}{2}G$.   (N.U.3)

17. Forces 3, 4, 2, 1 N act respectively along the sides AB, BC, CD, DA of a square ABCD of side 1 m. Reduce the system to: (i) a force at A and a couple; (ii) two parallel forces through B and C.

18. A, B are any two points in a lamina, on which a system of forces coplanar with it are acting, and when the forces are reduced to a single force at each of these points and a couple the moments of the couples are G_a and G_b respectively. Prove that when the reduction is made to a force at the mid-point of AB and a couple the moment of the couple is $\frac{1}{2}(G_a+G_b)$.   (C.W.B.)

19. Forces P, Q, P, Q act along the sides AB, BC, CD, DA respectively of a square; if these four forces and a fifth force R of given magnitude in the same plane have a resultant which passes through the centre of the square, prove that the line of action of R touches a fixed circle.   (I.S.)

20. A uniform square lamina ABCD, of side 0·6 m, and of mass 7 kg, can turn freely in a vertical plane about A, which is fixed and is kept in equilibrium with the diagonal AC horizontal by means of a pull exerted in a horizontal string through its highest point D, together with a couple. Find the moment of the couple when the magnitude of the reaction at A is 100 N.   (I.S.)

21. Forces P, $4P$, $2P$, $6P$ act along the sides AB, BC, CD, DA of a square ABCD of side a. Find the magnitude of their resultant, and prove that the equation of its line of action referred to AB and AD as coordinate axes is

$$2x-y+6a = 0. \qquad\qquad \text{(H.C.)}$$

22. Forces 1, 3, 5, 7, $9\sqrt{2}$ act along the sides AB, BC, CD, DA, and the diagonal BD of a square of side a, the senses being indicated by the order of the letters. Taking AB and AD as axes of x and y respectively, find the magnitude of the resultant and the equation of its line of action.   (H.C.)

23. ABC is an equilateral triangle; forces of 4, 2, and 2 units act along the sides AB, AC, BC in the directions indicated by the letters. Prove that if E is the point where the perpendicular to BC at B meets CA produced and if F bisects AB, the resultant is $2\sqrt{7}$ units acting along EF.   (H.C.)

24. ABCD is a quadrilateral in which AB = BC, CD = DA, A and C are right angles, B is 60°, D is 120°. Equal forces $\sqrt{3}P$ act along AD and DC; equal forces P act along CB and BA. Find the magnitude of their resultant, and the point in which it cuts BD produced.   (H.C.)

25. ABCD is a square; four forces, whose algebraic magnitudes form an arithmetical progression, act along the sides taken in order. Show that, if their resultant passes through a corner of the square, the progression is a diminishing one, in which, if the common difference is $2P$ the greatest force is $5P$ or $3P$. (H.C.)

3.28. Moment of a force about an axis

So far we have dealt only with coplanar forces, and have considered their moments about a *point* in their plane.

Suppose now that we have a rigid body which is free to rotate about some *axis* fixed in the body.

Any force (whose line of action is not parallel to, or does not pass through, this axis) will tend to turn the body about it. This introduces the idea of the *moment of a force about an axis*.

For the present we shall consider only cases where the force is perpendicular to the axis. In this case the moment of the force about the axis is defined as the product of the force and the perpendicular distance between the force and the axis. As earlier, we shall regard the moment as positive if the force tends to rotate the body about the axis in a counterclockwise direction.

It can be shown that the principle of moments holds for moments about a fixed axis.

3.29. EXAMPLE (i)

A square table stands on four legs placed at the mid-points of its sides; find the greatest weight which can be put at one of the corners of the table without upsetting it, the total weight of the table and legs being W.

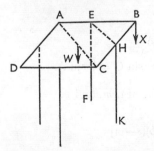

Fig. 3.48

Let ABCD (Fig. 3.48) represent the table and suppose the weight X be put at B.

This will tend to tilt the table about the line FK, joining the points of contact with the floor of the legs EF, HK. The weight W acts at the mid-point of AC, and its line of action is therefore at the same distance from FK as the line of action of the weight X.

Hence the greatest value of X is W.

EXAMPLE (ii)

A uniform circular plate is supported horizontally at three points in its edge, whose distances apart are l, m, n; find the proportion of the weight of the plate carried by each support.

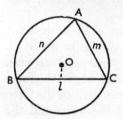

FIG. 3.49

Let A, B, C (Fig. 3.49) be the points of support, $BC = l$, $CA = m$, $AB = n$.

The weight of the plate acts at the centre of the circle O, which is the circumcircle of the triangle ABC.

The distance of O from BC is $R \cos A$, where R is the radius of the circle, and the distance of A from BC is $m \sin C$. Hence if P_A is the reaction at A, taking moments about BC, we get

$$P_A \times m \sin C = W \times R \cos A.$$

Also
$$R = \frac{l}{2} \operatorname{cosec} A$$

$\therefore$
$$P_A = W \frac{l}{2} \times \frac{\cos A}{m \sin C \sin A}$$

$$= \frac{Wl^2}{2mn} \times \frac{\cos A}{\sin^2 A}, \text{ since } \sin C = \frac{n}{l} \sin A.$$

This result should be expressed in terms of l, m, and n. Using the formula

$$\sin A = \frac{2}{bc} \sqrt{[s(s-a)(s-b)(s-c)]}$$

and remembering that $a = l$, $b = m$, $c = n$,

$$\sin^2 A = \frac{4}{m^2 n^2} \times \frac{(l+m+n)(m+n-l)(l+n-m)(l+m-n)}{16}$$

Also

$$\cos A = \frac{m^2+n^2-l^2}{2mn}.$$

$$\therefore \qquad P_A = \frac{Wl^2}{2mn} \times \frac{(m^2+n^2-l^2)}{2mn}$$

$$\times \frac{4m^2n^2}{(l+m+n)(m+n-l)(l+n-m)(l+m-n)}$$

$$= W\frac{l^2(m^2+n^2-l^2)}{(l+m+n)(m+n-l)(l+n-m)(l+m-n)}$$

and there are similar expressions for the reactions at B and C.

EXAMPLES 3.8

1. A uniform circular table, resting on four equal legs placed symmetrically round its edge, has a mass of 50 kg. Find the least weight which will just upset the table when hung from its edge.

2. A circular table stands symmetrically on three vertical legs distant 1 m from one another and attached to the table top at the vertices of the equilateral triangle ABC. The table-top has a mass of 40 kg, and a mass of 60 kg is placed at L (within the triangle ABC) distant 15 cm and 25 cm from BC and CA respectively. Calculate the pressures on each leg. (H.D.)

3. Show how to find the resultant of three parallel forces not in the same plane. A light table stands on three equal vertical legs, and a weight is placed at the centre of the circle inscribed in the triangle formed by the points of intersection of the legs. Show that the pressures on the legs are proportional to the opposite sides of the triangle. (I.S.)

4. A circular table of mass 20 kg is 1·2 m in diameter; it is supported by three equally spaced legs at its edge. Find the load that can be placed diametrically opposite one leg at the edge of the table, so that the whole weight of the table and the load is carried by the other two legs. (C.E.)

5. A round table 1·5 m in diameter has three symmetrically placed legs each 0·6 m from the centre. If the table has a mass of 25 kg, find the least weight which, placed on the edge of the table, will cause it to overbalance. What is the greatest weight which can be placed on the edge without overbalancing it?

6. A stool of mass 4 kg has a circular top of diameter 0·6 m. It is supported symmetrically in a horizontal position by three legs, 0·6 m long, each making an angle 60° with the ground, and fixed

into the circular top at the vertices of an equilateral triangle, the lengths of its sides being 0·3 m. Calculate the least weight which, when placed on the edge of the stool, will cause it to topple over.   (Q.E.)

REVISION EXAMPLES A

1. (i) Prove that the resultant of two forces P and Q which meet at an angle α is $\sqrt{(P^2+2PQ \cos \alpha+Q^2)}$.

   (ii) ABCDEF is a regular hexagon of centre O. Forces 2, 4, P, Q act from O along OA, OC, OE, OG respectively, where G is the mid-point of AB, and are in equilibrium. Find the values of P and Q.   (I.S.)

2. Particles of weights w_1 and w_2 lie at rest at the points A and B respectively on the upper half of a smooth circular wire whose plane is vertical, equilibrium being maintained by a light inextensible string which connects the particles and lies along the minor arc joining A to B. This arc subtends an angle α at the centre O of the circle. Prove that the tangent of the acute angle which OA makes with the vertical is

$$\frac{w_2 \sin \alpha}{w_1+w_2 \cos \alpha}.$$

   Show the heavier particle is nearer the highest point of the circle than the lighter particle.   (I.S.)

3. Prove that if three concurrent forces are in equilibrium each force is proportional to the sine of the angle between the other two. Three forces, P, Q, R are in equilibrium. P is given in magnitude and direction. The magnitude of Q is not given, but its direction makes an angle θ with the direction of P. Find the direction of R when its magnitude is least, and determine the corresponding magnitudes of Q and R.   (I.S.)

4. Two particles of weights $2W$, $3W$, fastened to the ends of a light inextensible string which passes over two smooth pegs on the same level distant a apart, are kept in equilibrium by a third particle of weight W' fastened to the part of the string between the pegs. If the angle between the oblique portions of the string is $120°$, prove that $W' = \sqrt{7}W$ and that the depth of W' below the level of the pegs is $2a/7\sqrt{3}$.   (I.S.)

5. Two rings of equal weight, connected by a string, can slide on two fixed rough rods which are in the same vertical plane and inclined to the downward vertical at equal angles of $45°$ on

opposite sides of it. If the coefficient of friction is $\frac{1}{3}$, prove that the maximum angle which the string can make with the horizontal, the rings remaining at rest, is θ, where $\tan\theta = \frac{3}{4}$. (I.S.)

6. Prove that the algebraic sum of the moments of a number of concurrent coplanar forces about a point in their plane is equal to the moment of their resultant about that point.

   (i) If three forces X, Y, Z, acting along the internal bisectors of the angles of the triangle ABC in order, are in equilibrium, prove that $X : Y : Z = \cos\frac{1}{2}A : \cos\frac{1}{2}B : \cos\frac{1}{2}C$.

   (ii) Forces P, Q, R act along the sides BC, CA, AB respectively, of a triangle ABC and their resultant acts along the line joining the incentre and the orthocentre. Prove that

   $P : Q : R = \sec B - \sec C : \sec C - \sec A : \sec A - \sec B.$
   (C.W.B.)

7. Obtain the conditions of equilibrium of a system of coplanar forces. A uniform square lamina ABCD is suspended by two strings attached to A and B which slope away from one another and make angles 30° and 45° respectively with the vertical. Find the inclination of AB to the horizontal. (C.W.B.)

8. A uniform square lamina ABCD of weight W is freely suspended at A. A weight w is attached to the lamina at B, and the system rests in equilibrium with AB inclined to the vertical at an angle of 30°. Find the ratio of w to W. If now an additional weight $2w$ is attached to the lamina at D, find an equation for the inclination of AD to the vertical in the new position of equilibrium. Prove that this inclination is greater than 30°.

9. To the end B of a uniform rod AB, of weight W, is attached a particle of weight w. The rod and particle are suspended from a fixed point O by two light strings OA, OB of the same length as the rod. Prove that, in the position of equilibrium, if T, T' are the tensions in OA, OB

$$\frac{T}{W} = \frac{T'}{W+2w}.$$

Prove also, that if OA makes angle α with the vertical,

$$\tan\alpha = \frac{(W+2w)\sqrt{3}}{3W+2w}.$$
   (H.C.)

10. A uniform rod AB of length $2a$ and weight W lies along a line of greatest slope of a plane inclined at angle θ to the horizontal, B being above A. The coefficient of friction between the rod and the plane is $\tan\lambda$, where λ is greater than θ. A cord, attached to A, passes over a small pulley at height a vertically above the mid-

point of the rod and supports a scale-pan in which gradually increasing weights are placed. Prove that if λ is greater than $45° - \frac{1}{2}\theta$ the rod will tilt before it slides.   (I.E.)

11. A thin rod of length a is in equilibrium with its ends resting on the inner rim of a smooth circular hoop, of radius a, fixed with its plane vertical. If the centre of gravity of the rod divides its length in the ratio 3 : 4, prove that its inclination to the vertical is $\tan^{-1} 7\sqrt{3}$.

Determine the ratio of the reaction on the lower end of the rod to that on the upper end.   (L.A.)

12. Two equal smooth cylinders of radius a and weight W lie in contact along generators on a horizontal table. A third equal cylinder is placed symmetrically upon them, and the system is kept in equilibrium by a band which passes round the cylinders in a plane perpendicular to the generators. Find the tension of the band if the lower cylinders are just about to separate. If the band is elastic and its natural length is $12a$, prove that the tension of the band at any extension x would be

$$\frac{Wx}{4\sqrt{3}(\pi-3)a}.$$ (H.C.)

13. Two uniform ladders, each of weight W and length $2b$, are hinged smoothly at their upper ends and stand on a smooth horizontal plane. A weight w is hung from a rung of one of the ladders at a distance d from its lower end, and the ladders are prevented from slipping by means of a light rope of length $2a$ attached to their lower ends. Find the pressure of each ladder on the ground and prove that the tension in the rope is

$$\frac{a(2Wb+wd)}{4b(4b^2-a^2)^{\frac{1}{2}}}.$$ (I.S.)

14. Three equal uniform rods, AB, BC, CD, each of length $2a$ and weight W, are smoothly jointed at B and C and rest with AB, CD in contact with two smooth pegs at the same level. In the position of equilibrium AB and CD are inclined at an angle α to the vertical, BC being horizontal. Prove that the distance between the pegs is $2a(1+\frac{2}{3}\sin^3 \alpha)$. If β is the angle which the reaction at B makes with the vertical, prove that

$$\tan \alpha \tan \beta = 3.$$ (H.C.)

15. A pentagon ABCDE of smoothly jointed uniform rods, each of weight W and of the same length, is supported symmetrically in a vertical plane with CD horizontal and AB and AE in contact with smooth pegs at the same horizontal level and at such a distance apart that the pentagon is regular. From the equilibrium

of the pentagon find the reactions on the pegs, and by considering the equilibrium of the rods BC, CD, DE, show that the horizontal components of the reactions at B, C, D, E are equal and of magnitude $W \cot (2\pi/5)$. Show further that the reaction at A is a horizontal force

$$\left(\frac{5}{2} \tan \frac{\pi}{5} - \cot \frac{2\pi}{5}\right) W. \qquad \text{(H.S.C.)}$$

16. A square ABCD is formed from four equal uniform rods, each of weight W, freely jointed at their ends. The square is suspended freely from A and a weight $3W$ is hung from the lowest point C, the square shape being maintained by a light horizontal strut joining the mid-points of AB and AD. Prove that the thrust in this strut is $10W$. (L.A.)

17. Two uniform bars AB and BC, each of length $2a$ and weight W, are smoothly jointed together at B. A light ring is attached to C and threaded on a fixed smooth horizontal rod, and A is freely jointed to a fixed point at depth $3a$ vertically below the rod. Prove that equilibrium is possible if, and only if, one of the bars is vertical. Find the reaction at C when the system is in equilibrium: (i) with AB vertical; (ii) with BC vertical. (L.A.)

18. Each of the two legs of a step-ladder is 2 m long, is uniform, and has a mass of 9 kg. The legs are smoothly hinged together at one end of each, and the mid-points are connected by an elastic cord of unstretched length 1 m and modulus λ. The ladder stands on smooth horizontal ground with its feet apart and supports a mass of 75 kg at the hinge. If, in the equilibrium position, the extension in the cord is 10 cm, calculate the value of λ.

19. ABCD is a plane quadrilateral with AB = 0·9 m, BC = 1·2 m, CD = 3·6 m, DA = 3·9 m and AC = 1·5 m, B and D being on opposite sides of AC. AB is vertical and B is below A. CD, DA, AC, and CB represent four light rods smoothly jointed together to form a framework, A and B being fixed points in a vertical wall. A load of mass 75 kg is suspended from D. Find the force in each of the four rods. (L.A.)

20. Two equal uniform rods AB, BC, each of weight W, are freely jointed together at B. They rest in contact with a smooth solid circular cylinder which is fixed with its axis horizontal, the plane of the rods being at right angles to this axis. In the position of equilibrium the rods are at right angles; show that the length of each rod is equal to twice the diameter of the cylinder. (L.A.)

21. Two uniform rods AC, BC, whose weights are proportional to their lengths, are freely jointed together at C, and the ends A

and B are freely hinged to two points in a vertical line. Show that the reaction between the rods at C acts along the bisector of the angle ACB. (L.A.)

22. ABCD is a rectangle with AB $= a$, BC $= b$. M is the mid-point of BC. Three forces are represented completely by k**AM**, k**MC**, k**CD**, where k is positive. Find the magnitude and direction of their resultant and the distance from A of its line of action.

Find the magnitude of the couple that must be combined with the three forces in order that the resultant of the system shall pass through the mid-point of AB. Indicate the sense of this couple in a diagram. (L.A.)

23. Forces P, Q act along the lines BA, CA in the directions indicated by the order of the letters. Write down the equivalent pair of forces acting along the internal and external bisectors of the angle A, where $\angle$ BAC $= 2\alpha$.

The triangle ABC is isosceles and right-angled at A; BCD is an equilateral triangle on the other side of BC. Forces of 3, 2, 3, 10, 14 N act along AB, BC, CA, BD, CD in the directions indicated by the order of the letters. Prove that the line of action of the resultant force is at a distance $(\sqrt{3}/15)$ AB from A. (L.A.)

24. Prove that a coplanar system of forces which is not in equilibrium may be reduced to a single force or to a couple.

ABCD is a square; forces of magnitude 3, 2, 4, 3, P units act along AB, CB, CD, AD, DB respectively in directions indicated by the order of the letters. If the system is equivalent to a couple, find the value of P. (L.A.)

25. A system of coplanar forces has anticlockwise moments M, $2M/3$, $3M/2$, respectively about the points $(0, 0)$; $(a, 0)$; $(0, 2a)$ in its plane. Calculate the magnitude of the resultant of the forces and prove that the equation of its line of action is

$$3y - 4x + 12a = 0. \qquad \text{(L.A.)}$$

26. Show that a system of coplanar forces, not in equilibrium, can be reduced either to a single force or to a couple.

Forces 4, 3, 3 N act respectively along the sides AB, BC, CA of an equilateral triangle ABC of side 0·6 m. Find the magnitude and direction of their resultant and obtain the perpendicular distance of its line of action from C.

An additional force is introduced at C in the plane ABC. If the system is now equivalent to a couple, find its moment and the magnitude and direction of the additional force. (L.A.)

27. Show that the moment of a couple is the same about all points

in its plane. Prove that a force F and a couple of moment M acting in the same plane are equivalent to a single force.

Forces of magnitudes 3, 4, 6, 7 units act along the sides AB, BC, CD, DA of a square ABCD of side a, the directions of the forces being indicated by the order of the letters. In addition, a couple acts in the plane of the square. If the whole system is equivalent to a force acting through the centre of the square, find the magnitude and sense of this couple. Find also the magnitude and line of action of this force. (L.A.)

28. ABCD is a square plate acted upon by the following forces in its plane: $5\sqrt{2}$ N at C along AC produced; $15\sqrt{2}$ N at B along DB produced; 10 N at D along a line making an angle of 30° with AD produced on the side remote from BC. A force P is to be placed at A so that the whole system reduces to a couple. Find the magnitude and direction of P, and the moment of the couple if the side of the square is of length 1 m. (I.E.)

29. Prove that two coplanar couples of equal and opposite moment are in equilibrium. Show that the resultant of any number of coplanar couples is a couple whose moment is the algebraic sum of the moments. ABCDE is a regular pentagon. Five forces each equal to P act along AE, ED, DC, CB, BA. Five forces each equal to Q act along AC, CE, EB, BD, DA. Prove that the ten forces will be in equilibrium if P and Q are in a certain ratio and find the ratio. (C.W.B.)

30. A force acting at the point (x, y) has components (X, Y) parallel to the axes of coordinates. Prove that it may be replaced by a force X along the axis of x, a force Y along the axis of Y, and a couple $xY - yX$. Forces of magnitudes 1, 5, 9, 11, 7, 3 act in the sides AB, BC, CD, DE, EF, FA of a regular hexagon of side a in the senses indicated by the order of the letters. Taking as axes OA and OH, where O is the centre of the hexagon and H is the mid-point of BC, find the forces along the axes and the couple by which the system may be replaced. Prove also that the system is equivalent to a single force which meets the axes at the points $(-9a/4, 0)$ and $(0, -9\sqrt{3}a/2)$. (H.C.)

31. ABC is a triangle in which AB $= a$, BC $= 2a$ and the angle B $= 90°$. The moments of a system of forces in the plane of triangle ABC are $+M_1$, $+M_2$, and $-M_3$ about A, B, and C respectively, anticlockwise moments being considered positive. Determine the magnitude and direction of the resultant of the system and the point at which it cuts BC.

If the resultant is perpendicular to AC, prove that
$$4M_1 = 5M_2 + M_3.$$
(L.A.)

32. ABC is an equilateral triangle of side $2a$. The moments of a system of forces, acting in the plane of the triangle ABC, about A, B, and C are M_1, M_2, and M_3 respectively, all moments being measured in the same sense. Prove that the magnitude of the resultant of the system is

$$\frac{\sqrt{(M_1^2 + M_2^2 + M_3^2 - M_2 M_3 - M_3 M_1 - M_1 M_2)}}{a\sqrt{3}}.$$

Find the moment of the system about the point D, where D is the reflection of C in AB. (L.A.)

33. ABCD is a square, H and K the mid-points of CD and BC respectively. A force P acts along AH, a force Q along KH, and a force R along KA, in the directions indicated by the order of the letters.

   (*a*) If the resultant acts along the diagonal BD, find the ratios $P : Q : R$.

   (*b*) Show that the forces can never be in equilibrium whatever the ratios $P : Q : R$, but that they could reduce to a couple if the force Q were reversed in direction and find the ratios $P : Q : R$ in that case. (L.A.)

34. A table has a uniform circular top of diameter 1·2 m and has a mass, together with its legs, of 15 kg. The three equal vertical legs are joined to the top at points A, B, and C, which are equally spaced round the rim of the table, and the table stands on a horizontal floor. What is the greatest weight which can be placed on the table without overturning it, no matter at what point the weight is placed?

   A weight of mass M kg is placed on the edge of the table at a point P, where PB $= 0·3$ m. Prove that the greatest value of M consistent with equilibrium is just over 21.

   If $M = 15$, calculate the reactions of the floor on each of the three legs. (L.A.)

35. Forces of magnitudes P, Q, R, S act along the sides AB, BC, CD, DA of a square ABCD. Find the two possible sets of values for Q, R, S given that $P = 5$, the resultant of all four forces has a magnitude $\sqrt{5}$, and its line of action goes through D and the mid-point of BC. (O.C.)

36. Three uniform rods OA, AB, and BC, each of length $2a$ and weight W, are freely jointed together at A and B and hang freely from a fixed pivot at O. A force P acting in a horizontal direction is applied at C to the rod BC and an equilibrium

position is reached when BC makes an angle of 45° with the horizontal.

(a) Find P in terms of W.

(b) Prove that the pull on the pivot at O is $\frac{1}{2}W\sqrt{37}$.

(c) Show that the distance of C from the vertical through O is about 2·44a. (L.A.)

37. A light framework consists of five rods forming the four sides of the square ABCD and one diagonal AC. The framework is suspended by the corner A and supports weights mW at B and nW at D. Find the thrusts in BC and CD and the inclinations of the rod AB to the vertical.

Prove that the tension in the rod AC is

$$Wmn \sqrt{\left(\frac{2}{m^2+n^2}\right)}. \qquad \text{(O.C.)}$$

38. A pin-jointed framework in the form of a regular pentagon ABCDE of equal uniform rods, each of weight W, is suspended from A. The framework is kept in shape by a light rod freely jointed at its ends to the mid-points of BC and DE. Show, by calculation, that the thrust in the light rod is approximately 6·16W. (L.A.)

39. Four uniform rods, AB, BC, CD, DE of equal length are freely jointed at B, C, D. The rods AB, DE each have weight w; the rods BC, CD each have weight W. When the rods hang suspended from the points A and E held at a distance $2a$ apart and at the same level, the points B, C, D are found to lie on the semicircle with AE as base. Prove that

$$w = (2\sqrt{2}+1)W,$$

and find the reaction at C in terms of W. (O.C.)

40. A hollow circular cylinder of thin material has diameter $2a$, length $2a \tan \alpha$, and weight W. It is open at both ends, and stands with one end resting on a smooth horizontal table. A uniform rod of weight w and length $2l$ ($2l > 2a \sec \alpha$) is placed inside the cylinder to rest against the upper rim with its lower end in contact with both the table and the wall of the cylinder. All the surfaces are smooth. Prove that, if equilibrium is possible, then

$$l < 2a \sec^3 \alpha \quad \text{and} \quad Wa > wl \sin^2 \alpha \cos \alpha. \qquad \text{(O.C.)}$$

CHAPTER FOUR

GRAPHICAL CONSTRUCTIONS

4.1. The magnitude and direction of the resultant of any number of coplanar forces can be found graphically by means of the polygon of forces. When the forces act at a point the resultant must pass through this point, so that we know its line of action. When the forces are acting on a rigid body we can still find the magnitude and direction of the resultant by drawing the polygon of forces, but we require a further construction to determine the line of action.

We shall now consider how this may be done, and also how to apply the graphical method to determine the stresses in a framework of light rods acted on by given forces.

4.2. To find, graphically, the resultant of any number of coplanar forces

Let the lines of action of the forces P, Q, R, S be as in Fig. 4.1A.

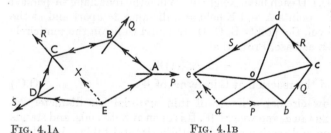

FIG. 4.1A   FIG. 4.1B

Draw the figure $abcde$ (Fig. 4.1B) having its sides ab, bc, cd, de parallel and proportional to P, Q, R, S respectively. Join ae.

By the polygon of forces ae represents the resultant of P, Q, R, and S in magnitude and direction. We will denote the resultant by X.

To find the line of action of the resultant, take any point o and join it to a, b, c, d, and e.

Take any point A on the line of action of P (Fig. 4.1A) and draw AE and AB parallel to *oa* and *ob*, and let AB meet the line of action of Q in B. P is equivalent to forces represented by *ao* and *ob* acting along EA and BA respectively.

Through B draw BC parallel to *oc* to meet R in C.

Q is equivalent to forces *bo*, *oc* acting along AB, CB respectively; the former of these balances the force *ob* along BA.

From C draw CD parallel to *od* to meet S in D.

R is equivalent to forces *co*, *od* acting along BC, DC respectively; the former balances the force *oc* acting along CB.

From D draw DE parallel to *oe* to meet AE in E.

S is equivalent to forces *do*, *oe* acting along CD and ED respectively; the former balances the force *od* acting along DC.

Hence the forces P, Q, R, and S are equivalent to forces *ao*, *oe* acting along EA, ED respectively, and since these intersect in E, their resultant must pass through E.

The resultant is therefore a force X passing through E. parallel to *ae* and represented in magnitude by *ae*.

We have therefore determined the line of action of the resultant as well as its magnitude and direction.

The figure *abcde* is called the *Force Polygon* for the given system, and the figure ABCDE is called a *Link* or *Funicular Polygon*.

If the link polygon were formed of strings it would be kept in equilibrium in the shape ABCDE by the forces P, Q, R, S acting at A, B, C, D and a force equal and opposite to their resultant acting at E.

This is the origin of the name funicular polygon, which means a rope polygon.

It is evident that by taking different positions for A and *o* the shape of the funicular polygon can be varied in any number of ways, but the final point of intersection of AE and DE will always lie on the same straight line, the line of action of the resultant.

4.3. If the point *e* of the force polygon coincides with *a* the polygon is said to close, and the resultant force vanishes.

This does not ensure, however, that the forces are in equilibrium.

For in this case *oe* and *oa* coincide, and AE, DE will be parallel, and the forces acting along them will be equal and opposite.

Hence, unless DEA is a straight line, i.e. unless the funicular polygon closes, we are left with a couple.

This means that *if the forces are in equilibrium, both the force polygon and the funicular polygon must close.*

4.4. EXAMPLE

ABC *is a triangle whose sides* AB, BC, CA *are respectively* 12, 10, *and* 15 *cm long, and* BD *is the perpendicular from* B *on* CA. *Find graphically the magnitude and line of action of the resultant of the following forces:* 8 *from* A *to* C, 4 *from* C *to* B, 3 *from* B *to* A, 2 *from* B *to* D.

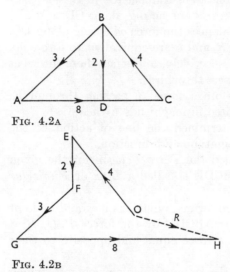

FIG. 4.2A

FIG. 4.2B

The forces are shown in Fig. 4.2A. Their resultant can be found in magnitude and direction by drawing a force-polygon, as in Fig. 4.2B, where the vectors OE, EF, FG, GH represent the forces 4, 2, 3, 8 respectively. The resultant of the forces, R, is represented by the closing side OH.

By measurement, $R = 3 \cdot 4$ and the angle it makes with GH, that is, with AC, equals 13° approximately.

To find the line of action of R, it is not necessary in this case to draw a funicular polygon. For, since three of the forces act through B, their resultant passes through B, and its magnitude and direction is represented by the vector OG in Fig. 4.2B.

If therefore we draw through B a line parallel to OG to meet CA produced at X, say, the resultant of the four forces must pass through this point.

4.5. Bow's Notation

Graphical constructions involving coplanar forces are often facilitated by the use of Bow's notation. According to this, the plane in which the forces act may be regarded as divided up into spaces or compartments by the lines of action of the forces. The spaces are lettered, for example, A, B, C, D, . . . and the force along the line dividing space B from space C is denoted by *bc* and so on.

It is specially convenient in dealing with frameworks, as will be shown later (4.10).

4.6. Resultant of any number of parallel forces

The case when the forces are parallel is a very important one in practice. The method of finding the resultant is exactly similar to that in 4.2, but now the force polygon is a straight line.

Let the forces be P, Q, R, S, T and their lines of action as in Fig. 13.3A.

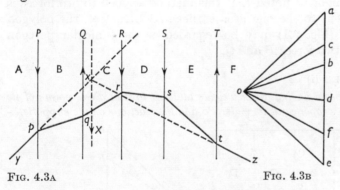

FIG. 4.3A   FIG. 4.3B

Letter the spaces A, B, C, D, E, F as shown, and on a line parallel to the direction of the forces mark off *ab* downwards to represent P to scale (Fig. 4.3B), *bc* upwards to represent Q, *cd, de* and *ef* to represent $R, S,$ and T.

Then *af* represents the resultant in magnitude and direction, and its line of action can be found as follows.

Take any point *o* and join *oa, ob, oc, od, oe,* and *of*.

Take any point *p* on the line of action of P and draw *py, pq* parallel to *ao, ob,* and let *pq* meet the line of action of Q in *q*.

P is equivalent to forces *ao, ob* along *py* and *pq*.

From q draw qr parallel to oc to meet R in r.

Q is equivalent to forces bo, oc acting along qp, qr, and the former balances the force ob along pq.

From r draw rs parallel to od to meet S in s.

R is equivalent to forces co, od acting along rq, rs, and the former balances the force oc acting along qr.

From s draw st parallel to oe to meet T in t.

S is equivalent to forces do, oe acting along sr, st, and the former balances the force od acting along rs.

From t draw tz parallel to of.

T is equivalent to forces eo, of, acting along ts, tz, and the former balances the force oe along st.

We are thus left with a force ao along py and a force of along tz.

Produce yp and zt to meet in x; then the resultant of the forces must pass through x.

The resultant is a force X represented in magnitude and direction by af and passing through x.

It might be noted that this method applied to *two* forces is equivalent to the one used earlier in 2.2. In fact, the polygon HABK (Fig. 2.1) may be regarded as the funicular polygon for the two forces P and Q.

EXAMPLE (i)

A light beam having given weights attached to it at given points of its length is supported at its ends. To find the reactions on the supports.

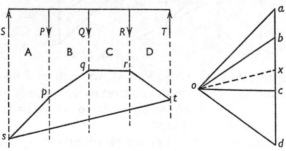

FIG. 4.4A   FIG. 4.4B

Let the weights be P, Q, R acting as shown in Fig. 4.4A, and let S, T be the reactions on the beam at the supports.

Letter the spaces A, B, C, D, as shown, and draw the force polygon $abcd$ (Fig. 4.4B). Take any point o and join oa, ob, oc, od.

Take any point p in the line of action of P, and draw ps parallel to oa to cut S in s and pq parallel to ob to cut Q in q.

From q draw qr parallel to oc to cut R in r, and from r draw rt parallel to od to cut T in t.

The forces P, Q, R are equivalent to forces represented by ao and od acting along ps and rt.

Join st and draw ox parallel to st to cut ad in x.

The force ao acting along ps is equivalent to a force represented by ax acting downwards along the line of S, and a force xo acting along ts.

The force od acting along rt is equivalent to xd acting downwards along the line of T and a force ox acting along st. This latter force balances the force xo acting in the direction ts, and we are left with vertical forces represented by ax and xd acting along the lines of S and T.

The reactions on the beam at the ends must be equal and opposite to these forces and are therefore equal and opposite to ax and xd.

EXAMPLE (ii)

PQRS is a light string attached to two fixed points P *and* S, *and carrying weights at* Q *and* R. *Determine the ratio of these weights if* PQ *is inclined at 30° to the vertical,* QR *at 60° to the vertical, and* RS *is horizontal.*

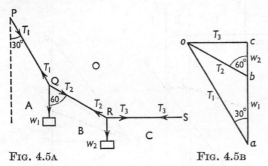

FIG. 4.5A   FIG. 4.5B

The form of the string PQRS is shown in Fig. 4.5A. It is the funicular polygon for the parallel forces W_1 and W_2 acting at Q and R respectively.

From this funicular polygon we can draw the force polygon. For in Fig. 4.5B let ba represent the weight W_1; draw lines from a and b parallel to the strings QP and QR respectively, and let them meet at o. Then this triangle abo is the triangle of forces for the forces W_1, T_1, and T_2 acting at Q.

If now we draw oc parallel to the string RS, that is, horizontal, then the triangle bco is the triangle of forces for the forces W_2, T_2, and T_3 acting at R.

Hence cb represents W_2 on the same scale as ba represents W_1, and therefore $W_1/W_2 = ab/bc$.

But $ac = oc \tan 60° = oc\sqrt{3}$

and $bc = oc \tan 30° = oc/\sqrt{3}$

$\therefore$ $ac = 3bc$

$\therefore$ $W_1 + W_2 = 3W_2$ or $W_1 = 2W_2.$

This result can also be obtained by resolving the forces acting at Q and R horizontally and vertically as follows:

For Q, $T_1 \cos 30° - T_2 \cos 60° = W_1$

and $T_1 \sin 30° - T_2 \sin 60° = 0.$

Also for R, $T_2 \cos 60° = W_2$

and $T_2 \sin 60° - T_3 = 0$

leading to $W_1 = 2W_2.$

EXAMPLE (iii)

Forces of 5, 9, −7, 3, and −10 N act along parallel straight lines, and their distances apart, in the order given, are 10, 5, 7, 3 cm. Find by means of the vector and link polygons the magnitude and sense of the resultant couple.

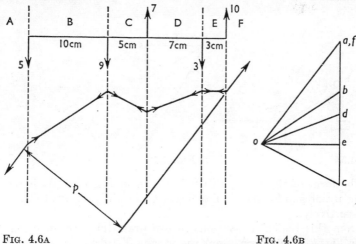

FIG. 4.6A   FIG. 4.6B

Draw the space diagram as in Fig. 4.6A, and letter it as shown.

Draw the force polygon as shown in Fig. 4.6B. In this figure $ab = 5$, $bc = 9$, $cd = 7$, $de = 3$, and ef or $ea = 10$, i.e. the polygon closes.

Construct the funicular by drawing in the compartments A, B, C, D, E, F of the space-diagram lines parallel to oa, ob, oc, od, oe, of respectively.

Each of the parallel forces can be resolved into two components acting along sides of the funicular polygon, as shown in Fig. 13.6A. These components balance in pairs except the first and the last, which are equal and parallel and in opposite directions and so form a couple.

The magnitude of these forces is represented by oa in the force-diagram, and their distance apart, p say, can be obtained from the space-diagram, in Fig. 4.6A.

The moment of the couple is therefore given by $oa \times p$, and is counterclockwise.

From the figures $oa = 12.7$ N and $p = 15.7$ cm. Hence the moment of the couple is $12.7 \times 15.7 = 199$ N cm.

This value can be checked by taking the moments of all the forces about some point, for example, a point on the line of action of the force 5. We get

$$-9 \times 10 + 7 \times 15 - 3 \times 22 + 10 \times 25$$

$$= -90 + 105 - 66 + 250$$

$$= 199 \text{ N cm.}$$

EXAMPLES 4.1

1. Three like forces of 3, 5, and 4 N respectively are parallel to one another, and the lines of action are at intervals of 1 m. Give a graphical construction for the position of their resultant. (I.S.)

2. A beam 9 m long is supported at its two ends, which are on the same level. Loads of 5, 3, 9, and 2 Mg rest on this beam at distances of 2.4, 3.6, 5.1, and 7.5 m from the left-hand end. Obtain the reactions at the supports graphically by link and vector polygons. Also calculate the reactions. (I.C.)

3. Find graphically the position of the resultant of four parallel forces of magnitudes $+7$, $+4$, -5, $+2$ N, the distances between them being 1, 0.5, and 1.2 m in the order given. (I.E.)

4. Loads of mass 2, 3, and 5 kg are placed on a beam 3 m long at distances of 0.3, 0.9, and 1.5 m from one end. Find graphically the line of action of the resultant.

5. A horizontal beam 20 m long is supported at its ends and carries loads of mass 2, 3, 6, and 4 kg at distances of 3, 6, 12, and 15 m respectively from one end. Find graphically the thrusts on the ends.

6. Weights of mass 8, 3, 2, and 6 kg are suspended at distances of 2, 3, 6, and 8 m from one end of a light beam 10 m long supported at its ends. Find graphically the thrusts on the supports.

7. Like parallel forces of 2, 4, 6, 8, and 10 N act at distances of 1 m apart. Find graphically the position of the line of action of their resultant.

8. The loads on the wheels of a locomotive are the weights of 10, 10, 18, 16, 8, and 8 Mg respectively, and their distances apart are 1·2, 3, 2·4, 1·8, and 1·2 m respectively. Find graphically the magnitude and position of the resultant thrust on the rails, and check the result by calculation.

9. A horizontal bar 10 m long carries weights of mass 4, 5, 6, and 7 kg at distances respectively 2, 5, 7, and 9 m from one end; give graphical constructions (i) for the resultant of the weights, (ii) for the pressures on the supports when the bar is supported at its two ends, neglecting the weight of the bar itself.   (Q.E.)

10. The spaces between and the loads taken by the axles of a locomotive are as follows, reading from front to rear:

Spaces		2·7		3		2·4		3		m
Loads	12		12		25		25		5 units

Find by a graphical construction the horizontal distance of the centre of gravity of the locomotive from its front axle.   (Q.E.)

11. A uniform beam, AB, of mass 50 kg and length 6 m, is supported at A and at a point 1·2 m from B. It carries loads of mass 30 kg at points 1·5 m and 2·4 m from A and a load of mass 40 kg at B. Find *graphically* the pressures on the supports.

12. A light string is fastened to a fixed point A and passed over a smooth pulley D at the same level as A, and carries a known weight W at its end. If unknown weights W_1 and W_2 are fastened to points B and C of the string between A and D, show how W_1 and W_2 can be found from the directions of the portions AB, BC, and CD of the string in the equilibrium position.

13. ABCD is a light string. The ends A, D are fastened to fixed points in the same horizontal line. Weights of mass 2·5 kg and P kg are attached at B and C. Determine graphically the value of P and the tensions in the portions AB, BC, CD if AB and CD make angles of 60° and 45° respectively with the horizontal and the angle BCD is 140°.

14. Forces 1, 3, −4 N act in order round the sides of an equilateral triangle ABC of side 5 cm drawn on a rigid lamina. Give a graphical construction for the magnitude, direction, and position of their resultant. (C.W.B.)

15. Forces, 3 N, 2 N, −3·5 N act in order round three consecutive sides of a regular hexagon of side 5 cm drawn on a rigid lamina. Give a graphical construction for the magnitude, direction, and line of action of their resultant. (N.U.3)

4.7. Frameworks

In most cases in which a graphical method is used it is known that the forces are in equilibrium. The forces are usually acting on some sort of framework, and the problem is to find the stresses in the various parts or 'members' of this framework. The method of modifying the construction of the preceding paragraphs for this purpose are illustrated in the following examples.

In considering the equilibrium of a *light* rod it is evident that, if the only forces acting on it are applied at the ends, these forces must be directed along the rod or they cannot balance one another; they must also, of course, be equal and opposite. The stress in a *light* rod, acted on by forces at the ends only, must therefore consist of a thrust or a tension along the length of the rod.

A rod which is in a state of thrust is called a *Strut*, and a rod in a state of tension is called a *Tie*.

4.8. *A closed polygon of light rods, freely jointed at their extremities, is in equilibrium under the action of a given system of forces applied at the joints; to find the stresses in the rods.*

Let AB, BC, CD, DE, EA represent the rods freely jointed at their ends, and let forces P, Q, R, S, and T act at the joints as in Fig. 4.7A.

Let the resulting forces acting on the joints along the rods be T_1, T_2, T_3, T_4, T_5 respectively. The forces on the rods are equal and opposite to these.

Draw the force polygon *abcde* (Fig. 4.7B) having its sides parallel and proportional to P, Q, R, S, and T respectively. Since the forces are in equilibrium, this polygon must close.

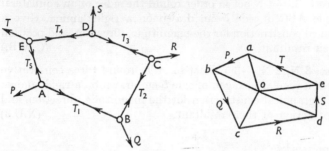

FIG. 4.7A   FIG. 4.7B

Through a draw ao parallel to AE, and through b draw bo parallel to AB. The triangle boa has its sides parallel to the forces P, T_1, and T_5 which act on the joint at A. Its sides are therefore proportional to these forces, and the sides bo and oa must represent T_1 and T_5 on the same scale that ab represents P.

Join oc, od, oe.

The sides ob, bc of the triangle obc represent two of the forces T_1 and Q, which act at B. Hence co represents the third force acting at B, i.e. T_2, and must therefore be parallel to BC.

Similarly do represents T_3 and eo represents T_4.

The lines oa, ob, oc, od, oe therefore represent, in magnitude and direction, the forces along the sides of the framework.

4.9. It is clear that, of the figures in the last paragraph, *abcde* is the force polygon and ABCDE a funicular polygon for the system of forces P, Q, R, S, T.

Also if *abcde* represents a jointed framework acted on by forces oa, ob, oc, od, oe, then ABCDE is the force polygon for this system of forces, *abcde* being the funicular polygon.

Hence either of these polygons may be the framework or funicular polygon, and then the other is the corresponding force polygon. For this reason such figures are said to be *Reciprocal*.

4.10. Bow's Notation

As explained in 4.5, there is another system of lettering the figures which may be used. It is known as Bow's notation. Consider the framework of 4.8.

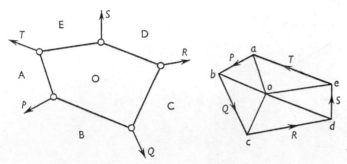

FIG. 4.8

Draw the framework and forces as in Fig. 4.8, and instead of lettering the corners of the framework, letter the *spaces* or *compartments* between the forces and bars, e.g. let the space between T and P be called A, that between P and Q be called B, that between Q and R be called C, and so on.

The line of action of P is the boundary between the spaces A and B, and in the force polygon the line representing P is called ab, that representing Q is called bc and so on.

The force polygon is then $abcde$.

Calling the pole of the force polygon o the space within the funicular is then called O.

A small letter attached to a vertex of the force polygon corresponds to a big letter attached to a space of the funicular.

EXAMPLE (i)

ABCD *is a quadrilateral formed of four light rods freely jointed at their extremities. The angles at A and B are each right angles, the angle ADC is 60° and AD = CD. It is stiffened by a rod AC, and at B and D act forces of 40 N, in such a manner that the frame is in equilibrium. Find the tensions or thrusts in the rods.*

Since AD = DC and ADC = 60°, the triangle ADC is equilateral and the angle CAB = 30°.

Draw AB as in Fig. 4.9, and AC, making BAC = 30° to cut the perpendicular to AB at B in C. Then construct the triangle ACD on AC.

Since the two forces of 40 N balance they must act in the same straight line, i.e. along BD, and in opposite directions. Take them as acting outwards.

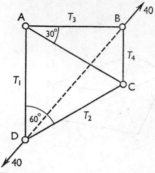

FIG. 4.9

Method (i)

Draw the triangle of forces for the corner D, and obtain the tensions in AD and DC, denoted by T_1 and T_2 ($T_1 = 15 \cdot 1$, $T_2 = 30 \cdot 2$).

Draw the triangle of forces for the corner B, and obtain the tensions in AB and BC, denoted by T_3 and T_4 ($T_3 = 26 \cdot 2$, $T_4 = 30 \cdot 2$). The thrust in AC is found by drawing the triangle of forces for A or C. (This thrust is $30 \cdot 2$.)

The triangles of forces for each joint can be combined compactly in one diagram as explained below.

Method (ii)

Using Bow's notation, we letter the compartments of the diagram (Fig. 4.10A) P, Q, R, S as shown.

To draw the force polygon we start at one of the corners, say B, where one of the known external forces acts, and draw pq to represent the 40 N force in magnitude and direction. We then draw ps parallel to the rod (AB) between the compartments P and S, and qs parallel to the rod (BC) between the compartments Q and S. Then psq is the triangle of forces for the forces acting at the joint B, and since the 40 N force is in the direction q to p, the directions of the other forces acting at B must be as shown in Fig. 4.10A.

Proceeding to the neighbouring joint A, we draw pr parallel to the rod (AD) between the compartments P and R, and sr parallel to the rod (AC) between the compartments S and R. Then prs is the triangle of forces for the forces acting on the joint A, and since the direction of the force *ps* is known, the direction of the others can be found and are as shown in Fig. 4.10A.

If we now join *qr* the triangle *pqr* is the triangle of forces for the forces acting on the joint D, and the directions of the forces are as shown in Fig. 4.10A.

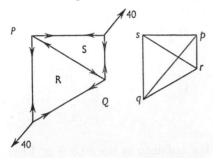

FIG. 4.10A   FIG. 4.10B

The forces acting on the joint C are represented by the triangle *rqs*, and the directions are as shown. It is clear that this triangle is equilateral, and hence the three forces acting on the joint C are equal.

The magnitudes of all the forces may be obtained by measuring the sides of the figure *pqrs*, and using the scale that *pq* represents 40 N.

The forces in the rods are equal and opposite to those acting on the hinges at their ends; consequently (reversing the directions shown in Fig. 4.10A) the rods AB, BC, CD, DA are in tension and the rod AC in thrust.

EXAMPLE (ii)

ABC *is a triangular framework of light rods freely jointed together and placed on supports at A and B so that AB is horizontal and C above AB, and the plane ABC is vertical. BC = 6·6 m, CA = 5·4 m, AB = 6 m. A weight of mass 200 kg is suspended from C. Find by the methods of graphical statics the thrusts or tensions in the three rods.* (H.S.D.)

Draw the framework to scale as in Fig. 4.11, with the lines of action of the external forces *outside the framework.* The external forces are the 200*g* N, and the vertical reactions of the supports *P* and *Q* at A and B.

The reactions *P* and *Q* may be found graphically as in 4.6.

In this simple case, however, there is no need to find P and Q initially: they may be found as follows from the triangle of forces for each joint.

Letter the spaces formed by the lines of action of the external forces and the bars of the framework, X, Y, Z, W, as shown.

The space Y extends indefinitely to the left, being separated from the other spaces by the 200g N force, the bar AC, and the force P.

Similarly, the space X extends indefinitely to the right.

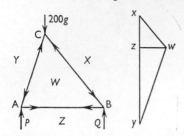

FIG. 4.11

Starting with the joint C; it is acted upon by the 200g N and the stresses along AC and CB. Now draw xy vertically downwards to represent 200g N; complete the triangle of forces for C by drawing yw parallel to AC and xw parallel to CB. Since the forces 200g N represented by xy is vertically downwards, the directions of the forces acting on the joint C are therefore as shown.

Considering next the joint B, we complete the triangle of forces xwz by drawing wz parallel to BA. zx must therefore represent the reaction Q and wz the stress in AB.

Similarly, yzw is the triangle of forces for the joint A and yz represents the reaction P. Again the directions of the forces acting at the joint A can be found from the triangle yzw, since the force P represented by yz is upwards.

Since the directions of the forces acting on the joints are as shown, the bar AB is in tension and the bars BC and CA in thrust.

The tensions and thrusts can be denoted by arrows, as shown, or by putting − for a tension and + for a thrust. Sometimes they are distinguished by putting one and two strokes through the bar. Great care must be taken to remember that in obtaining the directions of the forces from the triangle of forces we must go round the triangle in order, e.g. for A, since yz represents P, zw represents the force acting on the joint A along AB, i.e. it acts towards the right away from A. To find the nature of the stresses in the rods the arrows may be reversed.

EXAMPLE (iii)

A framework of seven rods is in the form of three equilateral triangles ABC, BCD, CDE. It rests on smooth vertical supports at A and E, with BD *and* ACE *horizontal, and* BD *above* ACE, *and carries loads of 5 units at* B, *5 units at* C, *and* 10 *units at* D. *Neglecting the weights of the rods, find the reactions at* A *and* E, *and determine. preferably by means of a stress diagram, the stresses in each of the rods, indicating which are thrusts and which are tensions.*

Draw the framework and insert the external forces as in Fig. 4.12.

The reactions P and Q acting on the framework at A and E can be found graphically, but in this case it is much easier to find them by taking moments.

Taking moments about A we get

$$4Q = 5\times1+5\times2+10\times3 = 45$$

∴ $$Q = 11\tfrac{1}{4} \text{ and } P = 8\tfrac{3}{4}.$$

We then letter the spaces as in Fig. 4.12 (all the external forces being drawn *external* to the framework), and draw the force polygon as described below.

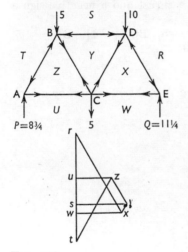

FIG. 4.12

On a vertical line with scale 2 cm = 5 units, mark off $rs = 10$, $st = 5$, and then $tu = 8\tfrac{3}{4}$ upwards to represent P. Now mark off $uw = 5$ to represent the weight of 5 units at C. wr represents Q the reaction at E.

Construct the triangle of forces for the forces acting on the joint A by drawing uz parallel to AC and tz parallel to AB. uz represents the stress along AC and zt the stress along AB. The directions of the forces acting on the joint A are shown. AC is in tension and AB in thrust.

Of the force polygon for B we have already st and tz, and to complete it we draw zy parallel to BC and sy parallel to BD.

zy represents the stress in BC (a tension) and ys the stress in BD (a thrust).

Of the force polygon for D we have already rs and sy and to complete it we draw yx parallel to DC and rx parallel to DE.

yx represents the stress in CD (a tension), and xr the stress in DE (a thrust).

Since wr represents Q and rx the thrust on E due to DE, wx must be the third side of the triangle of force for E, i.e. wx must be parallel to CE and xw represents the stress in CE (a tension).

We notice that we have also a closed polygon $yzuwx$ for the joint C. The directions of the forces acting on this joint can be checked since the force uw is downwards.

The stresses in the various rods can be found by measuring the corresponding sides of the force-diagram. They are approximately as follows, a positive sign denoting a thrust and a negative sign a tension.

Stresses in AB $= +10$, AC $= -5$, BC $= -4\frac{1}{4}$, BD $= +5$, CD $= -1\frac{3}{4}$, CE $= -6\frac{1}{4}$, DE $= +13$.

EXAMPLE (iv)

A crane, as in Fig. 4.13, is pinned at A and kept vertical by a horizontal pressure at B.

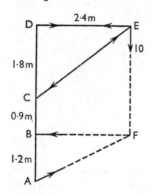

FIG. 4.13

If a load of mass 10 *Mg be suspended from* E, *show how to find the reactions at* A, B, *and the forces in* CE, DE *graphically. Also calculate the reactions and forces.*

Method (i)

Draw the figure to scale, say 1 cm = 1 m.

Since the rod AD is acted on by forces at points other than its ends, the stress in it will not be a simple thrust or tension. To find the reactions at A and B we consider all the external forces acting on the crane, i.e. these two reactions and the weight of 10 units. The unit is the weight of 1 Mg.

Draw BF horizontal to meet the vertical through E in F, then the reaction at A must pass through F.

Join AF. The triangle ABF will serve as a triangle of forces for the reactions and the load. The lengths of BA, AF, FB, are 1·2, $1·2\sqrt{5}$, 2·4 cm, and these represent 10 units, the reaction at A and the reaction at B respectively.

Hence the reaction at A is $10\sqrt{5}$ units, and the reaction at B is 20 units.

The triangle DCE is a triangle of forces for the corner E.

The lengths of DC, CE, ED are 1·8, 3, and 2·4 cm, and these represent 10 units, the stress in CE and the stress in ED respectively. Hence the stress in CE is $\frac{50}{3}$ units (a thrust), and the stress in ED is $\frac{40}{3}$ units (a tension).

Method (ii)

To *calculate* the reaction at B and A.

Let P units be the reaction at B, then, taking moments about A for the whole frame,
$$1·2P = 2·4 \times 10$$
$$\therefore \qquad P = 20.$$

Let X, Y units be the horizontal and vertical components of the reaction at A, then
$$X = P = 20$$
and
$$Y = 10.$$

If R is the resultant reaction,
$$R = \sqrt{[20^2 + 10^2]} = 10\sqrt{5} \text{ units.}$$

R is inclined to the horizontal at an angle $\tan^{-1} \frac{1}{2}$.

To calculate the stresses in CE, ED.

Let T_1 units be the thrust in CE.

Resolving vertically for the corner E,
$$T_1 \sin \text{CED} = 10$$
$$\therefore \qquad \tfrac{3}{5}T_1 = 10 \text{ or } T_1 = \tfrac{50}{3}.$$

Let T_2 units be the tension in ED.

Resolving horizontally for the corner E,
$$T_2 = T_1 \cos \text{CED} = \tfrac{50}{3} \times \tfrac{4}{5} = \tfrac{40}{3}.$$

EXAMPLES 4.2

1. A framework is built up of rigid bars AB = AD = 0·6 m., BC = CD = 37·5 cm, smoothly hinged together. The framework is placed on a horizontal table in such a manner that the angle BAD = 60°, and A, C are on opposite sides of BD. B and D are joined by a string capable of supporting a weight of mass 14 kg. Find the greatest forces which can be applied to A and C so that the framework is in equilibrium in the given position with the string intact, and determine the corresponding stresses in the members of the framework, stating whether they are tensions or compressions.   (H.S.D.)

2. A rod AB 3 m long, hinged to a wall at B and carrying a weight of mass 50 kg at A, is supported by a cord attached to the middle point C of the rod and carried to a point D of the wall 2·1 m above B, the length of CD being 1·8 m. Draw a diagram to scale, and from a triangle of forces estimate the tension in the cord and the magnitude and direction of the reaction of the hinge B.

3. A uniform beam AB, of mass 50 kg, is supported by strings AC and BD, the latter being vertical and the angles CAB and ABD, being each 105°. The rod is maintained in this position by a horizontal force F applied at B. Show that the value of F is about 122 N.

4. Five equal light rods are freely jointed together at their extremities to form a pentagon ABCDE. The system is suspended from the joint A, and a load w is suspended from the joint C. the shape of a regular pentagon being kept by rods BE, BD. Draw a diagram showing graphically the magnitudes of the stresses in the various rods, and find which of the rods are in compression and which in tension.   (I.S.)

5. An equilateral triangle ABC formed of three light rods, each 0·6 m long jointed at their ends, is supported at a point in AC which is 37·5 cm from A, and a weight of mass 6·5 kg is hung from a point in AB 17·5 cm from A. Find (by graphical or other method) the tension in the rod BC, and the action between the other two rods at A.   (I.A.)

6. Three light bars are jointed together to form a triangular framework ABC in which the angles A and C are each 30°. The framework can turn in a vertical plane about the point B, and is kept in equilibrium with AB horizontal by a weight of mass 50 kg hung at C and a vertical force P at A. Find, graphically or otherwise, the magnitudes of the force P and of the stresses in the rods.   (H.S.D.)

7. Five light rods form a parallelogram ABCD and its diagonal BD. The sides AD and BC of the parallelogram are twice as long as the other two sides, and the angles at A and C are 60°. Two equal and opposite forces of magnitude F are applied at A and C along the diagonal AC so that the parallelogram remains in equilibrium. Find the stresses in all the rods. (I.S.)

8. The framework shown in Fig. 4.14 is formed by smoothly jointed rods, the length of AB and AD is 0·9 m, of BC and DC 1·2 m, the distance AC is 1·5 m, and E is the point mid-way between A and C. Forces of 40 N act at C and E as shown. Find the tension or compression in each rod of the framework. (Q.E.)

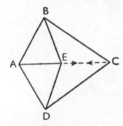

Fig. 4.14

9. A regular pentagon ABCDE (Fig. 4.15) jointed at the angles is stiffened by two jointed bars AC, AD. Two equal and opposite forces, each equal to 15 N, are applied at B and E. Find the stress in each bar of the framework, stating whether it is tensile or compressive. (Q.E.)

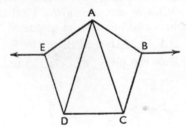

Fig. 4.15

10. A rectangle ABCD in which the sides AB, BC are respectively 8 and 12 cm is formed by smoothly jointed rods, and is made stiff by a diagonal BD; the frame is hung up from A and a weight of mass 10 kg is fastened at C; draw a stress diagram for the forces in the rods (which are supposed to be light), and hence find the stresses, distinguishing between thrusts and tensions. (I.S.)

11. Figure 4.16 represents a framework of nine smoothly jointed bars supported by ropes at A, B in the directions shown and carrying loads at X, Y. Find *graphically* the forces in each bar, indicating the results in your figure, and showing whether they are tensions or thrusts. (Neglect the weight of the bars.)

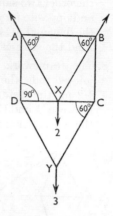

12. Four rods AB, BC, CD, and DE of a framework ABCDE form sides of a regular hexagon, and the framework is stiffened by rods joining AC, AE, and CE. Equal forces of 20 units are applied at the joints B and D from outside along directions perpendicular to AC and CE respectively, and the frame is kept in equilibrium by forces applied at A and E along AB and ED respectively. Find the magnitude of these forces and the stresses in the various members of the frame.   (N.U.)

FIG. 4.16

13. Three equal strings of negligible weight are knotted together to form an equivalent triangle ABC. A mass of 30 units is suspended from A, while the triangle and weight are supported by strings at B and C so that BC is horizontal. If the supporting strings are equally inclined at an angle of 135° to BC, find the tensions in the strings by a graphical method.   (N.U. 3 and 4)

14. Figure 4.17 represents the framework of a roof whose weight may be regarded as distributed in the manner shown. Find the stress in each of the nine members and indicate its nature.

$$AE = EC = ED = 2CF = 2FD = 2FB.$$

Weight at E = 200 units, at F = 100 units, at C = 150 units, and ACB is a right angle.

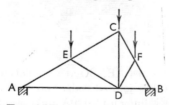

FIG. 4.17

15. The framework ABCDEF (Fig. 4.18) is composed of light rods, smoothly jointed; it is hung from smooth pins at A, B, and carries weights as shown. Find the stresses in all the rods to the

nearest unit, and mark with a double line the rods which are under a thrust.

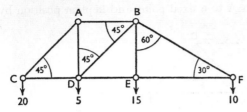

Fig. 4.18

16. Figure 4.19 represents a framework of nine light rods loaded as indicated, resting on vertical supports at A, B with AB horizontal. Find the reactions at A and B and the stresses in all the rods.

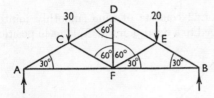

Fig. 4.19

17. The framework in Fig. 4.20 consists of four light bars, AB, BC, CD, DB freely jointed at B, C, D and attached to a vertical wall at A and D. A weight of 10 units is suspended from C. Find the stresses in all the bars and the reactions at A and D. Mark the struts with double lines.

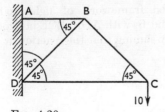

Fig. 4.20

18. A framework of nine equal light rods freely jointed together is formed by four equilateral triangles, ABC, CBD, CDE, EDF, the whole forming a parallelogram ABFE. The framework is placed in a vertical plane with the lowest rod AB horizontal, a weight of 20 units is suspended from F and vertical forces at A and B maintain equilibrium. Draw a force diagram to show the stresses in the rods.

19. Figure 4.21 represents a framework of light bars smoothly jointed so as to form three isosceles right-angled triangles. The frame is hinged at A to a fixed point and kept in position by a

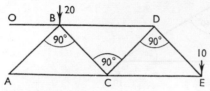

Fig. 4.21

light horizontal rod OB hinged to it at B and to a fixed point O; it is loaded as shown. Find the reaction at A and the tension in OB, and draw a stress diagram to give the stresses in all the bars.

20. The framework in Fig. 4.22 consists of nine smoothly jointed light rods, smoothly hinged to a fixed point at A, kept in position

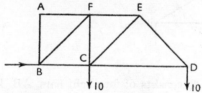

Fig. 4.22

by a horizontal reaction at B, and loaded with 10 units each at C and D, the angles in the figure are all 45° or 90°. Determine the stresses in the rods, marking the struts with a double line.

21. Figure 4.23 represents a loaded framework of light rods, attached to a vertical wall at A and B, with ADF, BCE horizontal. The angles are all 45° or 90°. Assuming that the reaction at B

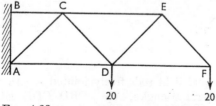

Fig. 4.23

is wholly along BC, find the magnitude and direction of the reaction at A, and find the stresses in all the rods, indicating which are in a state of thrust and which are in a state of tension. Measurements in a stress diagram will be sufficient.   (I.S.)

22. ABCDE (Fig. 4.24) is a light smoothly jointed vertical frame held so that ED is horizontal, the angles and lengths being as

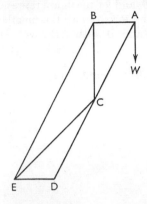

follows. ABED is a parallelogram, AB = 2 m, BC = 4 m, ABC is a right angle, and AC = CD. A weight W is hung at A. Find the stresses in the six rods, specifying their nature in each case.   (I.S.)

FIG. 4.24

23. Draw a triangle AET with AT = AE. Let AE represent to scale the horizontal, and AT the vertical part of a structure in which AT = 12 m. Divide AE into four equal parts, and letter the dividing points, B, C, D. Join TB, TC, TD, and let these lines with TE represent tie bars. Consider AB, BC, CD, DE to be rigid, weightless bars, and suppose loads of 1 unit to be suspended from the points B, C, D, and E. Obtain graphically or otherwise the tensions in the tie bars.   (I.E.)

24. Figure 4.25 represents a crane composed of four freely jointed, light bars, AC, BC, BD, CD in a vertical plane, supporting a

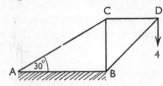

FIG. 4.25

weight of 4 units at D; AB and CD are horizontal, and BC vertical. Find graphically or otherwise, the force acting along each bar.

25. Figure 4.26 represents a bridge girder freely supported at A and E; CF and CH are squares, and AB = BF = DE. Loads of 10,

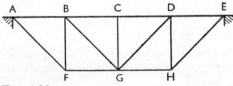

FIG. 4.26

15, and 12 units are applied at B, C, and D. Determine the stresses in AB, BF, BG, FG, and CG.   (I.E.)

26. Draw a rectangle ABCD with AB = DC = 15 cm, AD = BC = 5 cm. Bisect DC in E, and trisect AB in F and G, F being nearer to A than to B. Join DF, EG, FE, CG, and let the figure represent a bridge truss to scale. Determine the forces in all the members when the truss is supported at A and B with AB level, and loaded with 1 unit at F.   (I.E.)

27. In the frame ABCDE (Fig. 4.27) of freely jointed bars,

$$AD = DC = CE = EB = DE,$$

and $$AC = CB = 1{\cdot}8AD.$$

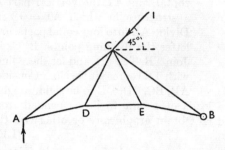

FIG. 4.27

It is freely supported at A, and B is hinged to a support at the same level. A force of 1 unit acts at C as shown; find the reactions at A and B, and show how to draw a stress diagram for the bars.   (I.E.)

28. The frame of smoothly jointed rods represented in Fig. 4.28 is supported at A and B, AB being horizontal, and weights of 8 and

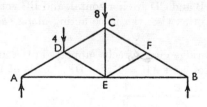

FIG. 4.28

4 units are suspended from C and D. The acute angles in the figure are 30° or 60°. Find by a stress diagram the stresses in each member of the frame.   (I.S.)

29. Three equal light bars are jointed together to form a triangular framework ABC. The framework is suspended from A, and

weights of 4 units and 5 units are attached at B and C. Find, graphically or otherwise, the angle which AB makes with the vertical and the force along BC. (H.S.D.)

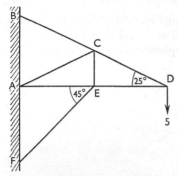

FIG. 4.29

30. Find the stresses in the bars of the wall crane shown in Fig. 4.29 due to a load of 5 units hanging from a chain at D. The chain passes over pulleys at D and E; it is fixed to the wall at F, and AE = ED. (I.E.)

31. The framework in Fig. 4.30 consists of seven light, smoothly

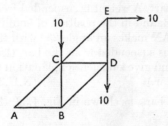

FIG. 4.30

jointed rods, the horizontal and vertical rods being equal. It is loaded at C and D with equal weights of 10 units, and a horizontal force of 10 units acts at E. The framework is hinged at B to a fixed support and anchored at A. Find the stresses in the rods and mark those rods which are ties. (H.C.)

32. A framework consists of seven light rods AB, BC, CA, BD, CD, DE, CE smoothly jointed together so that ABC, BCD, CDE are right-angled isosceles triangles, the right angles being at B, C, D respectively, and ACE are in one straight line, which is parallel to BD. It is pivoted to a fixed point at A and rests with AB vertical and B above A and the joint E against a support, which

exerts a pressure at right angles to AE. It carries loads of 10 units at each of the joints B, C, D. Show that there is no stress in CD, and find the stresses in the other rods, indicating which are thrusts and which are tensions.   (H.C.)

33. In the jointed frame, shown in Fig. 4.31, AB and CD are horizontal. The frame is supported at A and B and carries a load of 100 units at C. Find, either graphically or by calculation, the sup-

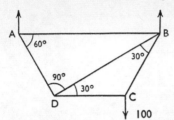

Fig. 4.31

porting forces at A and B and the forces in the five members of the frame.   (Q.E.)

34. A uniform bar AB is 0·75 m long and has a mass of 5 kg. It is suspended by strings AX, BY, respectively 0·6 m and 0·75 m long, which are tied to the ends of the bar and to two points X, Y at the same level and 1·2 m apart. A weight is suspended from the bar at a point 0·15 m from B, and in the position of equilibrium it is found that the string AX makes an angle 25° with the vertical. Find the weight which is suspended from the bar. (Use a graphical method if you can, and give sufficient explanation to make the details of the work clear.)   (N.U.3.)

35. A roof-truss is composed of nine bars, as shown in Fig. 4.32, the

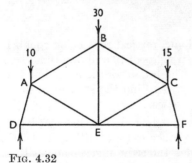

Fig. 4.32

bars, AB, BC, AE, EC, DE, EF being 4·5 m long and the bars AD, CF being 1·8 m long; it is supported at D and F with DEF horizontal. It may be assumed that the bars are smoothly jointed together and that their weights are negligible. If the roof-truss carries loads of 10, 30, and 15 units at A, B, and C, determine graphically the forces on the supports and the stresses in the bars.   (N.U.4.)

36. Draw a diagram to show the stress in each member of the frame in Fig. 4.33, loaded and supported as shown. Indicate by arrows whether the bars are in tension or thrust.   (C.W.B.)

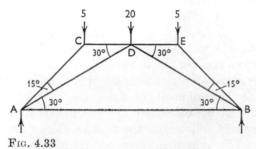

FIG. 4.33

4.11 Number of bars required to stiffen a framework

In the preceding paragraphs it has been assumed that the frameworks involved were *stiff*, i.e. that they were not deformed by the external forces acting on them.

We shall now consider how many bars are necessary to stiffen a frame with a given number of joints.

Suppose there are n joints A_1, A_2, . . . A_n, then assuming n to be not less than 3, we begin by stiffening two, A_1 and A_2, by means of one bar. The remaining $n-2$ joints have now to be connected to these. In order to connect rigidly a third joint A_3 to these two, it must be joined to both of them, and thus requires two more bars.

To connect a fourth joint A_4 to these it must be joined to any two of the first three, and this requires two more bars. Proceeding in this manner, we see that, except for the first two joints, each of the remaining $n-2$ joints requires two bars to stiffen it.

Hence to make a system of n joints rigid, a framework of $2(n-2)+1$, i.e. $2n-3$ bars is sufficient.

4.12 A system of joints made rigid by just the necessary number of bars is said to be *simply stiff* or *just stiff*.

When there are more bars than the necessary number the system is said to be *overstiff*.

When a simply stiff frame is subjected to a given system of external forces, acting on the joints, which keep it in equilibrium we can obtain sufficient equations to determine the reactions in all the bars.

For the conditions of equilibrium for the n joints, resolving in two directions for each, give $2n$ equations.

These equations, however, must involve the three conditions of equilibrium of the frame as a whole. This leaves therefore $2n-3$ independent equations, which in general are just sufficient to determine the unknown stresses in the $2n-3$ bars.

If there are more than the necessary number of bars we have not enough equations to determine all the stresses.

It must be noticed that because there are $2n-3$ bars to n joints it does not necessarily follow that the framework is stiff; one part may be stiffened by more than the necessary number of rods, and another part may not have a sufficient number. In addition to this there are what are called 'critical forms' where, although the frame has a structure which would be ordinarily sufficient for rigidity, it admits of very small deformation owing to some special relation between the lengths of the bars. These cases will not be considered.

4.13. In all the cases considered so far the framework has had at least one joint where only two bars meet (often called a 'single' joint). In these cases we must always start at such a joint where one of the external forces is acting, for we can then determine the stresses in the two bars hinged to that joint.

In some frameworks, however, there are no single joints where external forces are applied, and in such cases the ordinary method fails—we have no corner to start on.

There are various special methods for dealing with such cases, and we shall consider one of them which can often be applied. This is called the 'method of sections'.

4.14. Method of sections

This method can be applied whenever the frame can be regarded as made up of two distinct portions which are connected by *three* bars, i.e. if we can draw a line across the framework which cuts only three bars and divides it into two parts; the bars not being concurrent or parallel.

Thus suppose we want to find the stresses in the bars A_1A_2, A_2A_3, A_4A_5, of a framework of which a portion is shown in Fig. 4.34.

These bars can be cut by a line BCD which cuts no other bars. If we imagine the right-hand portion of the framework removed, then stresses P, Q, R in the three cut bars must be in equilibrium with the external force or forces acting on the left-hand portion. These external forces can be reduced to a single

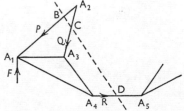

FIG. 4.34

force (or a couple) which can be resolved into three components acting along lines of the bars A_1A_2, A_2A_3, A_4A_5, since these are not all parallel or all concurrent, by the method of Example (iii), paragraph 3.27.

Hence the stresses P, Q, R are determinate.

In many cases the stresses can be obtained by taking moments.

Thus, if, in the figure above, the external force is a force F at A_1, and we take moments about A_2, for the *left-hand portion of the frame*, the moment of R is equal to the moment of F. Having found R, we take moments about A_1 and determine Q. P can then be found by taking moments about A_3.

The method of sections is, as mentioned in the last paragraph, especially useful when all the joints at which known forces act have three or more bars, as A_1 in the figure.

It may also be used in the case of a complicated framework when we require the stresses in certain bars only.

EXAMPLE

The symmetrical framework of smoothly jointed light rods shown in Fig. 4.35 is freely supported at A and B, and loaded with 10 units at C.

Show how to draw a stress diagram for the frame and determine the stress in FG.

It is clear, from symmetry, that the vertical reactions at A and B are each 5 units.

At each of the corners A, C, B where we have known forces acting there are, however, more than two bars; the ordinary method therefore fails, and we proceed by the method of sections.

We can cut the three bars AC, CD, FG, by a straight line meeting them in X, Y, Z respectively.

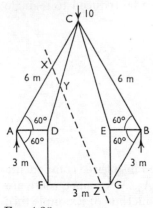

Fig. 4.35

Using these letters to denote the stresses in the respective bars, and considering the equilibrium of the part of the frame to the left of the section, we can obtain the three stresses by resolving the 5 units force at A along the lines of the cut bars.

We can also obtain these stresses by taking moments for the left-hand portion of the framework.

Taking moments about C, we have

$$Z \times 4 \cdot 5 \sqrt{3} = 5 \times 3$$

$$\therefore \qquad Z = \frac{10}{3\sqrt{3}}.$$

We can now find X and Y by taking moments about D and A in turn. We then apply the triangle of forces to the corners D and F, and so determine the stresses in AD and AF.

Otherwise having found Z, the stress in FG, we can start drawing the stress diagram in the ordinary way from the corner F as we now know one of the three forces acting there.

4.15. External forces acting on the bars

In the cases considered so far the external forces acting on the frame have been supposed to act only at the joints. When external forces are applied to the bars themselves the stress in each bar will no longer consist of a purely longitudinal thrust or tension. We shall see later (Chapter Eight) that in such cases the stress at any point of the bar in general consists of three parts: (i) a tension or thrust along the bar; (ii) a couple tending to bend the bar; (iii) a force acting across the bar perpendicular to its

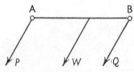

FIG. 4.36

length. All three of these vary from point to point of the bar, so that we cannot express the stress in a bar acted on by external forces as one fixed quantity in the way we have done for bars acted on by forces at their ends only.

We can, however, find the forces exerted by the bar on the joints at its ends in the following way.

Let a bar AB, joined at the ends, be acted on by a force W as shown in Fig. 4.36.

Resolve W into two parallel components P and Q at A and B. This transformation of W cannot affect the distribution of stress over the rest of the framework, so that when the forces P and Q are combined with the other forces acting at the joints A and B we can find the stresses in the bars in the ordinary way. The values found in this way for unloaded bars will be the actual values of the stresses in those bars, but the value found for a bar such as AB will give only the thrust or tension in it.

To find the force exerted by AB on A we compound with this thrust or tension a force equal to P. Similarly for the force exerted on B.

EXAMPLE

Fig. 4.37 represents a roof truss composed of seven equal bars, AE, EC, CD, DB, EF, FD, CF, *and two long horizontal bars,* AF, FB. *The length of* AB *is 3 m.*

The end B *is fixed, the end* A *supported on a roller. The dead weight of the roof is distributed as indicated* (1000 *units each at* A *and* B, 2000 *units each at* E, C *and* D). *Wind blowing from the right causes pressures as indicated perpendicular to* CDB (2500 *units at* B *and* C, 5000 *units at* D). *Neglecting the weights of the bars, draw the force diagram.*

What happens if both A *and* B *are fixed?* (C.S.)

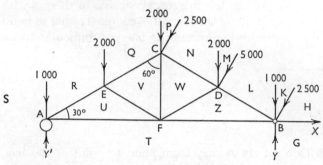

Fig. 4.37

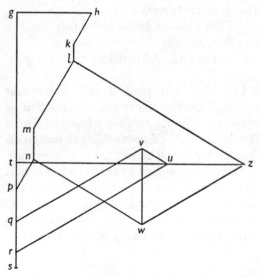

Fig. 4.38

There will be a vertical reaction Y' at A, and horizontal and vertical reactions X and Y at B.

These can be calculated by taking moments and resolving for the whole frame. The values are

$$X = 5000, \quad Y = 9800, \quad Y' = 6900.$$

Letter the spaces as shown, and draw the force polygon *ghklmnpqrstg* (Fig. 4.38), starting with *gh* to represent X, *hk* to represent the 2500 units force at B, *kl* the 1000 units force at B, and so on.

Now draw the force polygon *rstu* for the corner A.

Next draw the polygon *qruv* for E, *npqvwn* for C, *lmnwzl* for D. For B we have the polygon *ghklztg*, and *zt* gives the stress in BF.

The thrusts and tensions can be read off from the force diagram. If A is fixed as well as B the horizontal thrust X is indeterminate, as we have two unknown horizontal reactions in the same straight line, and cannot find them separately.

EXAMPLES 4.3

1. Sketch the force diagram for the Warren girder shown in Fig. 4.39, each triangle being isosceles and having its altitude equal to

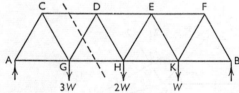

FIG. 4.39

   its base. Calculate the forces in the bars cut by the dotted line by considering the equilibrium of the frame to either side of this line.
   (I.C.)

2. Draw the force diagram for the symmetrical frame illustrated in Fig. 4.40, explaining how to proceed when the usual method fails.

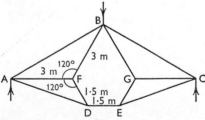

FIG. 4.40

3. The symmetrical framework shown in Fig. 4.41 is freely supported at A and C, and loaded with 10 units at B, 2 units at E,

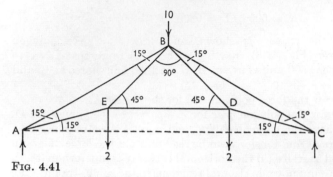

FIG. 4.41

and 2 units at D. Find by the method of sections the stresses in AB, EB, and ED.

4. Fig. 4.42 represents a roof truss on which the dead weight of the roof is distributed, as shown (1 unit each at A and B, 2 units each at E, C, D). The end A is fixed, the end B supported on a roller.

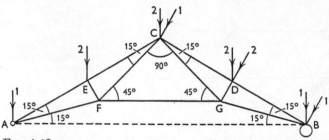

FIG. 4.42

Wind blowing from the right causes pressures as indicated perpendicular to BC. Neglecting the weight of the bars, draw the force diagram and find the tensions and thrusts in the various members.

CHAPTER FIVE

FRICTION

5.1. The laws of Friction and their application to the equilibrium of a particle on a rough surface were considered in Chapter Ten. We shall now consider cases where bodies other than particles are in equilibrium under frictional as well as other forces.

Before proceeding to discuss particular problems, there are one or two general points in connection with friction which are worth noticing.

5.2. Friction plays an important part in everyday mechanics. When a person is walking there is a tendency for the feet to slip backwards; this is prevented by the friction between the feet and the ground, which acts in the forward direction. The force of friction is really the propelling force; without friction it would be impossible to walk.

A railway engine cannot move forward, even without a train behind it, unless there is an external force acting in the forward direction. The engine can only cause its own driving wheels to rotate. If these are resting on smooth rails they will revolve without causing any forward motion. In practice, the friction between the driving wheel and the rail tends to prevent rotation or rather to prevent slipping at the point of contact C (Fig.

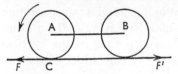

Fig. 5.1

5.1). The wheel is thus made to roll along the rail, and the friction force F called into play acts in a forward direction. This friction gives the magnitude of the tractive force of the engine. The other wheels of the engine or the wheels of a carriage of the train are drawn forward by a force applied to the axle as at B. Such wheels on smooth rails would simply skid forward *without turning*; the effect of friction which acts in the *backward direction*, as F' in Fig. 5.1, causes them to roll.

5.3. *Action of brakes*

The brakes on a motor car tend to stop the wheels rotating. If they actually lock the wheels (i.e. prevent them rotating altogether) sliding friction acts at the points of contact with the ground and reduces the car to rest.

In most cases, however, the brakes merely check the rotation of the wheels, which still keep rolling, and in this case it is not perhaps obvious how the force tending to stop the car is produced.

It is clear that checking the rotation of the wheel cannot, in itself, produce a retarding force on the car. If the ground were smooth, then even locking the wheels would have no such effect.

When a wheel, whose centre is A (Fig. 5.1) and point of contact C, is rolling freely there is no slipping at C. The backward velocity of C relative to A is equal to the forward velocity of A so that C is instantaneously at rest. Hence if the rotation of the wheel is checked the backward velocity of C is diminished, so that C tends to move forward with A. This brings into play a backward friction force at C, and it is this force which retards the forward motion of the wheel.

5.4. Problems involving friction can be classified roughly as follows:

(a) Those in which the body is rigid and equilibrium can only be broken by sliding, the direction of motion being obvious. For example, a ladder resting on rough ground against a rough wall in a vertical plane perpendicular to the wall. If no other forces act it is obvious that if motion ensues in the vertical plane it will be such that the lower end of the ladder slips away from the wall and the upper end slips downwards at the same time. One end cannot slip without the other.

(b) Those in which equilibrium may be broken either by sliding or tilting. For example, a block or cylinder resting on a rough plane which is gradually tilted. If the vertical through the centre of gravity comes outside the base before sliding commences the body will topple over before sliding.

(c) Those in which we have a non-rigid body such as two jointed rods AB, BC (Fig. 5.2) freely jointed at B and with the ends A and C resting on a rough horizontal plane. In this case it is not necessary for slipping to occur at *both* A and C. The friction may be limiting at A or at C without being limiting at both. It *may*, of course, be limiting at both. If there are no external forces other than the weights of the rod and weights placed on them, then, in equilibrium, the frictions at A and C must be equal and opposite, since they are the only horizontal forces acting.

Care must be taken, however, not to assume that either of them is limiting, i.e. equal to μR.

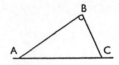

Fig. 5.2

4. More difficult problems where the direction of motion is not obvious; or where equilibrium may be broken by sliding or rolling, e.g. a ladder resting on rough ground and against a rough wall, but not in a vertical plane perpendicular to the wall; or a rough cylinder resting on two others, which in turn rest on a rough horizontal plane.

Examples of various kinds are illustrated in the following paragraphs.

5.5 *Problems involving slipping*

EXAMPLE (i)

One end of a uniform ladder, of weight W, rests against a smooth wall, and the other end on rough horizontal ground, the coefficient of friction being μ. Find the inclination of the ladder to the horizontal when it is on the point of slipping, and the reactions at the wall and ground.

Let AB (Fig. 5.3) be the ladder, G its centre of gravity, C the intersection of the wall and the ground. Since the wall is smooth, the reaction R at B must be perpendicular to the wall.

If S is the normal reaction on the ladder at A, then, since the ladder is on the point of slipping, the friction at A is μS towards the wall.

Method (i)

There are thus four forces acting on the ladder, and by resolving horizontally and vertically, and taking moments about C or A, we obtain three equations to find the unknown reactions R and S, and the angle θ.

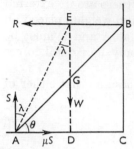

FIG. 5.3

Resolving horizontally $\qquad\qquad\qquad R = \mu S \qquad\qquad\qquad$ (i)

Resolving vertically, $\qquad\qquad\qquad S = W \qquad\qquad\qquad$ (ii)

Taking moments about C,

$$Rl \sin \theta + W\frac{l}{2} \cos \theta = Sl \cos \theta \qquad\qquad \text{(iii)}$$

where l is the length of the ladder.

From (i) and (ii) $R = \mu W = W \tan \lambda$, where λ is the angle of friction.

From (iii) $\qquad W \tan \lambda \sin \theta + \tfrac{1}{2}W \cos \theta = W \cos \theta$

$\therefore \qquad\qquad\qquad W \tan \lambda \tan \theta = \tfrac{1}{2}W$

$\therefore \qquad\qquad\qquad\qquad \tan \theta = \tfrac{1}{2} \cot \lambda.$

The resultant reaction at the ground is $\quad \sqrt{(S^2 + \mu^2 S^2)}$

$$= S\sqrt{(1 + \tan^2 \lambda)}$$

$$= W \sec \lambda.$$

Method (ii)

The angle of inclination of the ladder can also be found by considering the *resultant* reaction of the ground at A instead of the two components S and μS. This reduces the number of forces acting on the ladder to three, which must therefore meet in a point, viz. the point E, where the vertical through G cuts the line of action of R.

Also since the ladder is on the point of slipping, the resultant reaction at A makes an angle λ, the angle of friction, with the normal at A.

If (Fig. 5.3) the vertical EG cuts AC at D we have

$$\tan \theta = \frac{BC}{AC} = \frac{ED}{2AD} = \tfrac{1}{2} \cot \lambda.$$

The geometrical method is particularly useful when we require only the *position* of a body in limiting equilibrium, as we obtain the result without introducing the reactions and having to solve equations. Even when reactions are required, it is sometimes useful to use this method to find the position of the body and then find the reactions by resolving or taking moments.

EXAMPLE (ii)

One end of a uniform ladder, of weight W, rests against a rough wall, and the other end on rough horizontal ground, the coefficient of friction at the ground and wall being μ and μ' respectively. Find the inclination of the ladder to the horizontal when it is on the point of slipping, and the reactions at the wall and ground.

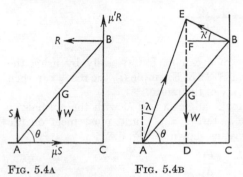

FIG. 5.4A   FIG. 5.4B

Let AB (Fig. 5.4A) be the ladder, G its centre of gravity, R and S the normal reactions acting on the ladder at the wall and ground. Since both ends are on the point of slipping, we have a force of friction $\mu' R$ acting *upwards* at B, and μS acting *towards* the wall at A.

Method (i)

Let θ be the inclination of the ladder to the ground, and $2l$ its length. Resolving horizontally,

$$\mu S = R. \tag{i}$$

Resolving vertically, $\quad S + \mu' R = W.$ (ii)

Taking moments about C,

$$Wl \cos \theta = S2l \cos \theta - R2l \sin \theta$$

or $\qquad\qquad W \cos \theta = 2S \cos \theta - 2R \sin \theta.$ (iii)

From (i) and (ii) we have

$$S(1+\mu\mu') = W,$$

$$\therefore \qquad S = \frac{W}{1+\mu\mu'}, \text{ and } R = \frac{\mu W}{1+\mu\mu'}.$$

Substituting μS for R in (iii),

$$W\cos\theta = 2S(\cos\theta - \mu\sin\theta)$$

$$= \frac{2W}{1+\mu\mu'},\ (\cos\theta - \mu\sin\theta)$$

$$\therefore \qquad 1 = \frac{2}{1+\mu\mu'},\ (1-\mu\tan\theta)$$

$$\therefore \qquad \frac{1+\mu\mu'}{2} = 1-\mu\tan\theta$$

$$\text{or} \qquad \mu\tan\theta = \frac{1-\mu\mu'}{2}$$

$$\text{or} \qquad \tan\theta = \frac{1-\mu\mu'}{2\mu}.$$

Method (ii)

In this case the value of θ can be found easily by using the resultant reactions at A and B. As in Example (i), the result can then be derived from the geometry of the figure.

These reactions must make angles λ and λ' with the normals at A and B (where $\tan\lambda = \mu$, $\tan\lambda' = \mu'$), and must meet on the vertical through G, at E say, as shown in Fig. 5.4B.

$$\text{Hence,} \qquad \tan\theta = \frac{BC}{AC}$$

$$= \frac{ED-EF}{AC}$$

$$= \frac{ED}{2AD} - \frac{EF}{2FB}$$

$$= \tfrac{1}{2}(\cot\lambda - \tan\lambda')$$

$$= \tfrac{1}{2}\left(\frac{1}{\mu} - \mu'\right)$$

$$= (1-\mu\mu')/2\mu.$$

We can then find R and S by resolving as above.

EXAMPLE (iii)

One end of a uniform ladder, of weight W, rests against a smooth wall, and the other end on rough ground, which slopes down from the wall at an angle α to the horizontal. Find the inclination of the ladder to the horizontal when it is on the point of sliding, and show that the reaction of the wall is then W tan (λ—α), *where λ is the angle of friction.*

Let AB (Fig. 5.5) be the ladder, G its centre of gravity, and θ the inclination to the horizontal.

Since the wall is smooth, the reaction R at B is perpendicular to the wall.

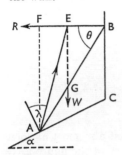

FIG. 5.5

When the ladder is on the point of slipping the resultant reaction at A makes an angle λ with the normal at A, and also passes through E, where the line of action of R meets the vertical through G.

Draw AF perpendicular to BE produced.

Then
$$\tan \theta = \frac{EG}{EB} = \frac{\frac{1}{2}AF}{FE}$$
$$= \tfrac{1}{2} \cot FAE$$
$$= \tfrac{1}{2} \cot (\lambda-\alpha)$$

which gives θ, the inclination of the ladder to the horizontal.

If 2l be the length of the ladder, then, taking moments about A,

$$R2l \sin \theta = Wl \cos \theta$$

∴
$$R = \frac{W}{2} \cot \theta = \frac{W}{2} \times 2 \tan (\lambda-\alpha)$$
$$= W \tan (\lambda-\alpha).$$

EXAMPLE (iv)

A uniform ladder rests with its lower end on rough horizontal ground and its upper end against a rough vertical wall, the ground and wall being equally rough, and the angle of friction λ. Show that the greatest inclination of the ladder to the vertical is 2λ.

When the ladder is in this position can it be ascended without slipping?

Let AB (Fig. 5.6) be the ladder, G its centre of gravity, θ the inclination to the vertical. When the ladder is on the point of slipping the resultant reactions at A and B make angles equal to λ

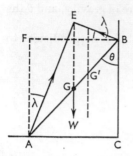

FIG. 5.6

with the normals at those points, and they must also meet at some point E on the vertical through G. The geometry of the figure gives

$$\angle AEG = \lambda, \quad \angle GEB = 90°-\lambda, \quad \angle AEB = 90°,$$

and hence the semicircle AFEB has centre G.

$$\therefore \qquad\qquad \theta = \angle EGB$$
$$= 2\angle AEG, \text{ since } AG = GE,$$
$$\therefore \qquad\qquad \theta = 2\lambda.$$

If an additional weight is placed anywhere on the ladder between A and G the centre of gravity of the ladder and added weight is moved to a point below G. The vertical through the new centre of gravity will then be to the left of that shown in the figure, and it is clear that the reactions at A and B can still meet on this vertical without their inclination to the normals at A and B being greater than λ; they will, as a matter of fact, be less than λ, so that the equilibrium is no longer limiting.

When the added weight reaches G the equilibrium is again limiting.

When the added weight gets above G the centre of gravity of the ladder and weight moves *above* G (say to G') and the reactions at A and B cannot meet on the vertical through G', as this would require one of the inclinations to the normal to be greater than λ. The ladder can therefore be ascended only as far as its centre.

If an additional weight is added at the foot of the ladder (as by someone standing on the bottom rung) the centre of gravity is brought below G. In this case the ladder can be ascended by another person above G, but only to such a height as to bring the centre of gravity of the ladder and the two added weights back to G.

EXAMPLE (v)

The upper end of a uniform ladder rests against a rough vertical wall and the lower end on a rough horizontal plane, the coefficient of friction in both cases being $\frac{1}{3}$. Prove that if the inclination of the ladder to the vertical is $\tan^{-1}\frac{1}{2}$, a weight equal to that of the ladder cannot be attached to it at a point more than $\frac{9}{10}$ of the distance from the foot of it without destroying equilibrium.

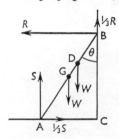

FIG. 5.7

Let AB (Fig. 5.7) be the ladder, G its centre of gravity, W its weight, and D the position of the extra weight W when the ladder is *on the point of slipping.*

Method (i)

If R, S are the normal reactions at the wall and ground the frictions at these points are $\frac{1}{3}R$ and $\frac{1}{3}S$.

Resolving horizontally $R = \frac{1}{3}S$ (i)

Resolving vertically, $S + \frac{1}{3}R = 2W$ (ii)

$\therefore$ $S + \frac{1}{9}S = 2W$ or $S = \frac{9}{5}W$

and $R = \frac{3}{5}W.$

Taking moments about A, if $2l$ is the length of the ladder, $AD = x$ and $\angle ABC = \theta$, where $\tan\theta = \frac{1}{2}$, we get

$\therefore$ $Wl\sin\theta + Wx\sin\theta = R2l\cos\theta + \frac{1}{3}R2l\sin\theta$

$\therefore$ $W\sin\theta + W\dfrac{x}{l}\sin\theta = \frac{6}{5}W\cos\theta + \frac{2}{5}W\sin\theta$

$\therefore$ $\dfrac{x}{l} = \frac{6}{5}\cot\theta + \frac{2}{5} - 1$

$$= \tfrac{12}{5} - \tfrac{3}{5} = \tfrac{9}{5}$$

$\therefore$ $x = \frac{9}{5}l = \frac{9}{10}$ of the length of the ladder.

Method (ii)

This problem can also be dealt with by the geometrical method.

For equilibrium it is necessary that the resultant reactions at A and B should meet on the vertical through the centre of gravity of the ladder and any added weight.

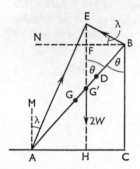

FIG. 5.8

If G (Fig. 5.8) is the centre of gravity of the ladder alone and G′ that of the ladder and added weight in the limiting position, then the reactions at A and B will meet the vertical through G′ in E so that

$$\tan \text{AEG}' = \cot \text{G}'\text{EB} = \tfrac{1}{3}$$

∴ $$\text{AH} = \tfrac{1}{3}\text{EH} \text{ and } \text{EF} = \tfrac{1}{3}\text{FB}.$$

Hence, $$\text{BC} = \text{EH} - \text{EF}$$

$$= 3\text{AH} - \tfrac{1}{3}\text{FB}.$$

But $$\text{BC} = 2\text{AC} \text{ since } \tan \theta = \tfrac{1}{2}$$

∴ $$2\text{AC} = 3\text{AH} - \tfrac{1}{3}\text{FB}$$

∴ $$2(\text{AH} + \text{HC}) = 3\text{AH} - \tfrac{1}{3}\text{HC}$$

∴ $$3\text{AH} = 7\text{HC}$$

∴ $$\text{AH} = \tfrac{7}{10}\text{AC}.$$

Hence, $$\text{AG}' = \tfrac{7}{10}\text{AB} = 7l/5.$$

Also, if x is the distance of the added weight W above A, since the centre of gravity of this and the ladder is at G′, we have, taking moments about A,

$$Wx + Wl = 2W \times \tfrac{7}{5}l,$$

∴ $$x = \tfrac{9}{5}l = \tfrac{9}{10} \text{ of the length of the ladder.}$$

EXAMPLE (vi)

A uniform rod rests in limiting equilibrium inside a rough hollow sphere, the rod being in a vertical plane through the centre of the sphere. Show that the rod makes with the horizontal an angle

$$\tan^{-1} \frac{\sin 2\epsilon}{\cos 2\alpha + \cos 2\epsilon}$$

where ϵ is the angle of friction, and 2α the angle subtended at the centre of the rod.

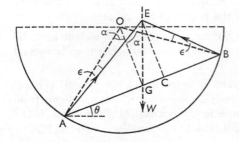

FIG. 5.9

Let AB (Fig. 5.9) be the rod, G its centre of gravity, O the centre of the sphere. Then the reactions at A and B meet at a point E on the vertical through G, and are inclined at the same angle ϵ to the radii OA, OB, since these are the normals to the sphere at A and B.

Draw EC perpendicular to AB. Then if θ is the inclination of AB to the horizontal, we get

$$\tan \theta = \frac{GC}{EC} = \frac{\frac{1}{2}(AC - CB)}{EC}$$

$$= \tfrac{1}{2}(\tan AEC - \tan CEB).$$

But,   angle $AEC = \alpha + \epsilon$ and angle $CEB = \alpha - \epsilon$

$\therefore$ $2 \tan \theta = \tan(\alpha + \epsilon) - \tan(\alpha - \epsilon)$

$$= \frac{\sin(a + \epsilon)}{\cos(\alpha + \epsilon)} - \frac{\sin(\alpha - \epsilon)}{\cos(\alpha - \epsilon)}$$

$$= \frac{\sin 2\epsilon}{\cos(\alpha + \epsilon)\cos(\alpha - \epsilon)}$$

$\therefore$ $\tan \theta = \dfrac{\sin 2\epsilon}{\cos 2\alpha + \cos 2\epsilon}.$

EXAMPLES 5.1

1. A uniform ladder rests with one end on horizontal ground and the other against a vertical wall, the coefficients of friction being respectively $\frac{3}{5}$ and $\frac{1}{3}$. Find the inclination of the ladder to the vertical when it is about to slip.

2. A uniform ladder rests with one end on rough horizontal ground, the coefficient of friction being $\frac{5}{8}$, and the other end against a smooth vertical wall. If the inclination of the ladder is $45°$, show that a man whose weight is equal to that of the ladder can only ascend three-quarters of the length of the ladder.

3. In the case given in the last example find what weight must be placed on the bottom of the ladder to enable the man to ascend to the top.

4. A uniform ladder of weight W leans against a smooth vertical wall, and its foot is on rough ground which slopes down from the wall at an inclination α to the horizontal. Prove that, if the ladder is in limiting equilibrium, its inclination to the wall is

$$\tan^{-1}[2 \tan (\epsilon-\alpha)],$$

where ϵ is the angle of friction. Prove also that the resultant reaction with the ground is then $W \sec (\epsilon-\alpha)$.   (I.S.)

5. A uniform stick of length l is placed in the rough ring of an umbrella-stand at a height h above the ground. It rests also on a smooth floor. Show that equilibrium is impossible unless the stick is vertical, if the coefficient of friction is less than

$$\frac{h}{\sqrt{(l^2-h^2)}}.$$ (I.E.)

6. A uniform rod of length l rests in a vertical plane against (and over) a smooth horizontal bar at a height h, the lower end of the rod being on level ground. Show that if the rod is on the point of slipping when its inclination to the horizontal is θ, then the coefficient of friction between the rod and the ground is

$$\frac{l \sin 2\theta \sin \theta}{4h - l \sin 2\theta \cos \theta}.$$ (I.E.)

7. A uniform thin heavy rod AB rests between two planes OA, OB inclined at $45°$ to the vertical and meeting in a horizontal line. The end A of the rod and the plane OA are rough, the other end B and the plane OB are smooth. Show that the rod will rest in any position in which it makes an angle θ with the smooth plane, provided that $\tan \theta$ lies between $1-2\mu$ and $1+2\mu$.

8. Two equal uniform rods AB, BC, of length $2l$, are rigidly connected at B, so that ABC is a right angle. Prove that if the rods rest in limiting equilibrium in contact with a fixed circular hoop radius a, so that AB is a horizontal, and BC a vertical tangent to the circle, then

$$2a(1-\mu) = l(1+\mu^2),$$

where μ is the coefficient of friction between the rods and the hoop. (I.S.)

9. A uniform ladder rests, with one end on the ground and the other end against a vertical wall, in a plane perpendicular to the line of intersection of wall and ground. The coefficients of friction at the wall and ground are both equal to tan 15°. Show that the inclination of the ladder to the ground cannot be less than 60°. Show that the ladder can just rest at an inclination of 30° if a weight $\frac{1}{2}(\sqrt{3}+1)$ of its own is attached to its lower end. (H.C.)

10. A uniform rod rests in limiting equilibrium equally inclined to two planes, the one horizontal and the other inclined at 120° to it. If the angle of friction between the rod and the inclined plane is 30°, show that the coefficient of friction between the rod and the horizontal plane is $\frac{1}{5}\sqrt{3}$. (H.S.D.)

11. A uniform plank AB, of length 2·4 m and mass 10 kg, rests with one end A on a rough floor, leaning at an inclination $\tan^{-1}\frac{4}{3}$ to the horizontal, against the edge of a smooth table of height 1·2 m. Show that the coefficient of friction μ between the plank and the floor is greater than $\frac{48}{89}$. If $\mu = \frac{3}{4}$, find what weight can be hung from the end B of the plank without it slipping. (H.S.D.)

12. A uniform heavy rod of weight W rests inclined at 45° to the horizontal with one end on the ground and the other against a vertical wall, the vertical plane through the rod being at right angles to the wall. The ground and wall are equally rough, the coefficient of friction between each of them and the rod being $\frac{1}{2}$. Show that the friction at the lower end of the rod may have any value between $\frac{1}{2}W$ and $\frac{1}{3}W$, and find the corresponding values of the frictions at the upper end. Discuss the case in which each coefficient of friction is $\sqrt{2}-1$. (H.S.C.)

13. A ladder of mass 50 kg rests with one end on the rough ground and the other against a smooth vertical wall. If the inclination of the ladder to the horizontal is 60°, determine the frictional force necessary to keep the ladder at rest. If the coefficient of friction between the ladder and the ground be $\sqrt{3}/4$, determine the greatest weight which can be suspended from the top of the ladder without causing it to slip.

14. A uniform rod is placed inside a rough cylindrical drum which is fixed with its axis horizontal. If the rod rests in a vertical plane perpendicular to the axis of the drum, show that its least possible inclination to the vertical is $\tan^{-1}$ (cot $2\lambda + \cos \gamma$ cosec 2λ), where λ is the angle of friction between the rod and the drum, and γ is the angle the rod subtends at the nearest point of the axis of the drum. (N.U.3)

15. The foot of a ladder, of length 9 m and mass 25 kg, rests on a rough horizontal surface, while the upper end rests in contact with a rough vertical wall, the ladder being in a vertical plane perpendicular to the wall. If the first rung of the ladder is 30 cm from the foot and the rest are at intervals of 30 cm, find the highest rung to which a man of mass 75 kg can climb without causing the ladder to slip, when the ladder is inclined at 60° to the horizontal and the coefficient of friction at each end is 0·25. (N.U. 3 and 4)

16. A uniform ladder of length $2a$ rests on a rough horizontal plane at a point O, and is held inclined at an angle α to the vertical by a rope tied at its top end and passing over a pulley fixed at a point distant $2a$ vertically above O. Prove that equilibrium is possible only if the coefficient of friction between the ladder and the plane is greater than $\tan \frac{1}{2}\alpha$. (C.W.B.)

17. Show that if the coefficients of friction with a vertical wall and a horizontal floor are respectively $\frac{1}{2}$ and $\frac{1}{3}$ the least inclination to the horizontal at which a uniform ladder can rest in a vertical plane with one end on the floor and the other against the wall is about 51° 20'. (C.W.B.)

18. A uniform rod rests in limiting equilibrium equally inclined to two planes, the one horizontal and the other inclined at an angle of 135° to it. If the angle of friction between the rod and the inclined plane is $22\frac{1}{2}$°, show that the coefficient of friction between the rod and the horizontal plane is $\frac{1}{17}(3\sqrt{2}+1)$. (H.S.D.)

19. A uniform rod AB rests with one end A in contact with a rough inclined plane, which makes 30° with the horizontal; the rod makes an angle 30° with the upward direction of the plane, and is in a vertical plane through the line of greatest slope. It is kept in equilibrium by means of a string attached to the other end B, and pulled parallel to the plane. Prove that the angle of friction at A must be at least $\tan^{-1}$ ($\frac{2}{3}\sqrt{3}$). (H.S.D.)

20. To one end A of a uniform heavy wire AB there is attached a light ring which can slide along a fixed rough horizontal rod. The other end B is joined by a light inextensible string to a fixed point C of the rod. If the wire is in limiting equilibrium when inclined at an angle α to the vertical and with the angle ABC $= 90°$, prove that the coefficient of friction μ between the ring and the rod is given by

$$\mu(1+\cos^2\alpha) = \sin \alpha \cos \alpha. \qquad \text{(L.A.)}$$

5.6. EXAMPLE (i)

A cone, of radius r and height h, rests on a rough plane, and the inclination of the plane to the horizontal is gradually increased; show that the cone will slide before it topples over if the coefficient of friction is less than 4r/h.

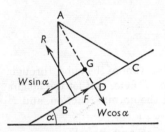

FIG. 5.10

Let W be the weight of the cone, G (Fig. 5.10) its centre of gravity, α the inclination of the plane. Let R denote the normal reaction, and F the frictional force acting on the cone. If the cone slides $F = \mu R$, and if the cone topples about B the reaction R must act at B.

The cone will slide if $W \sin \alpha > \mu W \cos \alpha$, i.e. if $\tan \alpha > \mu$.

Taking moments about B, the cone will topple over if

$$W \sin \alpha \times \frac{h}{4} > W \cos \alpha \times r$$

i.e. if $\qquad \tan \alpha > \dfrac{4r}{h}.$

If μ is less than $4r/h$, and α is gradually increased, $\tan \alpha$ reaches the value μ before it reaches the value $4r/h$, and the cone will slide.

If μ is greater than $4r/h$, $\tan \alpha$ reaches the value $4r/h$ first and the cone will topple over.

If $\mu = 4r/h$ the cone begins to slide and topple over at the same time.

EXAMPLE (ii)

A block in the form of a cube of side 2a stands on a horizontal plane, the coefficient of friction between the block and plane being μ. A gradually increasing horizontal force is applied to a vertical face of the cube, at right angles to it, and in a vertical plane through the centre of gravity of the cube. Show that if $\mu<\frac{1}{2}$ the cube will tend to slide without upsetting, but that if $\mu>\frac{1}{2}$ the cube will tend to upset without sliding, if the point of application of the force is at a height above the plane greater than a/μ.

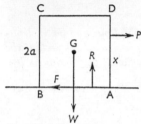

FIG. 5.11

Let ABCD (Fig. 5.11) be the vertical section of the cube, through its centre of gravity G, in which the force P acts, and let the height of P above the plane be x. Let R denote the normal reaction, and F the frictional force.

For sliding $\qquad P>\mu W$, since $F=\mu R$ and $R=W$.

For tilting, R must act at A and

$$Px>Wa, \text{ or } P>Wa/x.$$

Since the least value of a/x is $a/2a$ or $\frac{1}{2}$, as P increases it will reach the value μW first if $\mu<\frac{1}{2}$, so that in this case the cube slides without tilting.

If $\mu>\frac{1}{2}$ the cube will tilt or slide first according as

$$\mu> \text{ or } <\frac{a}{x}$$

i.e. $\qquad\qquad$ as $x>$ or $<\dfrac{a}{\mu}$.

Hence, if $x>a/\mu$ the cube will upset without sliding.

EXAMPLES 5.2

1. A uniform cylinder, of radius r and height h, is placed with its plane base on a rough plane whose inclination to the horizontal is gradually increased. Show that the cylinder will topple over before it slides if $2r/h$ is less than the coefficient of friction.

2. A right cone is placed with its base on a rough inclined plane; if 0·25 be the coefficient of friction, find the angle of the cone when it is on the point of slipping and turning over at the same time.

3. An equilateral triangle rests in a vertical plane with one side on a rough horizontal plane; a gradually increasing horizontal force acts on its highest vertex in the plane of the triangle. Prove that the triangle will slide before it tilts if the coefficient of friction be less than $\frac{1}{3}\sqrt{3}$.

4. A rectangle of sides a and h rests in a vertical plane with one of the sides of length a on a rough horizontal table. A gradually increasing horizontal force acts along the upper side in the plane of the rectangle. Find the condition that the rectangle shall tilt before it slides.

5. A right cone rests on a rough horizontal plane and is acted on by a gradually increasing horizontal force at its vertex. Show that the cone will turn over, or slide, according as the coefficient of friction is $>$ or $<r/h$, where r is the radius and h the height of the cone.

6. A triangular lamina ABC, right-angled at C, stands with BC on a rough horizontal plane. If the plane is tilted round an axis in its own plane perpendicular to BC, so that C is lower than B, show that the lamina will begin to slide, or tilt, according as the coefficient of friction is less, or greater than, tan A.

7. A uniform cubical block is supported on a rough inclined plane by a rope parallel to the line of greatest slope attached to the mid-point of the upper edge of the cube, which is horizontal. Show that the inclination of the plane must be less than $\tan^{-1}(1+2\mu)$, where μ is the coefficient of friction. (I.S.)

8. A cube rests on a rough plane of inclination α ($\alpha<\frac{1}{4}\pi$), with two of its upper and two of its lower edges horizontal. Show that it will not be possible to drag the cube up the plane without up-setting it, by means of a rope attached to the mid-point of the uppermost edge and pulled parallel to the greatest slope of the plane, if the coefficient of friction between the plane and the cube exceeds $\frac{1}{2}(1-\tan\alpha)$. (H.C.)

9. A uniform cube, of edge $4a$, stands on a rough horizontal plane. A gradually increasing horizontal force is applied to one of its vertical faces at a height a vertically above the centre of the face. Determine how equilibrium will be broken

    (i) when the coefficient of friction between the plane and cube is 0·5.

    (ii) when the coefficient is 0·7. (H.C.)

10. The triangle ABC, in which BC is horizontal and AB and AC are equal and greater than BC, represents the cross-section of a triangular prism standing on a rough horizontal plane on one of its rectangular faces, and the face represented by AB is subject to wind pressure. Show that when this pressure, assumed normal to AB, becomes sufficiently great the prism will topple over or slide according as the angle of friction is greater or less than $\pi - 2\alpha$, where α is the inclination of either sloping face to the base. (I.E.)

11. A cubical box, of edge a and weight W, stands on a rough horizontal plane, and a heavy bar of length b and weight w rests at an inclination of 45° with one end on the plane and the other against a vertical face of the box, the vertical plane through the bar passing through the centre of the box. If the lower end of the bar is prevented from slipping, find the least possible value of the coefficient of friction between the box and the plane in order that the box may not slide, friction between the bar and the box being neglected. Find also the ratio between the weights of the bar and the box if the latter is on the point of toppling over. (I.E.)

12. A uniform cube of edge $4a$ and weight W stands on a rough horizontal plane. A gradually increasing force P is applied inwards at right angles to a face F of the cube at a point distant a vertically above the centre of that face. Prove that equilibrium will be broken by sliding or by tilting according as the coefficient of friction between the cube and the plane is less than or greater than a certain value, and find this value.

If P has not reached a value for which equilibrium is broken, and $P = \frac{1}{2}W$, find how far from the face F of the cube the normal reaction acts. (H.C.)

5.7. EXAMPLE

Two uniform beams AB, BC, of equal length, are freely jointed at B, and rest in equilibrium in a vertical plane with the ends A and C on a rough horizontal plane. If the weight of AB is twice that of BC, show that there cannot be limiting friction both at A and C, and that if there is limiting friction at either of these points it is at C. Find also the coefficient of friction if the greatest angle that the rods can make with each other is a right angle.

The friction forces at A and C (Fig. 5.12) must be equal, since they are the only external horizontal forces acting on the beams. If

$\angle ABC = 2\theta$, and the reactions at A and C be R and S, then taking moments about A for *both* beams,

$$S \times 4l \sin \theta = W \times 3l \sin \theta + 2Wl \sin \theta$$

where $2l = $ length of AB or BC.

$\therefore \qquad\qquad S = \tfrac{5}{4}W$ and hence $R = \tfrac{7}{4}W$.

Now, if the friction is limiting at A its value must be μR or $\tfrac{7}{4}\mu W$, while if it is limiting at C its value must be μS or $\tfrac{5}{4}\mu W$. But the frictional force has the same value at A and C, and hence it cannot be limiting at both these points, for $\tfrac{5}{4}\mu W$ cannot equal $\tfrac{7}{4}\mu W$.

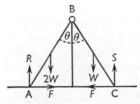

FIG. 5.12

It is also clear that F will reach the value $\tfrac{5}{4}\mu W$ before it reaches $\tfrac{7}{4}\mu W$, so that if it is limiting at either point it will be at C.

When $2\theta = 90°$ the point C will be on the point of slipping but A will not. The frictional force at both C and A will be $\tfrac{5}{4}\mu W$.

Hence, taking moments about B for the rod BC,

$$\tfrac{5}{4}\mu W \times 2l \cos \theta = S \times 2l \sin \theta - W \times l \sin \theta$$

$\therefore \qquad\qquad \tfrac{5}{2}\mu W = 2S \tan \theta - W \tan \theta.$

$\therefore \qquad\qquad \tfrac{5}{2}\mu W = 2 \times \tfrac{5}{4}W - W = \tfrac{3}{2}W$

$\therefore \qquad\qquad \mu = \tfrac{3}{5}.$

EXAMPLES 5.3

1. Two uniform rods AB and BC of the same thickness and material, and of length 0·9 m and 0·6 m respectively, are freely hinged together and rest in a vertical plane with the ends A and C on a rough horizontal plane. If the greatest possible value of $\angle ABC$ consistent with equilibrium is 90°, find the coefficient of friction between the rods and ground, and determine how equilibrium will be broken if the inclination of the rods is slightly increased beyond 90°. (H.S.C.)

2. Two equal uniform rods AB and BC of the same weight, freely jointed at B, rest in a vertical plane with the end A and C in a rough horizontal plane. If equilibrium is possible when ABC is any angle not exceeding a right angle, find the coefficient of friction between the rods and the plane. (H.S.C.)

3. Two uniform ladders, AB, BC of equal lengths and weights W, W' $(W>W')$ are hinged together at the top B and will stand on rough ground when containing an angle 2θ. Show that the total reaction at A makes a smaller angle with the vertical than at C. Assuming the coefficients of friction at A and C are each equal to μ, show that, as θ is increased, slipping will occur at C, and that

$$\mu = [(W+W')/(W+3W')] \tan \alpha$$

where α is the value of θ for which slipping first occurs. (H.S.D.)

4. AB and BC are two equal uniform rods of the same weight W, freely jointed at B. They rest in a vertical plane with the ends A and C in contact with a rough plane of inclination α, on a line of greatest slope of the plane with the rod BC horizontal. Find the pressures on the plane at the points A and C, and the friction at these points. Show that $\cos^2 \alpha$ must be $>\frac{1}{3}$, and that the friction at C is acting up or down the plane, according as $\cos^2 \alpha <$ or $>\frac{2}{3}$. If $\alpha = 30°$, and the friction at either of the points A and C is limiting, determine at which of these it is limiting and find the coefficient of friction. (H.S.C.)

5. Two equal uniform rods AB, BC, smoothly jointed at B, are in equilibrium with the end C resting on a rough horizontal plane and the end A held above the plane. Prove that, if α and β are the inclinations of CB and BA to the horizontal, the coefficient of friction must exceed

$$\frac{2}{\tan \beta - 3 \tan \alpha}.$$ (I.S.)

6. The two sides of a pair of steps are of the same length, but unequal weights, and they are freely jointed together at the top. If the steps are gradually opened out while standing on a rough horizontal plane, prove that the lighter side will tend to slip first, and that this will happen when the inclination of each side to the vertical is

$$\tan^{-1}\frac{(3w_1+w_2)\mu}{w_1+w_2}.$$

where w_1, w_2 are the respective weights of the two sides ($w_1<w_2$), and μ is the coefficient of friction at each point on the ground. (C.W.B.)

5.8. EXAMPLE (i)

A uniform sphere is held in equilibrium on a rough inclined plane of angle α by a force of magnitude $\frac{1}{2}W \sin \alpha$ applied tangentially to its circumference, where W is the weight of the sphere. Prove that the force must act parallel to the plane, and that the coefficient of friction must be not less than $\frac{1}{2} \tan \alpha$.

Let A (Fig. 5.13) be the point of contact of the sphere, C its centre, and a its radius. Then, if the sphere is not to roll about A the moment of the force $\frac{1}{2}W \sin \alpha$ about A must equal the moment of the weight about A, i.e. $Wa \sin \alpha$.

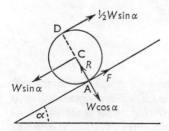

FIG. 5.13

Hence the force $\frac{1}{2}W \sin \alpha$ must be at a distance $2a$ from A, and must therefore act parallel to the plane at D, the other end of the diameter through A.

If there is to be no slipping, resolving parallel to the plane

$$F + \tfrac{1}{2}W \sin \alpha = W \sin \alpha$$

where $F \leqslant \mu R$ and $R = W \cos \alpha$

∴ $\qquad\qquad\qquad \frac{1}{2}W \sin \alpha \leqslant \mu W \cos \alpha.$

∴ $\qquad\qquad\qquad \mu \geqslant \frac{1}{2} \tan \alpha.$

EXAMPLE (ii)

A heavy chain is placed on a rough plane inclined to the horizontal at an angle α equal to the angle of friction, with a portion a m long along a line of greatest slope and the remainder of length b m hanging vertically over the top of the plane. If the chain is on the point of slipping, show that b is either zero or 2a sin α.

Let ABC (Fig. 5.14) represent the chain, and let w be its weight per unit length. The frictional force acting on AB is $\mu aw \cos \alpha$, and this equals $aw \sin \alpha$ since $\mu = \tan \alpha$.

If A is on the point of moving down

$$wa \sin \alpha - wa \sin \alpha = bw, \qquad \therefore b = 0.$$

If A is on the point of moving up,

$$bw = wa \sin \alpha + wa \sin \alpha,$$

$$\therefore \qquad b = 2a \sin \alpha.$$

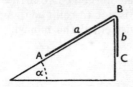

FIG. 5.14

EXAMPLE (iii)

Two rings of equal weight are free to move on a rough horizontal rod, the coefficient of limiting friction being μ. They are connected by a smooth string of length l, on which another ring of the same weight as the other two rings together can move freely. Prove that, in a position of equilibrium of the system, the two rings on the rod cannot be further apart than $2\mu(1+4\mu^2)^{-\frac{1}{2}}l$.

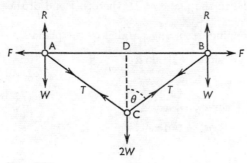

FIG. 5.15

Let W be the weight of each ring A and B (Fig. 5.15), $2W$ that of C.

Then, since C is free to slide on the string the tension is the same throughout the string and the vertical CD through C bisects the angle ACB. Let $\angle DCB = \theta$.

Resolving vertically for C,

$$2T \cos \theta = 2W.$$

The normal reaction R between A and the rod is

$$W + T \cos \theta = 2W,$$

and therefore the maximum friction at A is $2\mu W$.

For equilibrium $\qquad T \sin \theta = \mathrm{F} \not> 2\mu W$

or $\qquad\qquad\qquad\qquad W \tan \theta \not> 2\mu W$

or $\qquad\qquad\qquad\qquad\quad \tan \theta \not> 2\mu.$

If DB $= x$, DC$^2 = (l^2/4) - x^2$, and $\tan \theta = x/\mathrm{DC}$

$$\therefore \qquad\qquad \frac{x^2}{\frac{1}{4}l^2 - x^2} \not> 4\mu^2$$

$$\therefore \qquad\qquad x^2 \not> \mu^2 l^2 - 4x^2\mu^2$$

$$\therefore \qquad x^2(1 + 4\mu^2) \not> \mu^2 l^2$$

$$\therefore \qquad\qquad x \not> \frac{\mu l}{\sqrt{(1 + 4\mu^2)}}.$$

EXAMPLES 5.4

1. A heavy plank of length a lies on a rough horizontal plane. To one end is attached a rope which is kept inclined to the horizontal at an angle α. The end of the plank is gently raised. Show that the other end will slip at once if $\cot \alpha > \mu$, the coefficient of friction; and find the ratio of the tension of the rope to the weight of the plank, if $\mu \tan \alpha$ exceeds unity, when the slipping commences. (I.E.)

2. A body of weight W rests on a horizontal plane with which the coefficient of friction is μ. A horizontal force nW, which is $<\mu W$, is applied to the weight, and another horizontal force P, perpendicular to W is also applied. Find the least value of P which will just cause the weight to move, and find the inclination of the direction of motion to the direction of P. (I.E.)

3. A rigid square frame made of uniform heavy wire rests in a vertical plane on a rough horizontal cylinder of radius a, two of its sides being in contact with the cylinder. Show that the limiting angle of inclination of a diagonal to the vertical is given by the equation

$$b \sin \theta = a\sqrt{2} \sin \epsilon \cos (\theta + \epsilon),$$

where ϵ is the angle of friction and b is the distance of the centre of the square from the axis of the cylinder. (I.S.)

4. An equilateral triangle formed of wire is placed in a vertical plane with one side horizontal. On each side is strung a bead of weight W, and the beads are connected by an endless string, in tension, and passing through small smooth rings at the corners of the triangle. Prove that, if a gradually increasing horizontal force be applied to the bead on the horizontal side, the system will begin to move when the force is equal to $2\mu W$, where μ is the coefficient of friction between the beads and the wire.   (I.S.)

5. Two equal uniform rods, AB, CD, each of weight W, are freely jointed at their mid-points, and are placed in a vertical plane with the ends A and C on a rough horizontal plane of coefficient of friction μ. A string having weights, each equal to W, attached to its ends is passed over B and D. Prove that in the limiting position of equilibrium the rods are inclined to the horizontal at an angle $\tan^{-1} [3/(1+2\mu)]$.   (H.S.C.)

6. A wedge of weight W, lying on rough ground, has its thin end pushed against a smooth vertical wall, the sloping face of the wedge making an angle 2θ with the wall. A smooth right circular cylinder of weight W_1 lies between the wedge and the wall. Find the relation between W and W_1 when the wedge is just on the point of sliding, the coefficient of friction between the ground and wedge being μ.   (H.S.D.)

7. A light rod AD passes over a rough peg at B and under another rough peg at C in the same horizontal line, and has weights of 15 N and 9 N attached to it at A and D. The lengths AB, BC, CD are 0·9 m, 0·6 m, and 0·3 m. If the least horizontal force which will move the rod is 6 N, find the coefficient of friction, assuming the pegs equally rough.   (I.S.)

8. A wedge of mass M, whose faces are inclined at an angle α, is placed with one face in contact with a horizontal plane. A small object of mass m is placed in contact with the other face and is just prevented from sliding down by a horizontal force. If $\mu(< \tan \alpha)$ is the coefficient of friction for the object and the wedge, find the least value of the coefficient for the wedge and the plane in order that the wedge may remain at rest.

9. A rough circular cylinder of diameter d is fixed on an equally rough horizontal plane, and a uniform rod of length $2l$ rests tangentially against the cylinder in a vertical plane, which is perpendicular to the axis of the cylinder, one end of the rod being on the rough plane. If the friction is limiting at both ends of the rod, when the rod is inclined at 30° to the vertical, prove that the angle of friction is $\frac{1}{2} \sin^{-1} (l/d)$.   (H.S.D.)

10. Two equal particles, each of weight W, are placed on a rough horizontal table and connected by a taut inextensible string. Prove that the least horizontal force that can be applied to one of them in a direction making an angle θ with the string so as to cause them both to be on the point of motion is $2\mu W \cos \theta$, where μ is the coefficient of friction between either particle and the table.   (H.S.C.)

11. Figure 5.16 represents the central section ABCD of a child's box filled with sand and being dragged forward by a horizontal force applied at C. The box is attached to two crossbars at A and B,

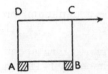

Fig. 5.16

and the whole is dragged across a rough floor whose coefficient of friction with the crossbars at A and B is μ. Given AB $= l$, BC $= h$, and the weight of the box and sand to be W, prove that the vertical reactions at A and B are

$$\frac{l-2\mu h}{2l}W \quad \text{and} \quad \frac{l+2\mu h}{2l}W$$

respectively.   (H.S.D.)

12. A heavy uniform rod is placed over a rough peg at A and under another rough peg at B, at a higher level than A, so that the rod lies in a vertical plane; show that the length of the shortest rod that will rest in this position is

$$a(1+\tan \alpha \cot \lambda),$$

where a is the distance between the pegs, α is the angle of inclination to the horizontal of the line joining the pegs, and λ is the angle of friction.

13. A uniform circular lamina of radius a and weight W rests with its plane vertical on two fixed rough planes each inclined at an angle of 45° to the horizontal, their line of intersection being perpendicular to the plane of the lamina. If the coefficient of friction at each contact is $\frac{1}{2}$, prove that the least couple required to rotate the lamina in its plane about its centre is of moment $2\sqrt{2}Wa/5$.   (L.A.)

14. A sphere is placed on a rough plane inclined at an angle α to the horizontal. A force is applied tangentially to the sphere at a point of its circumference and in the vertical plane containing the centre of the sphere and a line of greatest slope of the plane. The direction of this force makes an angle β with the horizontal (in the same sense as α). If the coefficient of friction is μ, prove that, for the sphere to rest in equilibrium,

$$\mu \geqslant \frac{\sin \alpha}{\cos \alpha + \cos \beta}.$$

If β is variable, prove that the least value of μ for which equilibrium is possible is $\tan \frac{1}{2}\alpha$.

If, however, the applied force is to be a minimum, prove that the least possible value of μ is $\frac{1}{2} \tan \alpha$.   (H.C.)

15. A wheel situated in a vertical plane is free to turn about its centre C. A uniform rod AB, weight W, is smoothly hinged at A, which is at the same level as C, and rests in contact with the wheel at D. Prove that to turn the wheel a couple of moment greater than

$$\frac{\mu W \times AB \times CD}{2\,AC}$$

is required, μ being the coefficient of friction between the rod and the wheel. Prove also that, when the wheel rotates so that the point in contact with the rod at D moves towards B, the reaction at A will be vertical, if the inclination of the rod to the horizontal is $\tan^{-1} \mu$.   (H.C.)

16. A circular cylinder of weight W, with its axis horizontal, is supported in contact with a rough vertical wall by a string wrapped partly round it and attached to a point of the wall, and making an angle α with the wall. Show that the coefficient of friction must not be less than cosec α, and that the normal pressure on the wall is $W \tan \frac{1}{2}\alpha$.   (H.C.)

17. A sphere of weight W is placed on a rough plane inclined at 45° to the horizon, and is kept in equilibrium, if possible, by a horizontal force P applied at the highest point of the sphere.
    (1) Show that equilibrium is impossible if μ, the coefficient of friction between the plane and the sphere, is less than $\sqrt{2} - 1$.
    (2) Show that equilibrium is possible if μ is equal to or greater than $\sqrt{2} - 1$; find the value of P, and determine whether equilibrium is limiting or not when μ is greater than $\sqrt{2} - 1$.
            (H.C.)

18. A thin hemispherical shell rests with its curved surface in contact with a rough horizontal plane, whose coefficient of friction is μ, and with a rough vertical plane, whose coefficient is μ'. If the shell is on the point of slipping when the plane of the rim makes $30°$ with the horizontal, find the relation connecting μ and μ'; and prove that if $\mu' < \frac{1}{3}$, μ must lie between $\frac{1}{5}$ and $\frac{1}{4}$. (The centre of gravity of a thin hemispherical shell bisects the radius.) (H.C.)

19. A uniform thin hemispherical bowl rests with its curved surface on a rough horizontal plane (coefficient of friction μ) and leans against a smooth vertical wall. Prove that when the bowl is on the point of slipping the inclination of the axis of the bowl to the vertical is $\sin^{-1} 2\mu$. (H.S.D.)

20. A uniform rod AB rests with one end A in contact with a rough inclined plane, which makes $30°$ with the horizontal; the rod makes an angle $45°$ with the upward direction of the plane, and is in a vertical plane through the line of greatest slope. It is kept in equilibrium by means of a string attached to the other end B, and pulled parallel to the plane. Prove that the angle of friction at A must be at least $\tan^{-1} \left(\dfrac{\sqrt{3}+1}{2\sqrt{3}} \right)$. (H.S.D.)

21. A uniform beam rests over two pegs, which are not at the same level, so that it is inclined at an angle θ to the horizontal. The angle of friction between the beam and the upper peg is λ_1, and that between the beam and the lower peg is $\lambda_2 > \theta > \lambda_1$. Show that the beam will just slip over the pegs when the ratio of the distances from its centre of mass to the lower and upper pegs respectively is

$$\frac{\sin (\lambda_2 - \theta) \cos \lambda_1}{\sin (\theta - \lambda_1) \cos \lambda_2}.$$ (I.E.)

22. A rectangular lamina ABCD has its plane vertical and perpendicular to a rough wall, AB being in contact with the wall. A string is attached to D and to a point K of the wall vertically above A, AK being $\frac{1}{2}$AD. Determine the least value of the coefficient of friction between the lamina and the wall that the lamina may rest in this position.

23. A uniform heavy beam AB of length $2l$ is standing vertically with the end A on the horizontal rough ground. To B is attached a light rope which passes under a small pulley fixed to the ground at a point C distant a from A. The end B is slowly lowered in the vertical plane through AC, towards the side remote from C, the

rope BC being kept taut. Show that the beam will slip when its inclination θ to the horizontal is given by

$$\cos \theta(a+2l \cos \theta) = 2\mu \sin \theta(a+l \cos \theta),$$

where μ is the coefficient of friction between the beam and the ground.   (H.S.D.)

24. A uniform rod AB of weight W rests in a horizontal symmetrical position with its ends resting on two planes inclined at angles $45°$ to the horizontal. The plane on which A rests is smooth, and a variable force P acts along the rod in the direction AB. Prove that so long as there is no slipping the reaction at A is constant, and find the frictional force at B. Find also the relation between P and W, and the coefficient of friction at B if there is no slipping.   (N.U.3.)

25. A hemisphere of radius r is fixed with its plane base on a rough horizontal table. A uniform rod ABC, of length $2a$ and weight W, is placed with one end A, resting on the table and the other end, C, free, a point B of the rod resting against the smooth curved surface of the hemisphere. If the rod makes an angle θ with the table, find the force of friction at A. If λ is the angle of friction at A, show that in the position of limiting equilibrium the angle θ satisfies the equation

$$a \sin \theta \sin (\theta+\lambda) = r \sin \lambda.$$ (N.U.3.)

26. A heavy uniform chain, of length l and weight lw, hangs from one end and carries a weight W at the other. Find the tension in the chain at a distance x from the top end. If the chain, instead of being fixed at its upper end, lies along a rough inclined plane for the upper half of its length and is still in equilibrium with the weight W attached at the lower end and hanging freely below the plane, prove that the coefficient of friction between the chain and the plane must not be less than

$$(2W/wl) \sec \alpha + \tan \alpha + \sec \alpha,$$

where α is the angle of the plane.   (N.U.3.)

27. An endless light string of length $(n+1)a$ passes across a rough table of width a and hangs below the table. To the part on the table is attached a weight W, and to the part below the table is attached a weight W/n. If the angle of friction on the table is λ, prove that, in limiting equilibrium, the parts of the string below the table makes angles with the vertical whose difference is 2λ, assuming that $n \tan \lambda < 1$.   (H.S.D.)

28. A uniform beam AB of weight W rests horizontally with A attached to a light rope which passes over a smooth fixed pulley

and supports a heavy particle hanging freely, while B is in contact with a rough plane of inclination α to the horizontal. The rope, beam and the line of greatest slope through B are in the same vertical plane. If λ is the angle of friction and B is about to slip downwards, find the weight of the particle and draw figures for the cases: (a) $\lambda < \alpha$; (b) $\lambda > \alpha$.

In case (a) show that, if $\alpha > \frac{1}{2}\pi - \lambda$, B cannot slip upwards, whatever the weight of the particle. (L.A.)

29. A circular cylinder of radius a is placed with its curved surface in contact with a horizontal plane and is fixed in position. A uniform rod AB of length $4a$ is laid across the cylinder in a plane perpendicular to the axis of the cylinder and has the end A on the horizontal plane. Contact between the rod and the cylinder is smooth and that between the rod and the plane is rough. The angle AB makes with the horizontal is 2α. If the rod is on the point of sliding down the cylinder, show that the coefficient of friction μ between the rod and the plane is given by

$$\mu = \frac{\sin 4\alpha}{\cot \alpha - 1 - \cos 4\alpha}. \qquad \text{(L.A.)}$$

30. A smooth right circular cylinder of radius a is fixed with its axis horizontal and touches a vertical wall along a generator. A rod of weight W rests in equilibrium across the cylinder at right angles to the axis and with one end against the wall.

(i) If the angle of friction between the rod and the wall is λ and the centre of gravity of the rod is vertically above the axis of the cylinder, show that the reaction between the rod and the cylinder is $W \cot \lambda$. Find the resultant reaction between the rod and the wall and the inclination of the rod to the vertical.

(ii) If the wall is smooth and the rod is inclined at 45° to the horizontal, show that the line of action of the weight of the rod will now be at a distance $a(\sqrt{2}-1)$ from the axis of the cylinder. (L.A.)

31. AB, BC, CD, and DE are four uniform rods of equal weights and equal length $2a$. They are freely jointed together by smooth hinges at B, C, and D and stand in equilibrium in a vertical plane, with the ends A and E on a rough horizontal plane, in the form of a symmetrical arch. The coefficient of friction between the plane and the rods is $\frac{1}{4}$. Prove that the maximum span AE of the arch is

$$\frac{2a(\sqrt{10+5\sqrt{2}})}{5}$$

and find the corresponding height of the arch. (L.A.)

32. A heavy uniform bar AB of length $(a+b)$ is bent at right angles at a point P, where AP $= a$, and rests with P on rough horizontal ground and B against a smooth vertical wall, the vertical plane containing the bar being at right angles to the line of intersection of the wall and the ground. The coefficient of friction between the bar and the ground is $\frac{1}{2}$. Prove that equilibrium is possible only if the acute angle between PB and the ground lies between α and β, where

$$\tan \alpha = \frac{b^2}{a^2} \text{ and } \tan \beta = \frac{b^2}{a^2+ab+b^2}. \qquad \text{(L.A.)}$$

33. Three uniform bars AB, BC, and CD, each of length a and weight W, are freely jointed at B and C. The system hangs symmetrically, being supported by two small identical light rings at A and D which can slide on a fixed rough horizontal rod. In the position of limiting equilibrium the length AD is $11a/5$. Find the coefficient of friction between a ring and the rod.

Obtain also the magnitude and direction of the action at B.

(L.A.)

34. A cotton reel rests with its axis horizontal on a rough plane inclined to the horizontal at an angle α. The coefficient of friction between the plane and the reel is μ, where $\tan \alpha > \mu$. The reel is supported by a thread wrapped round it and coming off it in the vertical plane through its centre of gravity at an angle θ to the horizontal. If a is the radius of each end of the reel and b the radius of the middle part, prove that the greatest value of θ consistent with equilibrium is given by

$$\cos \theta = \frac{b}{a}\left(\frac{\sin \alpha}{\mu} - \cos \alpha\right). \qquad \text{(C.A.)}$$

35. A uniform rod AB, of length $2a$, rests on a fixed smooth peg P, with the end A in contact with a rough vertical wall, AB making an acute angle θ with the upward vertical. If x is the distance of P from A and μ is the coefficient of friction between the rod and the wall, prove that, when the end A of the rod is about to slip downwards,

$$(x-a) \tan^2 \theta - \mu a \tan \theta + x = 0.$$

Deduce that, if $x = 3a/2$, then this position of limiting equilibrium is not possible unless $\mu > \sqrt{3}$. (O.C.)

36. Two small rings A and B, of weights w and $3w$ respectively, are threaded on to a rough horizontal wire. The coefficients of friction between A and the wire and between B and the wire are $\frac{2}{3}$ and $\frac{1}{2}$ respectively. The ends of a light inextensible string are

fastened to A and B, and at the mid-point M of the string a weight W is attached. The angle AMB is a right angle. If the weight W is gradually increased, find at which of the rings equilibrium is broken by sliding.   (C.A.)

5.9. More difficult examples

The following examples are of a more difficult nature.

EXAMPLE (i)

A bead, threaded on a weightless string, rests on a rough plane inclined at $\alpha°$ to the horizon; the ends of the string are fastened to two points A and B on the plane in the same horizontal line. If μ, the coefficient of friction, is less than $\tan \alpha$, show that, in a position of limiting equilibrium, the friction is inclined to the horizontal at an angle $\sin^{-1} (\mu \cos \alpha)$.

The bead must move in an ellipse whose foci are at A and B (Fig. 5.17). If C represents the bead, then it must move along the

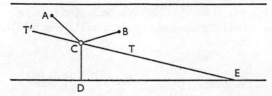

FIG. 5.17

tangent to the ellipse at C, which is equally inclined to CA and CB.

Let CT be the tangent at C and CD the line of greatest slope through C, and angle DCT $= \beta$.

The tension is the same throughout the string, so that the components along the tangent CT are equal and opposite.

Hence the component of weight along CT is balanced entirely by the friction $\mu W \cos \alpha$.

∴ $W \sin \alpha \cos \beta = \mu W \cos \alpha$ or $\cos \beta = \mu \cot \alpha$.

If CT meets the lower edge of the plane in E (Fig. 5.18) and F is

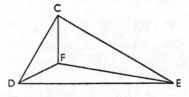

FIG. 5.18

the projection of C on the horizontal plane, the angle made by the friction with the horizontal is CEF.

Now $CF = CD \sin \alpha$

and $CE = CD \sec \beta$.

$\therefore \sin CEF = \dfrac{CF}{CE} = \dfrac{\sin \alpha}{\sec \beta} = \sin \alpha \cos \beta = \mu \sin \alpha \cot \alpha = \mu \cos \alpha.$

EXAMPLE (ii)

A cogwheel of radius 30 cm can turn on a fixed axle of radius 7·5 cm. If the angle of friction between the wheel and axle is 30°, determine the least vertical force applied to one end of a chain over the wheel which will just support a weight of mass 60 kg attached to the other end of the chain. Represent carefully in a diagram the line of action of the resultant pressure between the wheel and axle.

Let the least vertical force be P N.

We suppose that the wheel touches the axle only along a single horizontal line whose section is represented by the point C (Fig. 5.19).

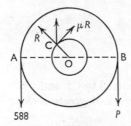

FIG. 5.19

This point must lie on the resultant of the weight $60 \times 9 \cdot 8 = 588$ N and the vertical force P N (neglecting weight of cogwheel).

The resultant reaction at C must be a vertical force equal to $(P+588)$ N, and this is composed of a reaction R N along the radius of the axle OC, and a force of friction μR N along the tangent to the axle at C in the direction shown in the figure.

$\therefore$ $R^2(1+\mu^2) = (588+P)^2$

$\therefore$ $R^2 \sec^2 \lambda = (588+P)^2$

$\therefore$ $R \sec \lambda = 588+P$

$\therefore$ $R = \dfrac{\sqrt{3}}{2}(588+P).$

Taking moments about O, we get

$$30P+7\cdot5\mu R = 30\times588$$

$\therefore \qquad 4P+\mu R = 2352$

$\therefore \qquad 4P = 2352-\tfrac{1}{2}(588+P)$

$\therefore \qquad \dfrac{9}{2}P = 2352-294 = 2058$

$\therefore \qquad P = \dfrac{4116}{9} = 457 \text{ approx.}$

EXAMPLE (iii)

A straight uniform pole AB *leans against a vertical wall. The lower end* A *is on the horizontal ground a m from the wall; the upper end* B *is on the wall b m above the ground and c m to one side of the vertical plane through* A *perpendicular to the wall. Assuming that the ground is rough enough to prevent slipping at* A, *prove that to prevent slipping at* B *the coefficient of friction between the pole and the wall must be not less than* $c(b^2+c^2)^{\frac{1}{2}}/ab$.

Let XY (Fig. 5.20) be the base of the wall, AC perpendicular to the wall.

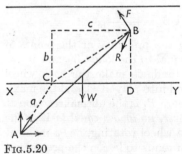

FIG.5.20

If the pole slips with A fixed it must describe a cone with vertex at A, that is, B must describe a circle on the wall with centre C.

The friction F therefore acts perpendicular to CB.

Let R be the normal reaction at B and $\angle CBD = \theta$. The components of friction are $F \sin\theta$ vertically and $F \cos\theta$ horizontally.

For equilibrium we have, taking moments about a vertical axis through A,

$$Fa \cos\theta = Rc \text{ and } F \not> \mu R.$$

Hence $\qquad \mu \not< \dfrac{c}{a} \times \dfrac{1}{\cos\theta}$, that is, $\mu \not< \dfrac{c}{a} \times \dfrac{(b^2+c^2)^{\frac{1}{2}}}{b}$.

EXAMPLE (iv)

The loads on the back and front axles of a motor lorry are W_1 and W_2 respectively. The height of the centre of gravity in h, the distance between the axles, is a, and the coefficient of friction between the wheels and the ground is μ. Find the maximum tractive force on starting when driven: (i) on the back axle; (ii) on the front axle. In the former case show that the reaction between the back wheels and the ground is increased in the ratio $a : a - \mu h$ when the lorry is just starting.

(N.B. *The tractive force = the friction between the ground and driving wheels.*)

Let A, B (Fig. 5.21) be the front and back axles, G the centre of

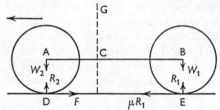

FIG. 5.21

gravity. The loads on A and B give us the horizontal distances of G from A and B. They are

$$\text{AC} = \frac{aW_1}{W_1 + W_2} \text{ and } \text{CB} = \frac{aW_2}{W_1 + W_2}.$$

When the engine is not working the pressures at the points of contact of the wheels D and E are equal to the loads W_2 and W_1.

(i) When the engine starts and tends to make the back wheel rotate a forward friction force is called into play at E, and the maximum value of this friction is μR_1, where R_1 is the normal reaction at E (Fig. 5.21). *This reaction is, however, no longer equal to W_1.* The friction μR_1 at E, by preventing the wheel rotating, has a moment about G on the lorry as a whole, and tends to lessen the pressure at D. The frictional force F called into play at D tends to make the front wheel rotate. The maximum tractive force on starting is therefore μR_1, corresponding to $F = 0$. In this case, if R_2 is the normal reaction at D, taking moments about G, we have

$$R_1 \times \text{CB} = \mu R_1 \times h + R_2 \times \text{AC}.$$

Also $R_1 + R_2 = W_1 + W_2$

$\therefore$ $R_1 \text{CB} = \mu R_1 h + (W_1 + W_2 - R_1)\text{AC}$

$\therefore$ $\dfrac{aR_1 W_2}{W_1 + W_2} = \mu R_1 h + aW_1 - \dfrac{aR_1 W_1}{W_1 + W_2}$

or $R_1 = \dfrac{a}{a - \mu h} W_1$

The reaction is therefore increased in the ratio of $a : a - \mu h$.

The maximum tractive force $= \dfrac{\mu a W_1}{a - \mu h}$.

(ii) When driven on the front axle, there will be a forward friction force μR_1 at D, and the friction at E will now tend to turn the back wheel.

Taking moments about G, we have

$$\mu R_2 h + R_2 \times \text{AC} = R_1 \times \text{CB} = (W_1 + W_2 - R_2)\text{CB}$$

$$\therefore \quad \mu R_2 h + \frac{R_2 a W_1}{W_1 + W_2} = a W_2 - \frac{a R_2 W_2}{W_1 + W_2}$$

$$\therefore \quad \mu R_2 h + a R_2 = a W_2$$

$$\therefore \quad R_2 = \frac{a}{a + \mu h} W_2.$$

The maximum tractive force is now $= \dfrac{\mu a W_2}{a + \mu h}$.

The relative values of these tractive forces depend on W_1 and W_2.

If $W_1 = W_2$, a greater tractive force is obtained by driving on the back wheels.

EXAMPLE (v)

Two equal cylinders rest in parallel positions on a horizontal plane. An isosceles triangular prism, whose vertical angle is α, rests between them in a symmetrical position, its base being horizontal. If all the surfaces are equally rough, show that equilibrium will be preserved if the coefficient of friction exceeds $\tan \frac{1}{4}(\pi - \alpha)$.

Considering a vertical section (Fig. 5.22), let A, B be the position of the axes of the cylinders, E and F the points of contact of the

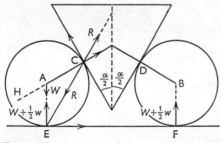

FIG. 5.22

cylinders with the ground, C and D the points of contact of the wedge with the cylinder. Let W be the weight of each cylinder, and w that of the wedge.

In problems of this kind it is very important to see clearly how equilibrium can be broken.

In this case, if equilibrium is broken, the wedge will descend vertically. To allow this to happen the cylinders must either: (i) roll apart about their points of contact E, F, the wedge slipping at C and D, or (ii) slide at E and F, so that they rotate inwards and let the wedge down.

Now for the equilibrium of the cylinder A, the resultant reaction at C must pass through E, where the weight and resultant reaction at E meet. The cylinder cannot then roll about E. Hence there can be no rolling if the angle of friction at C exceeds the angle ACE.

But $\angle\,\mathrm{HAE} = \frac{1}{2}(\pi-\alpha)$.

$\therefore$ $$\angle\mathrm{ACE} = \tfrac{1}{2}\,\angle\,\mathrm{HAE} = \tfrac{1}{4}(\pi-\alpha),$$

$\therefore$ $$\tan\lambda \text{ must exceed } \tan\tfrac{1}{4}(\pi-\alpha)$$

or μ must exceed $\tan\frac{1}{4}(\pi-\alpha)$.

This condition ensures only that equilibrium shall not be broken by rolling apart at E and F, and we must now find the condition that there shall be no slipping at E and F.

To balance the weight and the resultant reaction at C the resultant reaction at E must act in a line, between EA and EC, i.e. a smaller angle of friction is required at E.

We can find the value of this angle as follows.

The normal reaction at E is $W+\frac{1}{2}w$.

If the resultant reaction at C has the value R, on resolving vertically for the wedge, we get

$$2R\,\cos\tfrac{1}{4}(\pi-\alpha) = w.$$

The horizontal component of R is

$$\tfrac{1}{2}w\,\frac{\sin\frac{1}{4}(\pi-\alpha)}{\cos\frac{1}{4}(\pi-\alpha)}$$

and there can be no slipping at E if

$$\mu(W+\tfrac{1}{2}w) > \tfrac{1}{2}w\,\tan\tfrac{1}{4}(\pi-\alpha),$$

where μ is the coefficient of friction at E, i.e. if

$$\mu > \frac{w}{2W+w} \text{ times the coefficient of friction at C.}$$

This is clearly satisfied if the coefficient of friction at E equals that at C.

EXAMPLE (vi)

A plane is inclined to the horizontal at an angle greater than $\tan^{-1} \sqrt{2}$. *The coefficient of friction between this plane and a solid uniform cube lies between* 1 *and* $\sqrt{2}$. *Show that, if the cube is placed on the plane with a diagonal of the face in contact with the plane along a line of greatest slope of the plane, it will simply slide down the plane without toppling; but that, if the cube is placed on the plane with an edge of the face in contact with the plane along the line of greatest slope, it will topple as well as slide.* (H.C.)

Since the tangent of the angle of inclination of the plane is greater than the coefficient of friction, the cube will slide in either case directly it is placed on the plane. There is no question of the usual alternative of sliding *or* toppling.

Let ABCD (Fig. 5.23) represent a section of the cube through its

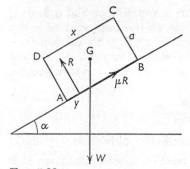

FIG. 5.23

centre of gravity G by a vertical plane through a line of greatest slope of the plane. Let the side of the cube be a and its weight W.

Let CD $= x$. In one case we have to consider $x = a\sqrt{2}$, and in the other $x = a$.

The forces acting on the cube in addition to its weight W are the normal reaction R acting at a distance y from A, and the friction μR acting along the direction AB. When the cube is on the point of toppling R will act at A, that is, y will be zero.

The cube will slide down the plane and not topple if

$$R = W \cos \alpha$$

and $$\mu R < W \sin \alpha$$

and, taking moments about G, if

$$R(\tfrac{1}{2}x - y) = \mu R \times \tfrac{1}{2}a \text{ and } y > 0,$$

that is, if $$x - \mu a > 0.$$

The first two conditions lead to $\mu<\tan\alpha$, which is satisfied, since $\tan\alpha=\sqrt{2}$ and μ lies between 1 and $\sqrt{2}$. The third condition involves x and so depends on the position of the cube.

When $x=a\sqrt{2}$ the third condition becomes $\mu<\sqrt{2}$, which is true, so that in this position the cube will slide and not topple.

When, however, $x=a$ the third condition becomes $\mu<1$, which is not true. In this position the cube will slide and topple. Indeed, the condition for toppling is obtained by putting $y=0$ (that is, R acting at A) and taking moments about G, which leads to

$$R\times\tfrac{1}{2}a<\mu R\times\tfrac{1}{2}a$$

that is, $\mu>1$, which is true.

EXAMPLES 5.5

1. A heavy rectangular block lies on a rough floor and a force is applied in a central vertical plane at the mid-point of a top edge, pulling upwards at an angle $\theta(<90°)$ with the top face. The vertical section has horizontal sides of length b and vertical sides of length a. Show that the block will begin to turn about a bottom edge if

$$\tan\theta>\frac{\tan\theta-2\mu}{\mu\tan\alpha},$$

where $\tan\alpha=b/a$ and μ is the coefficient of friction.   (H.S.C.)

2. A uniform rod AB of length $2a$ rests on a horizontal floor. It is raised by means of a rope attached to the end A and passing over a pulley at a height $2h$ vertically above B. Assuming that the end B does not slip on the plane, find the angle between the re-action of the plane and the vertical at B when the rod makes an angle θ with the vertical. Hence show that the rod may be raised into a vertical position without the end B slipping, provided that $h>a/\mu$, where μ is the coefficient of friction.

(H.S.C.)

3. A uniform cubical block, of side $2a$ and weight W, rests on a rough horizontal plane. A man pushes it at right angles to one of its faces. Prove that the least push that will move it is the smaller of the two forces Wa/h and μW, where h is the height of the man's hands above the ground and μ is the coefficient of friction between the block and the ground. If the coefficient of friction between the man and the ground is μ', what must be the least weight of the man that he may be able to move the block?

(H.C.)

4. A uniform rod AB lies on a rough horizontal plane. A cord, attached to the end B, passes over a small pulley fixed at a point D, vertically over a point C lying in the line AB produced beyond B. Prove that, if the cord is pulled with a gradually increasing force, the rod will begin to slide along the ground if μ, the coefficient of friction between the rod and ground, is less than cot DBC; but that, otherwise, it will begin to turn about the extremity A, and A will not slip until $\mu(\tan \beta - 2 \tan \alpha) = 1$, where α and β are the angles which AB and BD respectively then make with the horizontal. (H.C.)

5. A uniform solid cube stands on a rough horizontal plane, and an exactly similar cube is placed on it so that the faces coincide. The coefficient of friction between the two cubes is $\mu'(<1)$, and the coefficient between the lower cube and the plane is μ. A gradually increasing horizontal force is applied to the upper cube at right angles to one of its faces at the centre of that face. Prove that, when equilibrium is broken, the upper cube will slide on the lower, while the lower remains at rest, or both cubes will move together as a single rigid body according as 2μ is greater or less than μ'. (H.C.)

6. A uniform plank, length l, standing on rough ground, is to be lowered from the vertical position by paying out from a fixed point Q a rope tied to the plank at P. The upper end of the plank initially coincides with Q; μ is the coefficient of friction with the ground. Show that, if P is above the centre of the plank, it will not upset; and this condition being satisfied, that if $\mu<1$ the plank must slip; that if $\mu>1$ the plank will slip unless the distance of P from the free end is less than $(\mu-1)l/(2\mu-1)$. (H.C.)

7. Two equal uniform rods AB, BC, each of weight W, are freely hinged together at B and rest in a vertical plane with the angle ABC a right angle, and the ends A and C on a rough horizontal plane. A string is attached to the mid-point of BC and pulled parallel to AC away from AB with a gradually increasing horizontal force. Prove that if the coefficient of friction between the rods and the plane is $1\frac{1}{3}$, equilibrium will be broken by C slipping while A remains at rest, as soon as the force exceeds $2W$, but that, if the coefficient of friction is equal to $1\frac{2}{3}$, equilibrium will be broken by A slipping, while C remains at rest, as soon as the force exceeds $3\frac{1}{4}W$. (H.C.)

8. Two cylinders, each of radius a and mass 5 kg, rest on a rough table with their axes horizontal and parallel and at a distance $3a$ apart; a third cylinder of the same radius and weight rests

between them, friction being just sufficient to preserve equilibrium. Find the least coefficients of friction required between the upper and lower cylinders and between the lower cylinders and the table, and find the magnitude of the total reaction at each line of contact. (H.S.C.)

9. Forces of P each are applied symmetrically to opposite sides of a cone of weight W and vertical angle 2α, the forces being in a vertical plane through the axis of the cone and acting upwards at angles β with the lower slopes of the cone. Prove that it is impossible to raise the cone by the application of these forces unless $\mu > \cot \beta > \tan \alpha$, where μ is the coefficient of friction. If μ is greater than $\tan \alpha$, find the least possible value of P that is necessary to allow the cone to be raised in this manner.

(H.S.C.)

10. A wheel is turning on a fixed horizontal axle of radius a, the bearing being sufficiently loose for the contact to be only along a single horizontal line; show that, if the angle of friction between wheel and axle is λ, the resultant pressure between them is tangential to a circle of radius $a \sin \lambda$ concentric with the axle. If the weight of the wheel is W, and its radius is b, and the force applied is X, find the least value of X which will turn the wheel if X is applied downwards at the end of a horizontal radius. Show also that, if X is applied horizontally at the top of the wheel,

$$X = W \tan \theta$$

where $b \sin \theta = a \sin \lambda$. (H.S.D.)

11. A light string is attached to one end A of a heavy uniform beam AB of mass M, and passes over a smooth pulley, having a mass m hung from its free end. The other end B of the beam rests on a rough horizontal plane, the coefficient of friction between the plane and the beam being μ. Show that, if

$$\mu < \frac{M}{\sqrt{(M^2 - m^2)}}, \text{ and } m < \frac{M}{\sqrt{2}},$$

the beam is on the point of slipping when its angle of inclination to the horizontal has one of the values

$$\tan^{-1} \left[\frac{M^2 - 2m^2 \pm M \sqrt{\{m^2 - \mu^2(M^2 - m^2)\}}}{2\mu(M^2 - m^2)} \right]. \quad \text{(H.S.C.)}$$

12. C is the mid-point of a uniform plank of mass M and length $2l$. The plank is placed horizontally over a rough horizontal cylinder of radius R, so that its length is perpendicular to the generators of the cylinder; a point O of the plank is then in contact with the cylinder. The plank is then allowed gradually to assume an

inclined position. Show that the plank will rest in equilibrium on the cylinder without slipping if $OC < R\lambda$, where λ is the angle of friction. A mass m is attached to one end of the rod. Find the range of possible positions of O so that the rod can, if made to move slowly, assume a position of equilibrium.   (H.S.D.)

13. Two equal uniform right circular cylinders rest on a rough horizontal plane with their axes parallel. A uniform wedge in the shape of an equilateral triangular prism is laid symmetrically between them so that two of its faces are inclined at equal angles to the vertical and are each in contact with one of the cylinders. No part of the wedge touches the horizontal plane. The coefficient of friction is the same between all the surfaces in contact. Show that whatever the weight of the wedge the system will remain in equilibrium if the coefficient of friction exceeds $1/\sqrt{3}$, but that otherwise a very light wedge would cause the cylinders to separate by rolling on the plane.

14. Two uniform planks, AB, BD, of lengths $2a$, $2b$, and weights W and W' respectively, can turn freely about fixed hinges at A and B, which are at right angles to the lengths of the planks. The hinges are parallel, and in the same horizontal plane, and $BD > AB > AC$. Show that in the limiting position of equilibrium in which the end C of the first plank presses against the second plank, and the angle ACB is obtuse.

$$Wa \sin 2a \cos \lambda + W'b \sin 2\beta \cos (\alpha + \beta + \lambda) = 0,$$

where $\alpha = \angle\, CAB, \beta = \angle\, CBA$, and λ is the angle of limiting friction.

15. A cylinder of radius a rests with its curved surface on a horizontal floor. A uniform straight plank of length $2l$ lies symmetrically across it in such a position that its centre is in contact with the cylinder, and its lower end rests on the floor. The coefficients of friction at all three points of contact are equal. There is limiting friction at one of the three points of contact. Show that this point is never the point of contact of the cylinder with the floor, but is the point of contact of the plank with the floor or with the cylinder according as $3a^2 \lessgtr l^2$.   (C.S.)

16. A uniform rod, of weight W and length l, lies on a rough horizontal plane, the coefficient of friction being μ. A string is attached to one end, and is pulled horizontally in a direction perpendicular to the rod so that the tension gradually increases. Show that the rod begins to turn about a point $l/\sqrt{2}$ from the end to which the string is attached, and that the tension of the string

is then $(\sqrt{2}-1)\mu W$, assuming that the vertical reaction is distributed uniformly along the rod.   (C.S.)

17. A uniform ladder of weight W leans with one end against a wall and makes an angle θ with the floor. The angles of friction for floor and wall are respectively ϵ and λ. Explain why it is not in general possible to determine the reactions at the ends of the ladder. If a man of weight w slowly climbs the ladder, show that he can get to the top if

$$\tfrac{1}{2}\frac{W}{W+w} > \frac{\cos \lambda \cos (\epsilon+\theta)}{\cos (\lambda-\epsilon) \cos \theta}. \tag{C.S.}$$

18. A uniform circular hoop has a weight equal to its own attached to a point of its rim, and is hung over a rough horizontal peg. Prove that if the angle of friction is greater than $\tfrac{1}{6}\pi$ the system can rest with any point of the hoop in contact with the peg.

(C.S.)

19. Show that the greatest inclination to the horizon at which a uniform rod can rest in a rough sphere of radius a, and angle of friction λ, is

$$\tan^{-1}\frac{a^2 \sin \lambda \cos \lambda}{c^2-a^2 \sin^2 \lambda},$$

where c is the distance of the rod from the centre of the sphere.

(C.S.)

20. A uniform rod is placed over a rough horizontal rail and rests with one end against a rough vertical wall, the rail being parallel to the wall and perpendicular to the rod. If the rod is on the point of slipping up the wall, show that it will make an angle θ with the horizontal where

$$a \cos^2 \theta \cos (\theta+\lambda+\lambda') = c \cos \lambda \cos \lambda',$$

λ, λ' being the angles of friction at the rail and wall. It is assumed that the rod is sufficiently long to make such a position of equilibrium possible. The length of the rod is $2a$, and the rail is at a horizontal distance c from the wall.   (C.S.)

21. Two cylinders lie in equilibrium on a rough inclined plane, in contact with one another, with their axes horizontal. The upper cylinder, of radius a, is heavy, but the lower cylinder, of radius b, is of negligible weight. Prove that, if α is the inclination of the plane to the horizontal, $b>\tfrac{1}{4}a \tan^2 \alpha$, the coefficient of friction between the heavy cylinder and the plane must be at least

$$\frac{1}{2 \cot \alpha-(a/b)^{\frac{1}{2}}},$$

and the other coefficients of friction must be at least equal
to $(b/a)^{\frac{1}{2}}$. (C.S.)

22. A heavy particle of weight W is to be supported by a given force
equal to $\frac{1}{2}W$ on the upper portion of the outer surface of a fixed
rough circular cylinder. If the coefficient of friction is equal to
$1/\sqrt{3}$, find the greatest angular distance from the highest
generator of the cylinder at which the particle can be maintained
in equilibrium. (C.S.)

23. A uniform rod rests inside a rough vertical circle with its highest
point on a level with the centre of the circle. If the friction is
just sufficient to prevent sliding, show that the angle of friction
equals

$$\tfrac{1}{2}[\alpha-\sin^{-1}(\sin\alpha\cos 2\alpha)],$$

where α is the inclination of the rod to the horizontal. (C.S.)

24. A uniform beam AB lies horizontally on two rough parallel rails
at points A and C. Prove that the least horizontal force applied
at B in a direction perpendicular to AB, which is able to move the
beam, is the smaller of the two forces

$$\mu W\frac{b-a}{2a-b} \text{ and } \tfrac{1}{2}\mu W,$$

where AB is $2a$, AC is b, W is the weight of the beam, and μ is the
coefficient of friction at each point of contact. (C.S.)

25. The seat of a chair is a square of side 45 cm. The back and legs
are vertical, the latter being 45 cm long. The centre of gravity
is 15 cm from the back. The chair is drawn forward by a horizontal string attached to a point of the back distant 105 cm
from the ground. Show that the chair will slide or tilt according
as μ is less or greater than $\frac{2}{7}$. If the chair just slides in this
position, show that when an additional mass, equal to that of
the chair, rests in the centre of the seat, it will still slide forward
if the string is lowered through any angle less than $\tan^{-1}\frac{7}{10}$.
  (C.S.)

26. The distance between the axles of a railway track is $2a$, and the
centre of gravity is half-way between them and at a distance h
from the rails. With the lower wheels locked, the greatest incline
upon which the truck can rest is α. Show that the coefficient of
friction between the wheels and the rails is given by

$$\mu = \frac{2a\tan\alpha}{a+h\tan\alpha}.$$ (C.S.)

27. A homogeneous cube is supported, with a face flat against a rough
vertical wall and four edges vertical, by a force P applied at the

mid-point of the lowest edge, which does not meet the wall, in a plane perpendicular to that edge. Prove that if $\mu(= \tan \epsilon)$ is the coefficient of friction, the least value of P is (i) $\frac{1}{2} W \operatorname{cosec} \epsilon$;

(ii) $W \cos \epsilon$, or (iii) $W \cos \epsilon \sec \left(\tan^{-1}\dfrac{1}{\mu+2}-\epsilon\right)$, according as (i) $\epsilon > \frac{1}{4}\pi$, (ii) $\frac{1}{4}\pi > \epsilon > \frac{1}{8}\pi$ or (iii) $\epsilon < \frac{1}{8}\pi$. (C.S.)

28. Two cylinders, similar in all respects, of radius 15 cm, lie symmetrically in contact in a cylindrical trough of radius 54 cm, and a third cylinder, of radius 10 cm, lies on the two equal cylinders. The axes of all the cylinders are horizontal and parallel. Show that if the surfaces are smooth there will be equilibrium unless the weight of the upper cylinder is greater than $\frac{5}{2}$ of the weight of one of the others. Show also that however great the weight of the upper cylinder may be, there will still be equilibrium if all the surfaces are rough and no coefficient of friction is less than $\frac{1}{8}$. (C.S.)

29. A motor car of weight W is being retarded at rate f by application of the brakes. Determine the reactions between the wheels and the road, when the masses of the wheels may be neglected, and the brakes are applied only to the rear wheels. The centre of gravity of the car is at a height h above the road, and at horizontal distance a from the rear, and a' from the front wheels. Show that the maximum retardation, which can be obtained without skidding any wheel, is greater by the factor

$$(a+a'+h\mu)/a'$$

for a car braked on all four wheels than for a car braked on the rear wheels only, where μ is the coefficient of friction between the tyres and the road. If, for example, $a' = a = 2h$, $\mu = 0{\cdot}9$, four-wheel braking has the advantage in a factor of $2{\cdot}45$. Explain in general terms how it comes about that this factor can be greater than 2, when $a = a'$. (C.S.)

30. Two equal uniform ladders of weight w are rigidly fastened together at one end to form a step ladder, which stands on a plank. The angle between the ladders is 2α. A man of weight W stands on the top. The plank is slowly tilted about one end. Show that if β be the inclination of the plank to the horizontal the system will overturn if $\tan \beta$ reaches

$$\frac{W+2w}{W+w} \tan \alpha,$$

and will slip if $\tan \beta$ reaches μ, the coefficient of friction. (C.S.)

31. At points A, A′, A″ on a rough horizontal plane are placed weights

W, W', W'' ($W<W'$) connected by light inextensible strings AA', $A'A''$. The angle $AA'A''$ is obtuse, and equal to $\pi-\alpha$. Prove that the least force that can be applied to W'' so that all the weights may be just on the point of motion is $\mu(W''^2+x^2)^{\frac{1}{2}}$, where $x = W\cos\alpha+(W'^2-W^2\sin^2\alpha)^{\frac{1}{2}}$ and μ is the coefficient of friction.   (C.S.)

32. Three equal spheres rest in contact on a rough horizontal plane. An equal sphere of the same material is placed so as to rest symmetrically on them. Show that, if the coefficient of friction μ is greater than $\sqrt{3}-\sqrt{2}$, and all surfaces are equally rough, equilibrium will be maintained.   (C.S.)

33. A uniform rod ACB of length $2a$ is supported against a rough vertical wall by a light inextensible string OC attached to its mid-point C. The other end of the string is attached to a fixed point O on the wall. Show that the rod can rest with C at any point of a circular arc, whose extremities are at perpendicular distances a and $a\cos\lambda$ from the wall, where λ is the angle of friction.   (C.S.)

34. A weight W rests on a rough plane ($\mu = 1/\sqrt{3}$) inclined at $45°$ to the horizontal, and is connected by a light string passing through a smooth fixed ring A, at the top of the plane, with a weight $\frac{1}{3}W$ hanging vertically. The string AW makes an angle θ with the line of greatest slope in the plane. Prove that the greatest possible value of θ for equilibrium is $\sin^{-1}\frac{1}{3}$.   (C.S.)

35. A uniform rod rests in limiting equilibrium with its ends on a rough circular band whose plane is vertical. Prove that its inclination θ to the vertical is given by the equation

$$\sin\lambda\sin(\theta+\lambda) = \cos\theta\cos^2\alpha,$$

where λ is the angle of friction and 2α the angle subtended by the rod at the centre of the circle.   (C.S.)

36. A plank of breadth $2b$ and thickness $2c$ rests inside a horizontal cylinder of radius a with its long edges parallel to the axis of the cylinder and at such a height that it is just about to slip down. Show that the plank makes an angle θ with the horizontal given by

$$a\sin\lambda\cos(\theta-\lambda) = (a\cos\alpha-c)\sin\theta\cos\alpha$$

where λ is the angle of friction and $\sin\alpha = b/a$.   (C.S.)

37. A uniform cube of weight W and edge $2a$ is placed on a rough plane and a uniform sphere of weight W' and diameter $2a$ rests upon the plane, touching the cube at the centre of one of its faces F. The plane is gradually tilted round a line lying in the

plane and parallel to the face F of the cube. Show that if μ be the coefficient of friction for every contact and $\mu < 1$, then equilibrium will be broken by the cube slipping and the sphere rolling down the plane when the angle of inclination of the plane to the horizontal is α, where

$$\tan \alpha = \frac{\mu W}{W + (1-\mu)W'}.$$ (C.S.)

38. Two uniform rods AB, BC, of equal weight but different lengths, are freely jointed together at B and placed in a vertical plane over two equally rough fixed pegs in the same horizontal line. The inclinations of the rods to the horizontal are α, β, and they are both on the point of slipping. Prove that the inclination θ to the horizontal of the reaction at the hinge is given by

$$2 \tan \theta = \cot (\beta + \lambda) - \cot (\alpha - \lambda),$$

where λ is the angle of friction at the pegs. (C.S.)

39. A piece of uniform wire is bent into the shape of an isosceles triangle ABC, in which AB = AC. The triangle hangs in a vertical plane with BC in contact with a rough peg. Show that the triangle will rest in equilibrium whatever point of BC is in contact with the peg provided that the coefficient of friction is greater than $2 \tan\frac{1}{2}A(1+\sin \frac{1}{2}A)$. (C.S.)

40. Two uniform heavy rods AB, AC, each of length $2a$, are rigidly connected at A at right angles to each other, and rest on a fixed rough cylinder of radius c, the plane of the rods being perpendicular to the axis of the cylinders. Show that, if $c < \frac{1}{2}a$, $\tan \epsilon$ is the coefficient of friction between the rods and the cylinder, and α is the angle which the bisector of the angle BAC makes with the vertical in limiting equilibrium, then

$$c \sin (\alpha + 2\epsilon) = (a - c) \sin \alpha.$$ (C.S.)

41. On the radius OA of a circular disc as diameter a circle is described, and the disc enclosed by it is cut out. If the remaining solid rest in a vertical plane on two rough pegs on a horizontal plane subtending an angle 2α at the centre O, show that the greatest angle that OA can make with the vertical is

$$\sin^{-1} (3 \sin 2\lambda \sec \alpha),$$

where λ is the angle of friction at the pegs. (C.S.)

42. A thin uniform straight rod PQ of weight W rests partly within and partly without a uniform cylindrical jar of weight $4W$, which stands on a horizontal table. The rod rests in contact with the smooth rim of the jar, with its end P pressing against the rough

curved surface of the jar. If the rod is about to slip and the jar is about to upset simultaneously, prove that the rod makes with the vertical an angle

$$\tfrac{1}{2}\lambda + \tfrac{1}{2}\cos^{-1}(\tfrac{1}{3}\cos\lambda),$$

where λ is the angle of friction. (C.S.)

43. Two equal ladders are hinged at the top and rest on a rough floor forming an isosceles triangle with the floor of vertical angle 2θ. A man whose weight is n times that of either ladder goes slowly up one of them. Calculate the reactions at the floor when his distance from the top is x, and show that slipping begins when

$$\frac{nx}{l} = \frac{2\mu - \tan\theta}{\mu - \tan\theta} + n. \tag{C.S.}$$

44. Two rough planes intersect at right angles in a horizontal line and make angles

$$\alpha, \ \tfrac{1}{2}\pi - \alpha \ (\alpha < \tfrac{1}{4}\pi)$$

with the horizontal. Two equal rough cylinders with their axes in the same horizontal plane rest in contact with each other and each in contact with one plane. Prove that, if all the surfaces are equally rough, the coefficient of friction is not less than

$$\frac{\cos 2\alpha}{\sin\alpha + \cos\alpha + \sin 2\alpha}. \tag{C.S.}$$

45. Two supports P and Q are in the same horizontal plane and at a distance 0·9 m apart, and a uniform plank APQB, 2·4 m long, is in equilibrium resting on the supports. A horizontal force is applied to the plank at A in the direction perpendicular to its length, and is steadily increased in magnitude. Prove that, if the supports are equally rough, equilibrium will be broken by the plank slipping at P, provided that the length AP lies between 0·3 m and 0·9 m. (H.D.S.)

CHAPTER SIX

MACHINES

6.1. We shall now consider some simple examples of machines. A machine is a piece of apparatus in which work is done *on* the machine by applying a force, called the *Power or Effort*, at one part, and work is done *by* the machine in overcoming some external force, called the *Weight or Resistance*, at another part.

6.2. Work

The work done by a constant force is defined (Dynamics, Chapter Four) as the product of the force and the distance through which the point of application moves *in the direction of the force*. It is a scalar quantity.

Thus if a force P, acting on a particle at A in the direction AB (Fig. 6.1), move the particle from A to B, the work done by

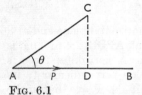

FIG. 6.1

the force is measured by $P \times AB$.

If, however, the particle moves from A to C, where angle $BAC = \theta$, the work done is *not* $P \times AC$, but the product of P and the projection of AC on AB, i.e. $P \times AD$ or $P \times AC \cos \theta$.

It is clear that some other cause must have been responsible for the motion in the direction DC, as P by itself can only cause motion in the direction of its own line of action.

It should be noticed that

$$P \times AC \cos \theta = P \cos \theta \times AC,$$

i.e. the work done by P is the same as that done by the component of P in the direction AC.

If the direction of the displacement is opposite to the direction of the component of the force we may say that the force does negative work, or that work is done against the force. Thus, when a weight descends under the action of gravity work is done by gravity; when the weight is raised work is done against gravity.

6.3. Units of work

If the force is in newtons and the distance in metres the work done is in joules.

Thus 1 joule is the work done by a force of 1 N in moving its point of application through 1 m in its own direction. The work done in lifting a mass of 1 kg vertically through 1 m is 9·8 J.

6.4. Work done by a number of forces

Consider a particle subject to a number of coplanar forces F_1, F_2, ... F_n. The total work done by the forces in any displacement of the particle equals the algebraic sum of the work done by F_1, F_2, ... F_n separately. We can show that this equals the work done by the resultant of the forces.

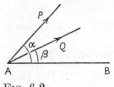

Fɪɢ. 6.2

We will first prove this result for two forces only. The work done by the resultant of two forces which are constant in magnitude and direction and acting on a particle is equal to the sum of the amounts of work done by the two forces separately.

Let P, Q be the two forces acting as shown on a particle at A (Fig. 6.2), and suppose the particle is displaced to B.

Let α, β, be the angles made by P and Q with AB.

The sum of the amounts of work done by P and Q is

$$P \cos\alpha \times AB + Q \cos\beta \times AB = (P \cos\alpha + Q \cos\beta)AB$$

But $P \cos\alpha + Q \cos\beta$ is the sum of the components of P and Q along AB, and this is equal to the component in this direction of the resultant of P and Q.

Hence the amount of work $(P \cos\alpha + Q \cos\beta)AB$ is also the amount done by the resultant.

It is clear that this result holds for any number of forces acting on a particle.

It also applies to any number of coplanar forces acting on a rigid body (see 9.2).

6.5. Work done by a couple

The work done by a couple acting on a rigid body (Dynamics, 4.22) in any small displacement equals the moment of the couple multiplied by the small angle through which the body turns.

In any finite displacement of the rigid body the work done by the couple is zero if there is no rotation of the body. If, however, the body rotates through an angle θ and the moment of the couple is M the work done is given by the expression $\int_0^\theta M \, d\theta$. This equals $M\theta$, if M is constant throughout the displacement.

6.6. Work done by a variable force

The work done by a variable force F acting on a particle (Dynamics, 4.15) can be expressed as the integral $\int F \cos \theta \, ds$, where θ is the angle between F and the element ds of the path of the particle.

It can be represented by an area under the graph of $F \cos \theta$ against s.

The work done by a set of variable forces acting on a particle or a rigid body in any displacement can be found by evaluating the work done by each of the forces and adding the results.

6.7. Simple machines

In most cases a machine is so arranged that, by applying a small force or effort, a larger force is overcome. Thus an inclined plane may be used as a simple machine.

Suppose we have a weight W resting on a smooth inclined plane AB (Fig. 6.3) connected to a weight P hanging freely by a light string passing over a smooth pulley at the top of the plane. P may be regarded as the *effort* used to raise a *weight W*.

If W rests in equilibrium on the plane, then the tension in the string T must equal $W \sin \alpha$. But $T = P$, and hence $P = W \sin \alpha$. Any force P slightly in excess of $W \sin \alpha$ will cause W to move up the plane. Thus the weight W can be raised vertically by applying a force less than W, which makes the inclined plane a useful machine. Indeed, historically it was one of the earliest machines used by man.

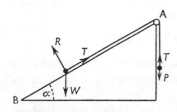

FIG. 6.3

We notice that when P moves downwards a distance x, W moves a distance x along the plane, that is, it is raised a distance $x \sin \alpha$ *vertically*. Hence the work done *by P* is Px, and this equals the work done *on W*, which is $Wx \sin \alpha$. (If the plane were rough Px would be greater than $Wx \sin \alpha$, since $P = W \sin \alpha + F$, where F is the frictional force called into play.) This result follows, of course, from the law of conservation of energy. In an ideal machine no energy is lost, and the work done *by P* is then equal to the work done *on W*, or in other words, the total work done *by P* and W is zero. This is sometimes known as the *Principle of Work*. In practice, however, the work done by P usually exceeds the work done on W, that is, more work has to be done as a result of using the machine (to overcome friction and to move certain parts of the machine). The great convenience of the machine is that usually P is less than W.

The following definitions are important.

6.8. If a force P applied to a machine causes it to exert a force W, the ratio W/P is called the **mechanical advantage** of the machine. The ratio of the distance moved by the effort P to that moved by the weight W is called the **velocity ratio**. These two ratios are related, as we shall now show.

If there is no friction, and the parts of the machine are weightless, then, by the principle of work,

$P \times$ distance through which P moves

$$= W \times \text{distance through which } W \text{ moves},$$

$$\therefore \quad \frac{W}{P} = \frac{\text{distance through which } P \text{ moves}}{\text{distance through which } W \text{ moves}}$$

$$= \text{velocity ratio},$$

i.e. in an ideal machine whose parts are weightless, and in which there is no friction,

Mechanical Advantage = Velocity Ratio.

Thus in all cases when the effort is less than the weight the effort has to move through a distance greater than the distance moved by the weight.

Occasionally a machine is arranged to give increased movement, and in this case the effort has to be greater than the weight moved.

In practical machines where there is friction the mechanical advantage is *less* than the velocity ratio.

6.9. The **efficiency** of a machine is measured by the ratio

$$\frac{\text{Useful work done by the machine}}{\text{Work supplied to the machine}}.$$

In an ideal machine the efficiency is therefore unity. The ratio giving the efficiency is often expressed as a percentage, and then the efficiency of an ideal machine is 100 per cent.

The efficiency may be determined experimentally by measuring the effort P required to raise a weight W and also the distances x and y moved through by P and W respectively.

$$\text{Efficiency} = \frac{Wy}{Px} = \frac{W/P}{x/y}$$

$$= \frac{\text{Mechanical advantage}}{\text{Velocity ratio}}.$$

In most cases the efficiency varies with the load.

It should be noticed that the only quantity connected with a machine which can be *calculated* from its dimensions is the velocity ratio.

The mechanical advantage and efficiency must be determined by experiment, except in the case of an ideal machine.

In actual machines it is often found that W and P are connected by a linear relation of the form

$$P = a + bW,$$

where a and b are constants.

This relation is often called the 'Law of the Machine'.

Clearly b is positive, but a may have any value. If a is zero, than $P = bW$ or $W/P = 1/b$, that is, the mechanical advantage, and therefore the efficiency, do not vary with W. This is the ideal case, and b is the reciprocal of the velocity ratio.

If a is positive, as it usually is (since even when $W = 0$, P is positive owing to friction), then the mechanical advantage increases as W increases. For

$$W/P = W/(a+bW) = 1/\left(\frac{a}{W}+b\right)$$

and this increases as W increases, if a is positive, and is less than $1/b$ for all values of W. Thus the efficiency increases with the load, but is always less than $1/br$, where r is the velocity ratio.

If a is negative W/P diminishes as W and P increase, i.e. the efficiency diminishes with the load. This is explained by the fact that with greater loads the friction between the various parts of the machine is increased.

EXAMPLE

A machine for lifting weights has a velocity ratio of 16, and it is found that efforts of 11 N and 19 N are needed to lift with it loads of 56 N and 112 N respectively. What is the efficiency in each case? Assuming a straight-line graph for load and effort, find the effort necessary to lift 224 N.

In the first case the mechanical advantage is $\frac{56}{11}$, and in the second case it is $\frac{112}{19}$.

Since $\qquad$ Efficiency $= \dfrac{\text{Mechanical advantage}}{\text{Velocity ratio}}$,

the efficiencies are $\dfrac{56}{11\times16}$ and $\dfrac{112}{19\times16}$, that is, $\frac{7}{22}$ and $\frac{7}{19}$.

Expressed as percentages, these are 31·8 per cent and 36·8 per cent.

Assuming that, since the relation between W and P is linear,

$$P = a + bW,$$

where a and b are constants, we have

$$11 = a+56b$$

and $$19 = a+112b$$

whence $$a = 3 \text{ and } b = \tfrac{1}{7}.$$

$\therefore$ $$P = 3+\tfrac{1}{7}W.$$

When $W = 224$, we get

$$P = 3+32.$$

$\therefore$ $$\textit{Effort} = 35 \text{ N}.$$

6.10. *System of pulleys with a single string*

In this system there are two blocks each containing pulleys, the upper block being fixed to a support and the lower block, which has the weight attached, being movable.

Figure 6.4 shows such a system when the number of pulleys in each block is the same, and Fig. 6.5 shows one where the number of pulleys in the upper block is greater than the number in the lower block.

In the first case one end of the string must be fastened to the upper block, and in the second case it must be fastened to the lower block.

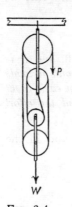

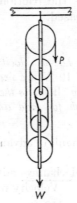

FIG. 6.4   FIG. 6.5

The relation between P and W may be obtained in two ways,

  (i) by considering the equilibrium or uniform motion of the lower block; or
  (ii) by the principle of work.

Let the weight of the lower block be w, and neglect friction.

(i) Since the pulleys are smooth, the tension is the same throughout the string and equal to P.

If there are n portions of string at the lower block the total upward force on this block is nP. Therefore for equilibrium or motion with uniform speed

$$W+w = nP.$$

We note that this is of the form $P = a+bW$, where a and b are positive constants. Further, $W/P = n-w/P$, which is always less than n, the velocity ratio of the system. We have assumed that all the portions of the string not in contact with the pulleys are vertical. If this is not the case the resultant of the two tensions P due to a string passing round one of the pulleys is not $2P$, but depends on the angle at which the two portions are inclined. In practice, the strings are not exactly parallel, but usually they are very nearly so.

(ii) If the load W and the whole of the lower block is raised a distance x, then in the first figure a length of string $2x$ will pass round the upper pulley of the lower block, a further length $2x$ round the lower one, and so on for any number of pulleys in the lower block.

Hence to keep the string taut P must move a distance $2x \times$ the number of pulleys at the lower block. But there are two portions of string to each pulley, and so the distance moved by P is nx, where n is the number of portions of string at the lower block. The velocity ratio is therefore n. By the principle of work

$$(W+w)x = Pnx,$$

$$\therefore \qquad W+w = nP,$$

as obtained above.

In the second figure a length of string $3x$ will pass round the upper pulley of the lower block, and an additional length of $2x$ for each other pulley. If p be the number of pulleys P must move $(2p+1)x$, but $2p+1 = n$, the number of strings, and we have the same relation between W, w, and P as before.

This system is often called the Block and Tackle.

EXAMPLE

A pulley system in which the same string passes round all the pulleys consists of six pulleys, the string being attached to one of the upper pulleys. When on the point of motion the tension of the string as it passes over each pulley is increased by 25 per cent. Find the force which will just lift a weight of 300 N, neglecting the weights of the pulleys themselves. Show also that the useful work done is about half that expended.   (H.S.D.)

The arrangement is shown in Fig. 6.6.

Let T be the tension in the end of the string attached to the upper block. The tension after passing round the first lower pulley is $(5/4)T$, and over the first upper pulley $(5^2/4^2)T$, and so on.

The tension at the free end is $\dfrac{5^6}{4^6}T$.

$$\therefore \qquad \frac{5^6}{4^6}T = P.$$

For the equilibrium of the weight

$$T\left(1+\frac{5}{4}+\frac{5^2}{4^2}+\frac{5^3}{4^3}+\frac{5^4}{4^4}+\frac{5^5}{4^5}\right) = 300$$

$$\therefore \qquad T\frac{\frac{5^6}{4^6}-1}{\frac{1}{4}} = 300$$

$$\therefore \qquad \frac{4^6}{5^6}\left(\frac{5^6}{4^6}-1\right)P = 75$$

$$\therefore \qquad \left(1-\frac{4^6}{5^6}\right)P = 75$$

$$\therefore \qquad P = \frac{5^6 \times 75}{61 \times 189} = 101 \cdot 6 \text{ N.}$$

If the lower block moves a distance x m, P moves a distance $6x$ m.

$$\therefore \qquad \text{Work expended} = 101 \cdot 6 \times 6x = 609 \cdot 6x \text{ J}$$

and   Useful work $= 300x$ J.

6.11. *System of pulleys when each string is attached to the support*

This arrangement is as shown in Fig. 6.7. The weight is attached to the lowest pulley.

The fixed pulley A_4 is usually inserted so that the effort P

may be applied in a downward direction; it does not affect the mechanical advantage.

As before, we suppose that all portions of the strings not in contact with the pulleys are vertical, and that there is no friction. We shall neglect the weights of the pulleys.

(i) Let T_1, T_2, ... be the tensions in the strings passing round A_1, A_2, ...

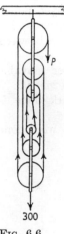

300

FIG. 6.6   FIG. 6.7

From the equilibrium of the pulleys, A_1, A_2, ... we have

$$W = 2T_1$$
and
$$T_1 = 2T_2$$
and
$$T_2 = 2T_3$$
$$\therefore \quad W = 2^3 T_3 = 2^3 P$$

and clearly for n movable pulleys,

$$W = 2^n P.$$

(ii) If P moves a distance x the upper movable pulley moves $x/2$.

The next moves $x/2^2$ and so on.

Hence with n movable pulleys the lowest one, and therefore the weight, moves a distance $x/2^n$.

From the principle of work

$$W \times (x/2^n) = Px$$
or
$$W = 2^n P, \text{ as before.}$$

EXAMPLE

A man of mass 70 *kg supports a weight of mass* 45 *kg by means of three movable pulleys arranged in the system where each pulley hangs in the loop of a separate string. The pulleys have masses* 2, 4, *and* 5 *kg respectively. What is the thrust of the man on the ground?*

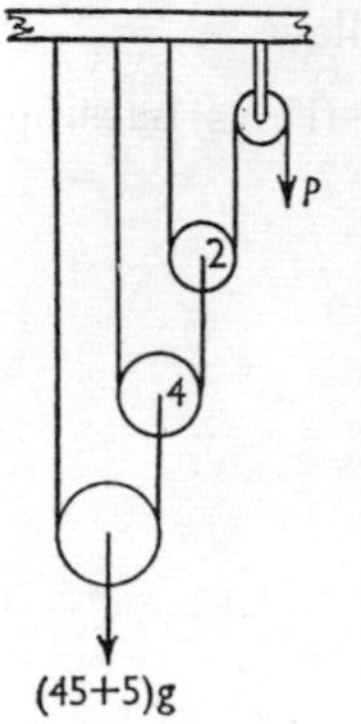

We shall assume that the end of the string on which the man pulls is passed over a fixed pulley as in Fig. 6.8, so that he pulls downwards. His thrust on the ground will be the difference between his weight and the pull he exerts.

If P moves a distance x the upper movable pulley moves $x/2$, and the work done is $2g \times x/2$.

The next moves $x/4$ and the work done is $4g \times x/4$.

The lower pulley moves $x/8$ and the work done is $50g \times x/8$.

$(45+5)g$

FIG. 6.8

Hence from the principle of work,

$$Px = \left(1+1+\frac{50}{8}\right)g\, x = 8{\cdot}25g\, x,$$

∴ $P = 8{\cdot}25g$ N.

Hence the thrust on the ground is $61{\cdot}75g$ N.

EXAMPLES 6.1

1. If there are four movable pulleys whose masses, commencing with the lowest, are 4, 3, 2, and 1 kg respectively, what force will support a weight of mass 500 kg?

2. If there are three movable pulleys whose masses, commencing from the lowest are 4, 4, and 2 kg respectively, what force will support a weight of mass 56 kg?

3. If there are four movable pulleys, each of weight w, and the effort be P, show that the stress on the beam is $15P-11w$ (P is supposed to act upwards, the string to which it is applied not passing over a fixed pulley).

4. When there are four movable pulleys find the effort necessary to support a weight of mass 400 kg, neglecting the weights of the pulleys. Find also the pull of each string on the beam; and if the sum is not equal to the weight, explain the difference.

6.12. *System of pulleys with each string attached to the weight*

This arrangement is shown in Fig. 6.9. The weight is suspended from a bar AB to which each string is attached.

The free portions of the strings are assumed to be parallel, and we shall neglect the weights of the pulleys.

(i) Let T_1, T_2, ... be the tensions in the strings passing round the pulleys A_1, A_2, ...

$$T_1 = P, \text{ since } A_1 \text{ is smooth.}$$

Also $\qquad T_2 = 2T_1 = 2P$, since A_1 is in equilibrium,

and $\qquad T_3 = 2T_2 = 2^2P$ since A_2 is in equilibrium,

and $\qquad T_4 = 2T_3 = 2^3P$ and so on.

If there are n pulleys,

$$\begin{aligned}
W &= T_1 + T_2 + \ldots + T_n \\
&= P(1 + 2 + 2^2 + \ldots + 2^{n-1}) \\
&= P\frac{2^n - 1}{2 - 1} = P(2^n - 1).
\end{aligned}$$

(ii) If AB is raised a distance x a length of string x passes over the uppermost pulley A_4; this lowers the next pulley A_3 a distance x. Since this pulley descends a distance x and the bar rises x, the string from the pulley to the bar shortens by $2x$, hence the next pulley A_2 descends a distance $2x + x = 3x$. The string from A_2 to the bar therefore shortens by $4x$, and the next pulley A_1 descends $4x + 3x = 7x$.

The string from A_1 to the bar therefore shortens by $8x$, and the weight P descends $8x + 7x = 15x$.

In the case shown there are four pulleys, and this distance is $(2^4 - 1)x$. With n pulleys the distance moved by P would be $(2^n - 1)x$, and from the principle of work,

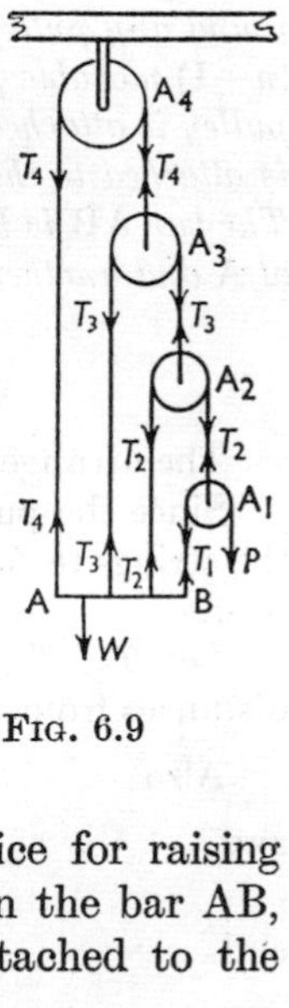

FIG. 6.9

$$Wx = P(2^n - 1)x$$

or $\qquad W = P(2^n - 1).$

The system of pulleys is not used in practice for raising weights. It is used to give a short strong pull on the bar AB, which is fixed, the uppermost pulley being attached to the

object, such as the top of a mast, on which it is required to exert
a pull. It is evident that as the tensions in the strings attached
to the bar are not equal, that if the bar is free, as shown in the
figure, it will not remain horizontal unless the weight is attached
at a particular point.

If the pulleys are of equal size the distances between the
points of attachment of the strings will be equal, and the condition for the bar to remain horizontal is obtained by taking
moments about one end.

In the case shown, $T_4 = 8P$, $T_3 = 4P$, $T_2 = 2P$, and
$W = 15P$.

If a is the distance between the strings the sum of the
moments of the tensions about B is

$$24Pa + 8Pa + 2Pa = 34Pa.$$

Hence, to keep the bar horizontal, W must be attached at a
point distant x from B, where

$$15Px = 34Pa$$

or $$x = \tfrac{34}{15}a.$$

EXAMPLE

*In a system of n equal weightless pulleys, in which the string passing
round any pulley has an end attached to a weightless bar* AB, *there are
$(n-1)$ movable pulleys and one fixed pulley; the string round the fixed
pulley is attached to the bar at* A *and that round the last movable pulley
is attached to the bar at* B *and has a weight P at the other extremity.
The bar* AB *is kept in horizontal equilibrium by a weight W attached
at* A *and another weight at* B: *prove that*

$$W = \frac{(n-2)2^n + 2}{n-1}P. \tag{H.S.C.}$$

The arrangement will be as in Fig. 6.10.

Since the pulleys are of equal size, the points of attachment of
the strings to AB will be equidistant; let the distance between them
be a.

Let T_1, T_2, ... T^n be the tensions in the strings, then their
distances from B will be 0, a, $2a$, ... $(n-1)a$ respectively.

Also $$T_2 = 2T_1 = 2P,$$

and $$T_3 = 2T_2 = 2^2T_1 = 2P^2,$$

                              $$\cdot \quad \cdot \quad \cdot \quad \cdot$$

and $$T_n = 2^{n-1}P.$$

Hence, taking moments for the bar about B,

$$W(n-1)a = 2P \times a + 2^2P \times 2a + 2^3P \times 3a \ldots + 2^{n-1}P(n-1)a$$
$$= Pa[2 + 2^2 \times 2 + 2^3 \times 3 + \ldots + 2^{n-1}(n-1)].$$

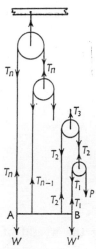

FIG. 6.10

To sum the series in the bracket, let

$$S = 2 + 2^2 \times 2 + 2^3 \times 3 + \ldots + 2^{n-1}(n-1)$$

$\therefore \quad 2S = \qquad 2^2 \times 1 + 2^3 \times 2 + \ldots + 2^{n-1}(n-2) + 2^n(n-1)$

$\therefore \quad -S = 2 + 2^2 + 2^3 + \ldots + 2^{n-1} - 2^n(n-1)$

$$= \frac{2(2^{n-1}-1)}{1} - 2^n(n-1)$$

$\therefore \qquad S = 2^n(n-2) + 2$

$\therefore \qquad W = \frac{2^n(n-2)+2}{n-1}P.$

6.13. *The Weston differential pulley*

The upper block A (Fig. 6.11) of this system has two grooves side by side, one of which has a slightly greater diameter than the other.

The weight is attached to the lower pulley B. An endless chain passes round the larger groove on the upper block, then round the lower pulley and the smaller groove on the upper block; the remainder of the chain hangs slack.

The chain is prevented from slipping by small projections or recesses in the groove.

The effort P is applied as in the figure.

(i) If T be the tension of the portions of the chain which support the lower pulley and the weight W we have (assuming

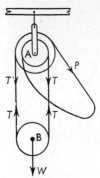

FIG. 6.11

that these portions are vertical and neglecting the weight of the chain and lower pulley)

$$2T = W.$$

If R and r be the radii of the larger and smaller grooves we have, taking moments about the centre of the upper block,

$$PR + Tr = TR$$

$$\therefore \qquad P = T\frac{R-r}{R} = W\frac{R-r}{2R}$$

$$\therefore \qquad \frac{W}{P} = \frac{2R}{R-r}.$$

By making R and r nearly equal a large mechanical advantange can be obtained.

(ii) Suppose the upper block is turned through an angle θ so that P moves downward; P moves a distance $R\theta$. The string on the left of B is shortened by $R\theta$, but an extra length $r\theta$ is let down on the right of B, owing to the smaller groove of A turning. Hence W rises $\frac{1}{2}(R\theta - r\theta)$ and, by the principle of work,

$$\frac{W}{2}(R-r)\theta = PR\theta$$

$$\therefore \qquad \frac{W}{P} = \frac{2R}{R-r}.$$

6.14. *The wheel and axle*

This consists of a wheel AB and an axle CD, which rotate together round the same axis as in Fig. 6.12. Figure 6.13 shows a sectional view.

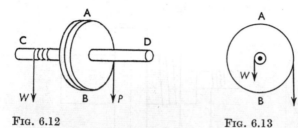

FIG. 6.12   FIG. 6.13

The effort P is applied to a string wound round the wheel, and the weight W is attached to a string wound round the axle in the opposite direction.

Let a, b be the radii of the wheel and axle respectively.

(i) Taking moments about the common axis,

$$Pa = Wb$$

$$\therefore \quad \frac{W}{P} = \frac{a}{b}.$$

(ii) If the wheel and axle be wound through an angle θ (in radians) so that P descends, P moves a distance $a\theta$, while W moves up a distance $b\theta$.

Hence, by the principle of work,

$$Wb\theta = Pa\theta$$

$$\therefore \quad \frac{W}{P} = \frac{a}{b}.$$

The mechanical advantage is limited in practice by the facts that if a is too large the machine is unwieldy, while if b is too small the axle may break.

A capstan and windlass are similar in action. In these there is only one cylinder, corresponding to the axle, the effort being applied at the ends of bars or at the end of a long handle perpendicular to the axle.

6.15. *The differential wheel and axle*

In this modified form of the wheel and axle the axle is made
of two parts, having different radii; the weight is attached to a
pulley, and the rope supporting this is wound in opposite directions on the two parts of the axle, the pulley hanging in the
loop between the two parts as in Fig. 6.14.

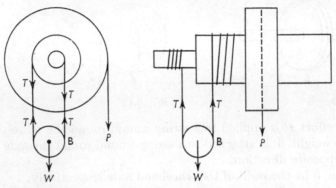

FIG. 6.14

As P descends, the rope round the pulley B is wound up on
the larger axle and unwinds on the smaller.

Let a, b, c be the radii of the wheel and the larger and
smaller portions of the axle respectively, and T the tension in
the rope supporting B (Fig. 15.10).

We assume that the portions of the rope round B are both
vertical, and neglect the weights of the rope and the pulley.

(i) From the equilibrium of W

$$2T = W, \text{ or } T = \tfrac{1}{2}W.$$

Taking moments about the common axis,

$$Pa + Tc = Tb$$

$$\therefore \qquad P = T\frac{b-c}{a} = W\frac{b-c}{2a},$$

$$\therefore \qquad \frac{W}{P} = \frac{2a}{b-c}.$$

By making b and c nearly equal a large mechanical advantage can be obtained without making the wheel too large or
the axle too thin.

(ii) If the wheel turns through an angle θ radians so that P descends, P moves down a distance $a\theta$.

The larger axle winds up a length $b\theta$ of rope, while the smaller axle unwinds a length $c\theta$.

Hence the weight rises a distance $\frac{1}{2}(b-c)\theta$, and from the principle of work,

$$\frac{W}{2}(b-c)\theta = Pa\theta$$

$$\therefore \qquad \frac{W}{P} = \frac{2a}{b-c}.$$

6.16. *Overhauling*

In many cases (as when raising heavy weights) it is important that the load shall not run back when the effort is removed.

If this happens the machine is said to 'overhaul'.

It is clear that, to prevent overhauling, the friction in the various parts of the machine must be too great for the load to overcome it, and this means that the efficiency cannot be very great.

Suppose an effort P just raises a load W, and that when P is removed W remains at rest.

When P is raising W the friction acts against P, and when P is removed it just supports W.

For small displacements x and y of P and W let the work done by the friction be F, then when the machine is worked

$$Px = Wy + F.$$

Now if the same displacement is given to W when F is supporting it, then

$$Wy = F.$$

Hence F must not be less than Wy, so that

$$Px \nless 2Wy,$$

$$\therefore \qquad \frac{Wy}{Px} \ngtr \tfrac{1}{2}$$

i.e. the efficiency must not be greater than $\tfrac{1}{2}$ or 50 per cent.

Any machine whose efficiency is less than 50 per cent will not overhaul. The usefulness of the Weston pulley, apart from its large velocity ratio, is due to the fact that its efficiency is usually less than 50 per cent.

EXAMPLES 6.2

1. The drum of a windlass is 10 cm in diameter, and the effort is applied to the handle 60 cm from the axis. Find the effort necessary to support a weight of mass 120 kg.

2. A wheel and axle is used to raise a load of mass 50 kg. The radius of the wheel is 50 cm, and while it makes seven revolutions the load rises 3·3 m. What is the smallest force that will support the load?

3. An anchor of mass 1 Mg is being raised by its chain being wound round a capstan of 22·5 cm diameter, which is turned by six men who work at the ends of capstan bars of 1·5 m effective length. Assuming that each exerts the same effort, find what that effort must be, if the efficiency is 56 per cent. (Neglect the weight of the chain.)   (N.U.)

4. A differential pulley, the two parts of which have respectively twenty-four and twenty-five teeth, is used to raise a weight of mass 500 kg. Show by a sketch how the apparatus is used, and determine its velocity ratio. Find also what effort must be exerted if the efficiency is 60 per cent.   (N.U.)

6.17. *The screw*

A screw consists of a bolt of circular section with a projecting *thread* running round it in a spiral curve, the inclination of this thread to a plane perpendicular to the axis of the bolt being the same at all points.

The distance between two consecutive threads, measured parallel to the axis, is called the *pitch* of the screw.

The screw works in a nut or fixed support, along the inside of which is cut out a hollow groove of the same shape as the thread of the screw and along which the thread slides.

The only movement possible for the screw is a rotation about its axis, and at the same time a motion parallel to the axis due to the thread sliding in its groove. In one complete revolution the screw moves parallel to its axis a distance equal to the pitch.

If the thread and groove are smooth and the axis of the screw inclined to the horizontal its weight will tend to cause it to move downwards and rotate. In practice, of course, friction is sufficient to prevent this.

In a screw-press or screw-jack one end of the screw is placed against the body on which a force is to be exerted, and the screw is driven forward in its support by means of a bar attached to the other end (Fig. 6.15).

Let a be the length of the arm at which the effort P is applied, and p the pitch of the screw.

In one revolution P moves a distance $2\pi a$, while the screw moves forward a distance p.

$$\therefore \qquad \text{the velocity ratio} = \frac{2\pi a}{p}.$$

If the screw is smooth this is also the mechanical advantage. Theoretically the velocity ratio can be made very large by making a large and p very small. The former, however, causes the machine to be unwieldy, while a very small pitch means a thin thread and consequent weakness.

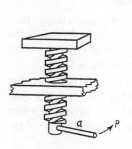

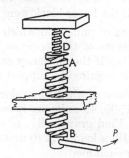

FIG. 6.15   FIG. 6.16

6.18. *The differential screw*

This machine gives a large velocity ratio without the draw-backs of a very long arm or small pitch.

One screw AB (Fig. 6.16) works in a fixed block.

The inside of this screw is hollow and a second screw DC of smaller pitch works inside it. The smaller screw is fastened to a block at C so that it cannot rotate, but can only move in the direction of its length.

When the effort arm has made one revolution the screw AB has advanced a distance equal to its pitch p_1, while the smaller screw goes into AB a distance equal to its pitch p_2.

Hence the smaller screw, and therefore the weight, has advanced a distance $p_1 - p_2$.

If a is the length of the effort arm the velocity ratio is

$$\frac{2\pi a}{p_1 - p_2}.$$

This can be made very large by making p_1 and p_2 nearly equal.

EXAMPLES 6.3

1. In the system of pulleys in which there is only one string it is found that weights of 5 and 6 units respectively will support weights of 18 and 22 units at the lower block. Find the number of strings and the weights of the lower block.

2. In the system of pulleys in which there is only one string it is found that a weight of 3 units supports a weight of 15 units and a weight of 5 units supports a weight of 27 units. Find the weight of the lower block, and also what would be the mechanical advantage if the lower block were weightless.

3. In the system of pulleys in which there is only one string there are five portions of the string at the lower block. What is the velocity ratio? If the efficiency of the apparatus is 50 per cent, what force is required to support a weight of 60 N?

4. With a machine of which the velocity ratio is 60, it is found that efforts of 21, 35, 49 units are necessary to lift loads of 400, 800, 1200 units. Find graphically, or otherwise, the probable effort required to lift a load of 2240 units, and find the efficiency of the machine for each load.

5. Find the condition of equilibrium in the system of pulleys in which each string is attached to a bar carrying the load, assuming that there are three movable pulleys, each of weight w, an effort P, and that the weight of the bar (including the attached weight) to which the four strings are attached is W. If the radius of each pulley, including the fixed pulley, is a, show that for the bar to remain horizontal and the strings vertical, the horizontal distance of the centre of gravity of the bar and weight from the point of attachment of the longest string must be equal to $\dfrac{11P + 5w}{W}a$.

6. If there are four pulleys in the system in which each string is attached to the weight, and each pulley weighs 2 N, what weight can be raised by an effort of 20 N?

7. In a system of pulleys in which each string is attached to the support there are three movable pulleys, each of mass 1 kg. The effort required to support a certain weight is twice that which would be required if these pulleys were weightless. Find the weight.

8. Find the condition of equilibrium for a system of pulleys in which each pulley hangs in the loop of a separate string, the strings being all parallel, and each string attached to the beam. The weights of the pulleys are to be taken into account. If there are five pulleys and each has a mass of 1 kg, what weight will a force of $5g$ N support on such a system, and what will be the total pull on the beam? (I.S.)

9. In a system of n pulleys in which the same string passes round all the pulleys, show that, if the weight of the pulleys is neglected, the mechanical advantage is n. A man of mass 80 kg uses such a system con sisting of seven pulleys, each of mass 1·5 kg, to raise a weight of mass 200 kg. If he pulls vertically downwards, what pressure does he exert on the ground? (H.S.D.)

10. A string with one end attached to a fixed point A passes under a heavy pulley P, then over a fixed pulley B, then under a heavy pulley Q, and has its other end attached to the centre of the pulley P, all the hanging parts of the string being vertical. By means of the principle of virtual work, or otherwise, find the ratio of the weights of P and Q when the system is in equilibrium. (H.S.D.)

11. A Weston differential pulley consists of a lower block and an upper block which has two cogged grooves, one of which has a radius of 10 cm and the other a radius of 9 cm; the efficiency of the machine is 40 per cent. Calculate the effort required to raise a load of 1500 N. (H.S.D.)

12. In a differential wheel and axle the radius of the wheel is 30 cm, and the radii of the axle are 5 cm and 7·5 cm. It requires an effort of 600 N to lift a weight of mass 1 Mg; calculate the efficiency of the machine for this load. (I.E.)

13. A man of mass 65 kg sits in a seat of mass 7 kg, which is suspended from a smooth pulley supported by the two parallel portions of a rope which is coiled in opposite directions round the two drums of a differential wheel and axle of radii 38 cm and 30 cm respectively. He raises himself by pulling one side of the rope. State which side, and show that to raise himself he must exert a pull exceeding $8·5g$ N. (I.E.)

14. Find the force necessary to sustain three movable pulleys, each of weight W, in that system of pulleys in which each movable pulley hangs in the loop of a separate string, the hanging parts of the strings being vertical, and one end of each string being attached to a fixed point. A, B are two such movable pulleys, each of weight W; a third pulley, also of weight W, hangs in the loop of a string whose ends are attached one to the axle of A and the other to the axle of B. If B is the upper of the pair A, B, what force must be applied to the free end of the string which passes under B so as to maintain equilibrium?　(I.A.)

15. If the fixed pulleys of a Weston differential block have ten and eleven teeth, and the effect of friction in the machine is to increase the effort by an amount which is a fixed proportion of the load, find that proportion when the efficiency is $\frac{1}{3}$.　(I.E.)

16. It is found that a couple of 35 N cm applied to a screw which has two threads to the cm can produce a force of 100 N. What is the efficiency of the screw?　(H.S.D.)

17. A copying press has a lever 50 cm long fastened at its centre to a screw, the thread of which makes 12 complete turns per 10 cm of length. Forces equal to 48 N (in opposite directions) are applied at the two ends of the lever; by the principle of virtual work, or otherwise, find the force exerted by the press.　(I.A.)

18. If an effort of 50 N, acting at the end of an arm 0·6 m long, produces in a screw press a thrust of 10^4 N, what is the pitch of the screw?

19. If the two screws in a differential screw have pitches of 0·8 cm and 1·2 cm respectively, and a couple of moment 3000 N cm applied to the larger screw produces a thrust equal to the weight of a mass of 500 kg, calculate the efficiency of the machine.

　(I.S.)

20. A bucket is lowered into a well by means of a windlass and a pulley. The end of the rope of the windlass (which in the more usual arrangement is attached to the bucket) is here attached to the frame of the windlass, and the pulley, with bucket attached, slides in the loop of the rope, the hanging parts of the rope being vertical. Neglecting friction and the weight of the rope, determine by the principle of virtual work, or otherwise, the force that must be applied at the arm of the windlass to maintain the bucket in equilibrium, having given the weight W of the bucket and its load, b the diameter of the barrel of the windlass, and a the length of the arm. In a practical machine friction cannot be neglected. What is the *efficiency* of this machine if the force that will just raise the bucket is nW?　(I.S.)

21. In a pair of pulley blocks there are three sheaves in each block, the mass of each block is 12 kg, and a weight of mass 132 kg is hung from the lower block. The efficiency is 60 per cent. Find the effort required to raise the weight and the reaction on the hook supporting the apparatus, neglecting the weight of the rope.

22. Two wheels of a radii a and b ($a>b$) are fixed to the same axle which is rotated by an arm of length c at right angles to the axis. A light cord has one end fastened to a point on the circumference of the larger wheel and the other to a point on the circumference of the smaller, so that when the arm is turned the cord is wound on one wheel and off the other. The cord passes under a smooth light pulley supporting a load W.

    Equilibrium is maintained by a force P applied at the end of the arm and at right angles to it in the plane in which it moves. Assuming that there is no friction and that the hanging parts of the string are vertical, find the relation between P and W.

    If, on the other hand, there is friction and the efficiency of the machine is 45 per cent, find the load W which can be just raised by a force P of 120 N, given that a, b, c are respectively 15 cm, 10 cm, and 40 cm.   (H.C.)

CHAPTER SEVEN

CENTRE OF GRAVITY

7.1. In Chapter Two we found the position of the centre of gravity of some bodies of simple form. We shall now show how to find the position of this point in a few other cases, and then how to obtain general formulae by means of which the centre of gravity may be found in more difficult cases.

7.2. Three rods forming a triangle

Let AB, BC, CA be three uniform rods of the same thickness and material forming a triangle ABC (Fig. 7.1).

Let D, E, F be the mid-points of BC, CA, AB.

Join DE, EF, FD.

The centres of gravity of the rods are at D, E, and F, and their weights, which are proportional to their lengths a, b, and c, may be taken to act at these points.

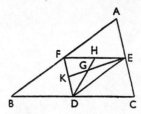

FIG. 7.1

The centre of gravity of the rods AB and AC is therefore at a point H in EF such that

$$c \times FH = b \times HE,$$

or
$$\frac{FH}{HE} = \frac{b}{c}.$$

But
$$b = 2DF \text{ and } c = 2DE$$

∴
$$\frac{FH}{HE} = \frac{DF}{DE}$$

∴ DH bisects the angle FDE.

Also the centre of gravity of the weight a at D and $b+c$ at H must lie on DH.

Hence the centre of gravity of the three rods must lie on DH. Similarly, the centre of gravity must lie on EK, which bisects the angle DEF.

The centre of gravity is therefore at G, where these bisectors intersect, and this point is the centre of the circle inscribed in the triangle DEF.

7.3. Tetrahedron

Let ABCD (Fig. 7.2) be a tetrahedron made of uniform material, E the mid-point of BD, and G_1 the centre of gravity of the base BCD.

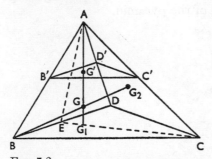

FIG. 7.2

Let B'C'D' be any section of the tetrahedron parallel to BCD. Then, from known results in geometry, we see that AE passes through the mid-point of B'D' and AG_1 passes through the intersection of the medians of the triangle B'C'D', i.e. through its centre of gravity G'.

Hence by considering the tetrahedron as built up of triangular slices parallel to BCD, it follows, since the centre of gravity of each slice lies in AG_1 that the centre of gravity of the whole lies in AG_1.

Similarly, it may be shown that the centre of gravity lies on the line joining B to the centre of gravity G, of the opposite face ACD.

Now it is also known that these lines intersect in a point G such that

$$G_1G = \tfrac{1}{4}G_1A$$

and

$$G_2G = \tfrac{1}{4}G_2B.$$

Hence the centre of gravity lies on the line joining the centre of gravity of any face to the opposite angular point at a distance equal to one-quarter of this line from the face.

Note. The centre of gravity of a tetrahedron is the same as that of four equal particles placed at its vertices.

For equal weights w placed at the vertices of the triangle BCD are equivalent to a weight $3w$ at G_1, the centre of gravity of BCD. Also $3w$ at G_1 and w at A are equivalent to $4w$ at G_1, since $G_1G = \frac{1}{4}G_1A$.

7.4. Pyramid on any base. Solid cone

Let OABCDE (Fig. 7.3) represent a uniform pyramid on any rectilinear base ABCDE. Let H be the centre of gravity of the base, and h the height of the pyramid.

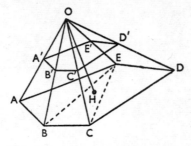

FIG. 7.3

Any plane parallel to the base will cut the pyramid in an area A′B′C′D′E′ similar to the base, and its centre of gravity will be similarly placed to that of the base and lie on OH.

By considering the pyramid to be made up of thin slices parallel to the base, since the centres of gravity of all these slices lie on OH, the centre of gravity of the whole pyramid must lie on OH.

By dividing the base into triangles such as ABE, BEC, CED, we can divide the pyramid into tetrahedra.

The centre of gravity of each of these is at a height $h/4$ above the base, and therefore the centre of gravity of the whole pyramid is at a height $h/4$ above the base.

Hence the centre of gravity is on the line joining the vertex to the centre of gravity of the base and one-quarter of the way up that line.

Since a right circular cone may be considered as the limiting case of a pyramid when the base is a regular polygon and the number of sides is increased indefinitely, the result just obtained applies to a solid cone.

The centre of gravity of a solid right circular cone is in the line joining the vertex to the centre of the base and at a distance from the base equal to one-quarter of the height of the cone.

For *any* solid cone the centre of gravity is in the line joining the vertex to the centre of gravity of the base and one-quarter of the way up this line from the base.

7.5. Curved surface of a right circular cone

By joining the vertex to points on the edge of the base indefinitely close together, we can divide the surface into an infinite number of parts, each of which is very approximately a triangular lamina. The centres of gravity of all these triangles lie on a plane parallel to the base at a distance from the vertex equal to two-thirds of the height of the cone, and therefore the centre of gravity of the whole surface must lie on this plane.

We can see, by symmetry, that the centre of gravity must lie on the axis of the cone.

Hence the centre of gravity of the curved surface is a point on the axis distant two-thirds of the height from the vertex.

7.6. Centre of gravity of a number of particles

Let a number of particles of weights w_1, w_2, w_3, ... w_n be placed in a plane at points A_1, A_2, A_3, ... A_n (Fig. 7.4), and

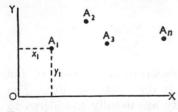

Fig. 7.4

let the coordinates of these points referred to two rectangular axes OX, OY in the plane be (x_1, y_1), (x_2, y_2), (x_3, y_3), ... (x_n, y_n).

Suppose the plane placed horizontally, so that the weights act perpendicularly to it, i.e. in the figure they are acting perpendicular to the plane of the paper.

The resultant of the weights is a weight $w_1+w_2+w_3+$ ... $+w$, or $\sum w$, and we know that the moment of this resultant about either of the axes OX or OY is equal to the sum of the moments of the separate weights about that axis.

Now the sum of the moments of the weights about OY is

$$w_1x_1+w_2x_2+ \dots +w_nx_n \text{ or } \sum wx.$$

Hence if $\bar{x}$ is the distance of the line of action of the resultant from OY,

$$\bar{x}\sum w = \sum wx$$

$$\therefore \quad \bar{x} = \frac{\sum wx}{\sum w}.$$

Similarly, if $\bar{y}$ is the distance of the resultant from OX,

$$\bar{y} = \frac{\sum wy}{\sum w}.$$

The line of action of the resultant weight therefore passes through a point G in the plane whose coordinates are $\bar{x}$ and $\bar{y}$, and this point must be the centre of gravity of the particles, which we know is a point in the plane.

This formula for the position of the centre of gravity will clearly apply also when w_1, w_2, ... w_n are the weights of bodies whose centres of gravity are at A_1, A_2, ... A_n.

Since $w = mg$, where m is the mass of the particle, the above formulae can also be written.

$$\bar{x} = \frac{\sum mx}{\sum m} \quad \text{and} \quad \bar{y} = \frac{\sum my}{\sum m}.$$

The point thus found by considering the *masses* of the particles instead of their weights is the *Centre of Mass*. The centre of mass and centre of gravity are usually considered to be the same (as they are in the case of bodies small in comparison with the earth). The centre of mass is a definite point, and its position is independent of the size of the body, depending only on its shape if the density is uniform throughout.

7.7. When we are given a finite number of particles of known weight and position the summations involved in finding the values of $\sum wx$ and $\sum wy$ are effected by ordinary addition.

In the case of a rigid body the number of particles is infinite. We then imagine the body as made up of strips or slices, the positions of whose centres of gravity are known. In a few simple cases, such as those already considered, we can by taking the strips or slices in two different directions, show that the centre of gravity of the whole body must be at the point of intersection of two straight lines.

In many cases, however, although it is easy to see one line in which the centre of gravity must lie, we have to determine the position in this line by using the formula,

$$\bar{x} = \frac{\sum wx}{\sum w}$$

where w now represents the weight of a strip or slice, and x the distance of its centre of gravity from some fixed point in the line. The value of $\sum wx$ must be obtained by integration, as we are dealing with an infinite number of strips.

In the case of a plane surface of area A, if x and y are the distances of any element δA of its area from two fixed axes, the

$$\bar{x} = \frac{\sum x \delta A}{A} \quad \text{and} \quad \bar{y} = \frac{\sum y \delta A}{A}$$

formulae give the position of the *Centroid* of the surface.

In the case of a thin *uniform* lamina (in which the mass is proportional to the area) the centroid is the asme as the centre of mass. If the lamina is *not* uniform these points will not be the same.

The centre of gravity, centre of mass, and centroid are usually the same, and the three terms are frequently used as if they really meant the same point. It must be remembered, however, that this is not true except in the case of uniform bodies in which the directions of the weights of the particles are considered parallel.

EXAMPLES 7.1

1. Particles of mass 2, 3, 6, and 9 kg are placed in a straight line AB at distances 1, 2, 3, 4 cm respectively from A. Find the distance of their centre of gravity from A.

2. A uniform rod AB is 4 m long and has a mass of 6 kg, and masses are attached to it as follows: 1 kg at A, 2 kg at 1 m from A, 3 kg at 2 m from A, 4 kg at 3 m from A, and 5 kg at B. Find the distance from A of the centre of gravity of the system.

3. Weights of mass 3, 4, and 5 kg are placed at the corners A, B, C respectively of an isosceles triangle in which AB = AC = 12 cm, BC = 8 cm. Find the distance of the centre of gravity of the weights from BC and from AD, the perpendicular from A to BC.

4. Weights of mass 1, 2, 3, 4 kg are placed at the corners A, B, C, D respectively of a square ABCD of side 8 cm. Find the distance of the centre of gravity of the system from AB and AD.

5. Weights of mass 1, 2, and 3 kg are placed at the corners of an equilateral triangle of side 9 cm. Find the distance of their centre of gravity from the first weight.

6. Weights of mass 5, 6, 9, and 7 kg are placed at the corners A, B, C, D of a square of side 27 cm. Find the distance of their centre of gravity from A.

7. ABC is an equilateral triangle of side 4 m. Weights of mass 5, 1, and 3 kg are placed at A, B, and C respectively, and weights of mass 2, 4, and 6 kg are placed at the mid-points of BC, CA, and AB. Find the distance of their centre of gravity from B.

8. Weights of mass 1, 5, 3, 4, 2, and 6 kg are placed at the angular points of a regular hexagon taken in order. Show that their centre of gravity is at the centre of the hexagon.

9. Weights of mass 5, 1, 3, 2, 4, and 15 kg are placed at the angular points of a regular hexagon taken in order. Find the distance of their centre of gravity from the 15 kg weight.

10. Weights of mass 1, 2, 3, 4, 5, and 6 kg are placed at the angular points of a regular hexagon taken in order. Find the distance of their centre of gravity from the centre of the hexagon, the length of the side being 14 cm.

11. ABC is an isosceles triangular lamina in which AB = AC = 15 cm, BC = 24 cm. The mass of the lamina is 24 kg, and weights of mass 6, 6, and 4 kg are placed at the corners A, B, and C respectively. Find the distance of the centre of gravity of the system from BC.

12. Masses of 1, 2, 3, 4 kg are placed respectively at the corners A, B, C, D of a rectangle; AB = 6 m, BC = 12 m. Find the perpendicular distances of the centre of gravity from AB and BC.

7.8. Centre of gravity of a compound body

If we know the weights and the centres of gravity of each of two parts of a body we can find the centre of gravity of the whole as follows.

Let G_1, G_2 (Fig. 7.5) be the centres of gravity of the two parts, W_1 and W_2 their weights.

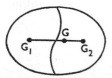

FIG. 7.5

These weights are like parallel forces acting at G_1 and G_2, and their resultant is equal to $W_1 + W_2$ and acts at a point G in $G_1 G_2$ such that

$$W_1 \times G_1 G = W_2 \times GG_2$$

$$\therefore \quad \frac{G_1 G}{W_2} = \frac{GG_1}{W_1} = \frac{G_1 G_2}{W_1 + W_2}$$

$$\therefore \quad G_1 G = \frac{W_2}{W_1 + W_2} G_1 G_2 \quad \text{and} \quad GG_2 = \frac{W_1}{W_1 + W_2} G_1 G_2.$$

This result may also be obtained by applying the general formula, or the method employed in proving the formula, which is really an application of the principle of moments.

We know that the resultant of W_1 at G_1 and W_2 at G_2 is $W_1 + W_2$, acting at some point G in $G_1 G_2$.

Taking G_1 as origin and moments about this point, we have

$$(W_1 + W_2)G_1 G = W_2 \times G_1 G_2$$

$$\therefore \quad G_1 G = \frac{W_2}{W_1 + W_2} G_1 G_2.$$

7.9. Centre of gravity of a remainder

If we know the weight and centre of gravity of a body, and the weight and centre of gravity of a part of it which is removed, the centre of gravity of the remainder is obtained as follows.

In the figure of the last paragraph let G be the centre of gravity of the whole body, W its weight, G_2 and W_2 the centre of gravity and weight of the part removed.

The centre of gravity of the remainder must obviously be in the same straight line as G_2 and G.

Also the sum of the moments of the weights of the two parts about any point in the line must equal the moment of the weight of the whole about that point.

Hence the moment of the remainder is equal to the moment of the whole, less the moment of the part removed.

If G_1 is the centre of gravity of the remainder, taking moments about G_2, we have

$$(W-W_2)G_2G_1 = W \times G_2G$$

$$\therefore \qquad G_2G_1 = \frac{W}{W-W_2}G_2G.$$

7.10. If a compound body is made up of several parts, or several parts are removed from a body, we take two axes and proceed as in obtaining the general formulae.

It is better to use the principle of moments directly, rather than quote the formulae.

For a body made up of several parts we have, about either axis,

Moment of whole = Sum of moments of parts.

Similarly for the remainder, after removing parts of a body, we have

Moment of remainder = Moment of whole
                    —Sum of moments of parts removed.

These methods are illustrated in the following examples.

It should be noticed that the position of the centre of gravity depends only on the *relative* values of the weights. It is often more convenient to use quantities *proportional* to the weights than the actual weights themselves.

7.11. EXAMPLE (i)

In a circular disc of 18 cm diameter a circular hole of 6 cm diameter is cut, the centre of the hole being 4 cm from the centre of the disc. Find the position of the centre of gravity of the remainder of the disc.

Let G, G_1 (Fig. 7.6) be the centres of the disc and hole. It is clear that the centre of gravity of the remainder must be in G_1G, on the side of G opposite to G_1.

It is convenient here to take moments about G; the moment of the whole disc is then zero, and the moments of the part removed and the remainder are equal and opposite.

FIG. 7.6

Tabulating weights and distances, we have

	Weight	Distance of C.G. *from* G
Disc	81	0
Part removed	9	4 cm
Remainder	72	x cm

Taking moments about G, we have

$$72x = 36$$
$$\therefore \qquad x = \tfrac{1}{2}.$$

Note. The squares of the radii are taken to represent the weights. This avoids introducing π and the density of the disc.

EXAMPLE (ii)
A sheet of metal is in the shape of a square ABCD with an isosceles triangle described on the side BC; if the side of the square be 12 cm, and the height of the triangle be 9 cm, find the distance of the centre of gravity of the sheet from the line AD.

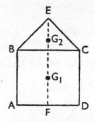

FIG. 7.7

Let ABECD (Fig. 7.7) represent the sheet, and draw EF perpendicular to AD. The centre of gravity of the square is at G_1, where $FG_1 = 6$ cm. The centre of gravity of the triangle is at G_2, in EF, where $EG_2 = \frac{2}{3} \times 9 = 6$ cm, so that $FG_2 = 15$ cm. The weights are proportional to the areas, i.e. to 144 and 54. Tabulating the weights and distances of the centres of gravity from the axes, we get

	Weight	Distance of C.G. from AD
Square	144	6 cm
Triangle	54	15 cm
Whole figure	198	x cm

Taking moments about AD,

$$198x = 144 \times 6 + 54 \times 15$$
$$= 93 \times 18,$$
$$\therefore \quad x = 8\tfrac{5}{11}.$$

EXAMPLE (iii)

The radii of the faces of a frustum of a cone are 2 m and 3 m, and the thickness of the frustum is 4 m. Find the distance of the centre of gravity from the larger face.

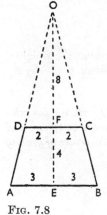

FIG. 7.8

Let ABCD (Fig. 7.8) represent a section of the frustum through its axis EF.

Produce AD, BC, EF to meet in O, the vertex of the cone from which the frustum is cut.

By similar triangles OFC, OEB, we have

$$\frac{OE}{OF} = \frac{EB}{FC} = \frac{3}{2}.$$

$$\therefore \quad \frac{OF+4}{OF} = \frac{3}{2}$$

$$\therefore \quad \frac{4}{OF} = \frac{1}{2} \text{ or } OF = 8 \text{ m.}$$

Now the volumes of similar solid figures are proportional to the cubes of corresponding dimensions,

$$\therefore \quad \frac{\text{Volume of cone ODC}}{\text{Volume of cone OAB}} = \frac{8^3}{12^3} = \frac{8}{27}.$$

The weights of the whole cone, the upper cone which is removed, and the frustum are therefore proportional to 27, 8, and 19 respectively.

The centres of gravity of all three lie in OE.

Tabulating weights and distances of C.G.s from O, we get

	Weight	Distance of C.G. from O
Whole cone	27	9 m
Part removed	8	6 m
Frustum	19	x m

Taking moments about O, we have

$$19x = 27 \times 9 - 48 = 195$$

$$\therefore \qquad x = \tfrac{195}{19}.$$

Hence the distance of the centre of gravity from E is

$$12 - \tfrac{195}{19} = \tfrac{33}{19} \text{ m.}$$

EXAMPLE (iv)

A sheet of paper is in the shape of a rectangle, 9 cm wide and 12 cm long; one of the shorter sides is folded down, so as to lie entirely along one of the longer sides. Find the position of the centre of gravity of the whole sheet thus folded.

Let AECFD (Fig. 7.9) represent the folded sheet, so that the triangular portion EFC is double.

It will be convenient to take DC, DA as axes of x and y.

The mass of ADFE is proportional to 27, that of EFC to 81.

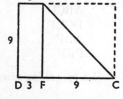

FIG. 7.9

Tabulating masses and distances of centres of gravity from the axes, we have

	Mass	Distance of C.G. from DA	Distance of C.G. from DC
ADFE	27	$\tfrac{3}{2}$ cm	$\tfrac{9}{2}$ cm
EFC	81	$3+3 = 6$ cm	3 cm
Whole figure	108	x cm	y cm

Taking moments about DA, we have

$$108x = 486 + \tfrac{81}{2} = \tfrac{1053}{2}$$

$$\therefore \qquad x = \tfrac{1053}{216} = 4\tfrac{7}{8}.$$

Taking moments about DC, we have

$$108y = 243 + \tfrac{243}{2} = \tfrac{729}{2}$$

$$\therefore \qquad y = \tfrac{729}{216} = 3\tfrac{3}{8}.$$

EXAMPLE (v)

ABCD *is a trapezium in which* AB, CD *are parallel, and of lengths* a, b. *Prove that the distance of the centre of mass from* AB *is* $\frac{1}{3}h\frac{a+2b}{a+b}$ *where* h *is the distance between* AB *and* CD.

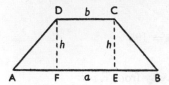

FIG. 7.10

Method (a)

Draw CE, DF perpendicular to AB (Fig. 7.10).
We then have

	Area	Distance of C.G. from AB
ABCD	$\frac{1}{2}(a+b)h$	x
DCEF	bh	$\frac{1}{2}h$
ADF	$\frac{1}{2}\text{AF}\times h$	$\frac{1}{3}h$
CEB	$\frac{1}{2}\text{EB}\times h$	$\frac{1}{3}h$

Hence, taking moments about AB,

$$\tfrac{1}{2}(a+b)hx = b\times\frac{h^2}{2}+\tfrac{1}{2}\times\text{AF}\times\frac{h^2}{3}+\tfrac{1}{2}\times\text{EB}\times\frac{h^2}{3}$$

$$= \frac{h^2}{2}\Big(b+\frac{\text{AF}+\text{EB}}{3}\Big).$$

But

$$\text{AF}+\text{EB} = a-b$$

$$\therefore \qquad \tfrac{1}{2}(a+b)hx = \frac{h^2}{2}\Big(b+\frac{a-b}{3}\Big) = \frac{h^2}{2}\times\frac{2b+a}{3}$$

$$\therefore \qquad x = \tfrac{1}{3}h\frac{a+2b}{a+b}.$$

Method (b)

The position of the centre of mass can also be found by supposing the trapezium divided into two triangles ADC, ACB, whose areas are $\frac{1}{2}bh$ and $\frac{1}{2}ah$ respectively, and then replacing them by particles of masses $\frac{1}{6}bh$ and $\frac{1}{6}ah$ at the vertices.

This gives $\frac{1}{6}(a+b)h$ at A and C, $\frac{1}{6}ah$ at B, and $\frac{1}{6}bh$ at D.

Taking moments about AB,

$$\tfrac{1}{2}(a+b)hx = [\tfrac{1}{6}bh+\tfrac{1}{6}(a+b)h]h$$
$$= \tfrac{1}{6}[a+2b]h^2$$

∴
$$x = \tfrac{1}{3}h\frac{a+2b}{a+b}.$$

EXAMPLE (vi)

A uniform rectangular board ABCD has AB = 10 cm, AD = 8 cm. Two square holes, each of side 2 cm, are cut in the board, and these are filled to the original thickness with metal whose specific gravity is nine times that of the material of the board. If the coordinates of the centres of the holes referred to AB, AD as axes of x and y and measured in centimetres are (4, 3), (7, 4) respectively, find the coordinates of the centre of gravity of the loaded board.

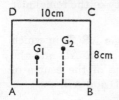

FIG. 7.11

Let G_1, G_2 (Fig. 7.11) be the centre of the holes.

The area of each hole is 4 cm², and its relative weight is 36. It is convenient to consider the loaded board as made up of the uniform board with added weights of 32 at G_1 and G_2, we then have,

	Weight	*Distance of* C.G. *from* AD	*from* AB
Unloaded board	80	5 cm	4 cm
Load at G_1	32	4 cm	3 cm
Load at G_2	32	7 cm	4 cm
Loaded board	144	x cm	y cm

Taking moments about AD, we have

$$144x = 400+128+224 = 752$$
∴
$$x = \tfrac{752}{144} = 5\tfrac{2}{9}.$$

Taking moments about AB, we have

$$144y = 320+96+128 = 544.$$
∴
$$y = \tfrac{544}{144} = 3\tfrac{7}{9}.$$

EXAMPLES 7.2

1. A sheet of paper in the shape of a rectangle ABCD, with an isosceles triangle described on the side BC, has the following dimensions: AB = 12 cm, AD = 8 cm, and the height of the triangle is 12 cm. Find the distance of the centre of gravity of the sheet from AD.

2. In a circular disc of 12 cm radius a circular hole of 2 cm radius is cut, the centre of the hole being 6 cm from the centre of the disc. Find the position of the centre of gravity of the remainder of the disc.

3. In a circular disc of 12 cm radius two circular holes of 2 cm radius are cut, the centres of the holes lying on two perpendicular diameters of the disc at a distance of 6 cm from the centre of the disc. Find the position of the centre of gravity of the remainder of the disc.

4. A sheet of paper is in the shape of a rectangle ABCD in which AB = 12 cm, AD = 8 cm. E is the mid-point of AD and the triangle CED is removed. Find the distance of the centre of gravity of the remainder of the sheet from AD and AB.

5. ABCD is a square board of side 12 cm, and E is a point in AD such that DE = 3 cm. The triangle CED is removed. Find the distance of the centre of gravity of the remainder of the board from AB and AD.

6. In a triangle ABC the mid-point of BC is D. A straight line is drawn through the centre of mass of the triangle parallel to BC, cutting the sides AB, AC in E and F respectively. Prove that the centre of mass of the quadrilateral BEFC lies in AD at a distance $\frac{7}{45}$AD from D.   (I.A.)

7. ABCD is a square lamina of side 9 cm. E, F are points in BC, CD such that EC, CF are each 3 cm. Find the centroid of the part ABEFDA of the lamina.   (I.A.)

8. ABCDEF is a sheet of thin cardboard in the form of a regular hexagon. Prove that if the triangle ABC is cut off and superposed on the triangle DEF the centre of gravity of the whole is moved a distance $\frac{2}{9}a$, where a is the side of the hexagon.   (I.S.)

9. ABCD is a rectangular plate; AB = 8 cm, BC = 12 cm, and E is the mid-point of BC. If the triangular portion ABE is removed from the plate and the remainder is then suspended from A, find the inclination of the side AD to the vertical.   (I.E.)

10. ABCD is a rectangular lamina, such that $AB = CD = 2a$, $AD = BC = 2b$; P and Q are the mid-points of BC and CD respectively. The triangular portion PCQ of the lamina is cut off. Prove that the perpendicular distances of the centre of gravity of the remainder from the sides AD, AB are $19a/21$ and $19b/21$ respectively. (H.S.D.)

11. A triangular plate is made up of two metals whose dividing line is parallel to a side and one-third up the median of that side. The ratio of the densities of the triangular and trapezoidal parts being $5 : 4$, find the position of the mass centre of the compound plate. (I.E.)

12. Three rectangular areas, 60 cm by 5 cm, 90 cm by 5 cm, 30 cm by 4 cm, are fitted together to form an I figure, the longest and shortest areas forming the cross-pieces. Find the distance of the centroid of the figure from the outer edge of the smallest area. (I.E.)

13. A uniform square plate of 30 cm side has two circular holes punched in it, one of radius 2 cm, and coordinates of centre (10, 13) cm referred to two adjacent sides of the plate as axes, the other of radius 1 cm, and coordinates of centre (20, 3) cm, find the coordinates of the centre of mass of the remainder of the plate. (I.S.)

14. Prove that, if ABCD is a trapezium with parallel sides AB, CD, its centre of gravity divides the line joining the mid-points of AB and CD in the ratio $AB+2CD : 2AB+CD$. (I.S.)

15. ABC is a triangle, in which $AB = AC$; D, E, F are the mid-points of BC, CA, AB respectively, and G is the centre of gravity of the triangle. If the portion AFGE is removed, find the distance from A of the centre of gravity of the remainder of the triangle given that $AD = 0.9$ m. (I.S.)

16. A uniform rectangular plate ABCD has a triangular portion removed by a straight cut through the mid-points of CB and CD. Show that the centre of gravity of the remaining portion is at a distance $\frac{1}{21}$ AC from the centre of the rectangle. (I.S.)

17. Two right circular cones have the same base, and their axes in opposite directions. Show that the centroid of the spindle-shaped solid formed of these two cones is midway between the centre of their base and the point bisecting the distance between the vertices. (I.S.)

18. ABC is a uniform equilateral triangular lamina of side 20 cm and mass 240 g, and masses of 30, 40, 50 g are placed at the corners A, B, C respectively. Find the distance of the centre of gravity of the whole system from the side BC. (H.S.D.)

19. The radii of the circular ends of a frustum of a solid circular cone are in the ratio 2 : 3. Prove that the distances of the centroid of the frustum from the ends are in the ratio 43 : 33. (H.S.D.)

20. A piece of cardboard is in the form of a rectangle, ABCD, where AB = 5 cm and BC = 8 cm. A piece of the cardboard ABE in the form of an equilateral triangle on AB as base is cut out. Calculate the distance of the centre of gravity of the remaining cardboard from DC. (H.S.D.)

21. A haystack has the form of a right circular cone standing on a circular cylinder. If the diameter of the base is 10 m, the height of the cylinder 4·8 m, and the length of the slant side of the cone 6 m, calculate the height of the centre of gravity of the haystack above the ground. (I.E.)

22. A closed regular tetrahedron is made of thin metal plate. Find the position of the centre of gravity when: (i) empty; (ii) filled with liquid whose weight is 3 times that of the tetrahedron. (I.E.)

23. A frustum of a cone has its circular ends of radii r_1 and r_2 and at a distance h apart. Its *curved surface* is covered with thin uniform material. Show that the height of the centre of gravity of the covering material above the end of radius r_2 is

$$\frac{h(2r_1+r_2)}{3(r_1+r_2)}.$$ (H.S.D.)

24. A uniform wooden triangular lamina in the form of an isosceles triangle of sides 12, 8, and 12 cm, has a thin metal band round it. If the weight of the metal band is twice that of the lamina, find the position of the centre of gravity of the lamina and band together. (H.S.C.)

25. A uniform wire is bent in the form of a triangle ABC and is suspended at A. Prove that a plumb-line hung from A will cross BC at the point D, where

$$\frac{BD}{DC} = \frac{a+b}{a+c}$$

the sides of the triangle being a, b, c. (H.S.D.)

26. The diameters of the plane ends of a frustum of a right circular cone are 30 cm and 90 cm respectively; the height of the frustum is 65 cm. Prove that its centre of gravity is 22·5 cm from the larger plane end. (I.S.)

27. A line is drawn through the centre of gravity of a triangle, parallel to one of its sides. Prove that the centre of gravity of the quadrilateral portion divides the median of the triangle in the ratio 7 : 38. (H.S.D.)

28. A tin can is in the form of right circular cone of semi-vertical angle 30°, with base all made of uniform thin sheet metal. It is filled with liquid, whose total weight is twice that of the can. Find the ratio of the height of the centre of gravity above the base to the height of the cone.   (H.S.D.)

29. A solid frustum of a uniform cone is of thickness h, and the radii of its end-faces are a and b $(a>b)$. A cylindrical hole is bored through the frustum. The axis of the hole coincides with the axis of the cone, and the radius of the hole is equal to that of the smaller face of the frustum. Show that the centre of gravity of the solid thus obtained is at a distance from the larger face of the frustum equal to

$$\frac{h(a+3b)}{4(a+2b)}.$$

30. Prove that if the radius of one plane end-face of a frustum of a right circular cone is n times the radius of the other end-face the centre of gravity divides the axis of the frustum in the ratio

$$(3n^2+2n+1) : (n^2+2n+3).$$

31. Find the centre of gravity of a hollow right circular cone with a circular base formed of the same material as the curved surface. If the height and diameter of the base are equal, find the angle the axis makes with the vertical when the cone is suspended from a point on the rim of the base.

32. A hollow conical vessel made of thin sheet metal is closed at its base; if it is cut across by a plane parallel to the base at half the perpendicular height, and the upper cone removed, prove that the distance from the base of the centre of gravity of the remainder of the vessel is

$$2lh/3(3l+4r),$$

where h is the perpendicular height of the original vessel, l its slant height, and r the radius of the base.   (I.S.)

33. Find the centre of gravity of an L-shaped uniform lamina, the outside height and breadth of the vertical arm being 10 cm and 1 cm, and the outside length and breadth of the horizontal arm being 6 cm and 2 cm respectively.   (I.A.)

34. A corner is cut off from a solid wooden cube by means of a saw-cut through the mid-points of the edges that meet in that corner. Find the perpendicular distances of the centre of gravity of the remainder of the cube from the three uncut faces.   (H.S.D.)

35. A regular tetrahedron of edge a is cut by a plane parallel to the base, and bisecting the perpendicular height of the tetrahedron. Find the position of the centre of mass of the frustum thus cut off.   (I.S.)

36. A four-sided plane lamina has the shape of a rectangle ABCD surmounted by a triangular part BEC, the side BE of which is a prolongation of AB. If the lengths of AB, AD, BE are a, b, c respectively, find how far the centre of gravity of the lamina is from AB and AD. Also prove that the centre of gravity will lie on BC if $c = a\sqrt{3}$.

37. There are four similar planks, each 3·6 m long, in a pile. The second plank projects 0·6 m beyond the first, the third 0·9 m beyond the second, and the fourth 1·8 m beyond the third; the sides are flush with each other. Find the centre of gravity of the four planks.   (I.A.)

38. A tank of sheet metal is a cube of side a, having in its base a central circular hole of diameter $a/4$; a cylinder of height $a/4$ is brazed externally to the edge of the hole and closed by a circular plate. The cube, cylinder, and plate being cut from the same sheet, find the mass centre.   (I.E.)

39. A solid cylinder of mass M and radius a has two small heavy discs, each of mass m, keyed to its ends so that the line joining the centres of the discs is parallel to the axis of the cylinder and at a distance c from it. Find where a cylindrical portion of radius b of the cylinder, with axis parallel to that of the cylinder, must be cut out in order that the centre of gravity of the remainder and the discs may be in the axis of the cylinder. What is the greatest value of the ratio m/M in order that this solution may be possible?   (I.E.)

40. A framework in the shape of a tetrahedron is made of six uniform rods, each rod being equal to its opposite rod. Show that the centre of gravity is at the same point as if the tetrahedron were solid.   (I.A.)

41. A sheet of metal is in the form of an isosceles triangle ABC, in which AB = AC = 20 cm, BC = 32 cm. The corner A is folded over until it comes to the middle of BC, and the metal is flattened down so that the fold is parallel to BC. Find the centre of gravity of the folded sheet.   (N.U.3.)

42. ABC is a triangular lamina. D is taken in BC, E in CA, and F in AB, such that BD : DC = CE : EA = AF : FB. Prove that the centre of gravity of the triangle DEF is the same as that of the original triangle ABC.   (I.S.)

43. (i) ABC is a triangle of rods. The rods are all uniform and of equal weight, but not necessarily of equal length. Find the centre of gravity.

    (ii) The bulb of a mercury thermometer is a sphere of internal diameter 1 cm. The tube of the thermometer is a circular cylinder of internal diameter $\frac{1}{4}$ mm. At freezing temperature the bulb and 2 cm of the tube are occupied by the mercury. At the boiling-point of water the bulb and 20 cm of the tube are occupied by the same quantity of mercury. Ignoring the expansion of the thermometer itself, find by how much the centre of gravity of the mercury rises between freezing and boiling temperatures.   (H.S.D.)

44. A thick conical shell open at the base, is bounded by two right circular cones, each of semi-vertical angle α, the base radii of which are r and $r-t$. Show that the height of the centre of mass above the base is $\frac{1}{6}(2r-t)$ cot α, where the square of t/r is neglected.   (H.S.D.)

45. ABCD is a uniform square plate whose edges are 8 cm long. Points Q, R are taken in BC, CD such that BQ = CR = 2 cm, and the portion CQR of the plate is cut off. Find the distances from AB, BC of the mass centre of the remainder.   (N.U.3.)

46. A uniform plate ABCD consists of two congruent triangles on opposite sides of AC, each having an obtuse angle at C. If C is a distance 3 cm from the line joining B and D, find the distance AC if the centre of gravity of the lamina coincides with C.
                                                                          (N.U.3.)

47. ABC is a uniform lamina right-angled at C and AC = b, BC = a. The mid-points of AC, BC are E and D respectively, and triangles are cut away by cuts EK and DH perpendicular to AB. Prove that the centre of gravity of the remainder CEKHD is at a distance $7ab/18(a^2+b^2)^{\frac{1}{2}}$ from KH.   (N.U.3.)

48. A uniform wire ABCD is bent at right angles at B and C in such a way that BA and CD are in the same sense, and the lengths of the parts AB, BC, CD are 6, 4, 2 cm, respectively. Find the distance of the centre of gravity of the wire from AB and BC. Show that the wire can be suspended with each part equally inclined to the vertical by a string attached at a point P, and give the length of BP.   (H.C.)

49. A window-cleaner's ladder is made up of two equal side pieces joined at one end and connected by ten equispaced rungs. The length of the lowest rung is one-eighth of the length of each side,

and the weight per foot run of the rungs is one-third that of the sides. Find the centre of gravity of the whole. The ladder is placed against a vertical wall in the limiting position of equilibrium. Find the position if the coefficient of friction at each end is one-fifth.   (I.E.)

50. A regular pyramid whose base is a regular hexagon is made of uniform thin sheet metal. If a is the radius of the circumcircle of its base, and $2a$ is the length of the slant edges, find the position of the centre of gravity of the total surface of the pyramid, and state its distance from the base.   (N.U. 3 and 4.)

51. An open rectangular tank of uniform thin material has length $3a$ m, breadth $2b$ m, and height $2c$ m, and can be closed by two lids of lengths a m and $2a$ m, respectively, hinged along each of the upper edges whose length is $2b$ m. Find the position of the centre of gravity when each lid is opened through an angle of 30°.   (C.W.B.)

52. A thin uniform wire is bent into the shape of the perimeter of a regular polygon of $2n$ sides. Find the position of the centre of gravity of one-half of this polygon. Hence deduce or otherwise obtain the position of the centre of gravity of a uniform semi-circular wire.   (C.W.B.)

7.12. Quadrilateral lamina

The position of the centre of gravity of a quadrilateral is obtained most easily by dividing it into two triangles by drawing one of the diagonals, and considering each of these triangles to be replaced by three particles equal to one-third of its weight placed at the three vertices.

There are several ways of expressing the position of this centre of gravity, one of which is illustrated below.

Let ABCD (Fig. 7.12) represent any quadrilateral lamina, AC and BD its diagonals intersecting at E.

Let W be the weight of the whole quadrilateral and W_1, W_2 the weights of the triangles ABD, BCD.

Since the areas ABD, BCD are in the ratio of AE to EC.

$$\frac{W_1}{W_2} = \frac{AE}{EC}.$$

Replacing ABD by equal particles $\frac{1}{3}W_1$ at A, B, D and replacing BCD by equal particles $\frac{1}{3}W_2$ at B, C, D we have $\frac{1}{3}W_1$

at A, $\frac{1}{3}W_2$ at C, and $\frac{1}{3}(W_1+W_2)$ at each of the corners B and D.

Now the centre of gravity of $\frac{1}{3}W_1$ at A and $\frac{1}{3}W_2$ at C is at a point F in AC such that

$$W_1 \times AF = W_2 \times CF$$

$$\therefore \qquad \frac{CF}{AF} = \frac{W_1}{W_2} = \frac{AE}{EC}.$$

Hence $\qquad\qquad CF = AE.$

The weights at A and C are equivalent to $\frac{1}{3}W$ at F.

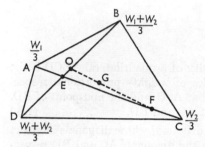

FIG. 7.12

The weights at B and D are equivalent to $\frac{2}{3}W$ at O, the mid-point of BD and the centre of gravity of the whole quadrilateral is therefore at G in OF, where $2OG = GF$.

EXAMPLE

G *is the centre of gravity of a uniform quadrilateral plate, G' is the centre of gravity of four equal particles placed at its corners, and O is the intersection of its diagonals. Prove that O, G, and G' are in the same straight line, and that* $OG' = 3GG'$. (H.S.D.)

Let ABCD (Fig. 7.13) be the quadrilateral.

Let W_1, W_2 be the weights of the triangles ABD, BCD respectively. Replace ABD by weights $\frac{1}{3}W_1$ at A, B, D, and replace BCD by weights $\frac{1}{3}W_2$ at B, C, D.

G is therefore the centre of gravity of these particles, and the whole weight W_1+W_2 acts there.

Now G' is the centre of gravity of these particles, together with particles $\frac{1}{3}W_2$ at A and $\frac{1}{3}W_1$ at C, which are equivalent to $\frac{1}{3}(W_1+W_2)$ at O (since $W_1/W_2 = AO/OC$).

Hence G' is the centre of gravity of W_1+W_2 at G and $\frac{1}{3}(W_1+W_2)$ at O, and must therefore lie in OG, and OG' $= 3$GG'.

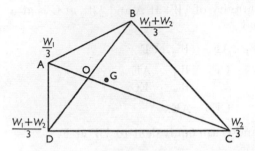

FIG. 7.13

EXAMPLES 7.3

1. Show that the centre of gravity of a quadrilateral ABCD is the same as that of three particles of weights proportional respectively to AO, OC, 2AC placed at A, C, and the mid-point of BD, where O is the intersection of AC and BD.   (I.S.)

2. ABCD is a uniform plane quadrilateral whose diagonals intersect in L; M and N are points on the diagonals AC and BD respectively, such that AM = CL and BN = DL. Show that the centre of mass of the quadrilateral ABCD coincides with that of the triangle LMN.   (H.S.C.)

3. ABCD is a uniform plane quadrilateral lamina, in which AB is parallel to DC. The length of AB is a and the length of DC is b. Prove that the centre of gravity of the lamina coincides with that of four particles of weights proportional to a, $a+b$, b, $a+b$, placed at A, B, C, D respectively. If AD = BC = c, and $b>a$, find the coordinates of the centre of gravity referred to rectangular axes, of which one is DA and the other is the perpendicular to DA, drawn through D.

4. Calculate, to two decimal places, the coordinates of the centre of gravity of a uniform quadrilateral lamina whose angular points are $(5, 0)$, $(3, 7)$, $(-2, 5)$, and $(-5, 0)$.   (Ex.)

5. (i) Prove that the centre of gravity of a uniform triangular plate ABC coincides with that of three equal particles at points X, Y, Z on the sides BC, CA, AB respectively, where BX : XC = CY : YA = AZ : ZB.

   (ii) ABCD is a trapezium the lengths of whose parallel sides AB, DC are a and b respectively; E, F are the mid-points of AB,

DC; H is the mid-point of EF. Prove that the centroid of the trapezium is the same as that of masses proportional to a, b, $2a+2b$, at E, F, H respectively. (H.C.)

6. A uniform plate in the form of a quadrilateral ABCD has its diagonals AC, BD crossing at right angles at a point O, so that AO = 5 cm, OC = 10 cm, BO = 2·5 cm, OD = 7·5 cm. Find the distance of the centre of gravity of the quadrilateral from each of the diagonals.

7. AD, BE, CF are the medians of a triangle ABC, and G is their point of intersection. If the portion AFGE is removed, show that the centre of gravity of the remainder is on GD at a distance $\frac{7}{12}$GD from D. (N.U.3)

8. ABCD is a quadrilateral, and it is 'reduced' to a triangle ABX of equal area by the usual construction (C being on the line BX). Prove that the centres of gravity of ABCD and of ABX are at the same distance from AC. (Ex.)

9. ABCD is a plane quadrilateral lamina whose diagonals meet at O; AO is less than OC and DO than OB. S is the mid-point of BD, T the mid-point of AC, and OSKT is a parallelogram. Prove that the centre of gravity of the lamina ABCD coincides with that of three equal particles at S, T, and K. (Ex.)

10. X, Y are the mid-points of two parallel sides of a quadrilateral whose lengths are a, b respectively. Show that G, the centroid of the quadrilateral, divides XY in the ratio $2b+a : 2a+b$. A uniform regular hexagonal area of side a has a part removed by a cut made parallel to one of the sides, and at a distance $\frac{1}{2}ka\sqrt{3}$ from it, where $k < 1$. Prove that the centre of gravity of the remaining portion is at a distance from the centre of the complete area equal to

$$\frac{k(3-k^2)}{3(6-2k-k^2)}a\sqrt{3}. \qquad \text{(C.W.B.)}$$

11. The mass centre of a uniform triangular plate ABC is G, and triangular pieces are cut from it by drawing parallels to BC, CA, AB cutting GA, GB, GC in the same ratio. Prove that the mass centre is unaltered. (C.W.B.)

12. A heavy quadrilateral lamina, weight $3W$, has a particle of weight W placed at the intersection of the diagonals. Prove that it will be in equilibrium in a horizontal position when four equal vertical forces support it, one at each corner. (Ex.)

7.13. *When a body is placed with its base in contact with a plane (rough enough to prevent sliding if inclined) it will be in equilibrium if the vertical line through its centre of gravity meets the plane within the area of the base.*

The forces acting on the body are its weight, acting vertically through its centre of gravity, and the reaction of the plane. If the plane is horizontal the reactions of the plane on the different parts of the base are all like parallel forces, and their resultant must obviously act at a point within the area of the base.

If the plane is inclined the resultant of the frictional forces on the different points of the base and the normal pressures of the plane must also have a resultant which acts at a point inside the area of the base.

Hence in both cases, if the resultant reaction of the plane is to balance the weight, the vertical through the centre of gravity must fall within the area of the base.

It should be noticed that if the base has re-entrant angles as in Fig. 7.14 the area of the base must be taken to mean the

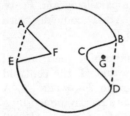

Fig. 7.14

area included in the figure which would be obtained by drawing a piece of thread tightly round the figure. Thus a point such as G would be within the area of the base.

7.14. *Stability of equilibrium*

If a rigid body be in equilibrium when one point only of the body is fixed it is clear that the centre of gravity of the body must be in the vertical line passing through the fixed point.

For let O (Fig. 7.15) be the fixed point, and G the centre of gravity of the body.

The only forces acting on the body are its weight acting vertically through G and the reaction R at the point O.

If these are to balance they must be in the same straight line, i.e. G must be in the vertical line through O.

This condition is satisfied in two cases, (i) when G is vertically below O, and (ii) when G is vertically above O.

Both these positions are positions of equilibrium, but there is a difference in the two cases.

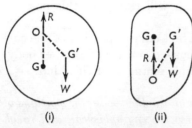

(i)   (ii)

Fig. 7.15

In the first case, if the body is slightly displaced it will tend to return to its equilibrium position; the moment about O of the weight through the displaced centre of gravity G' tends to bring the body back to its original position. In this case the equilibrium is said to be *Stable*.

In the second case, if the body is slightly displaced the moment about O of the weight acting through G', Fig. 7.15 (ii), tends to increase the displacement, and the body does not tend to return to its original position. In this case the equilibrium is *Unstable*.

A body is said to be in stable equilibrium when it returns to its original position after a *small* displacement. The extent to which it can be displaced without upsetting is a measure of its *amount* or *degree* of stability, and this we shall not consider.

7.15. A right circular cone with its base resting on a horizontal plane is in stable equilibrium; if slightly displaced it tends to return to its original position.

If placed with its vertex in contact with a horizontal plane and its axis vertical it will be in unstable equilibrium; if slightly displaced it will fall over.

A right circular cylinder with a plane end in contact with a horizontal plane is in stable equilibrium; if placed with its curved surface in contact with a horizontal plane it will rest in *any position*.

In this case the equilibrium is said to be *neutral*.

A uniform sphere resting on a horizontal plane is also in neutral equilibrium.

In most cases it is quite easy to see whether a position of equilibrium is stable or unstable. In certain cases, however, it is a matter of considerable difficulty to determine whether a body in a position of equilibrium will return to this position when slightly displaced.

Such cases arise when we have a curved portion of the surface of one body resting on the curved surface of another body, and we shall now consider a simple example of this kind.

7.16. *A body rests in equilibrium on another body, the portions of the surfaces near the point of contact being spherical, and rough enough to prevent sliding, and the line joining the centres of the two spheres being vertical. To find the conditions that the equilibrium shall be stable.*

Let O (Fig. 7.16) be the centre of the surface of the fixed

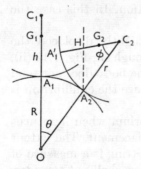

Fig. 7.16

body, C_1 that of the upper body. Since the surfaces are spherical, OC passes through the point of contact A_1, and since there is equilibrium, the centre of gravity, G_1, of the upper body must be vertically above A_1, and therefore in A_1C_1.

Let $OA_1 = R$, $A_1C_1 = r$, and $A_1G_1 = h$.

Suppose the upper body displaced through a small angle, such that if A_2 is the new point of contact the angle $A_1OA_2 = \theta$.

Let A_1', C_2, G_2, be the new positions of A_1, C_1, G_1.

O, A_2, and C_2 will be in a straight line, and the arc A_2A_1' equals the arc A_1A_2.

Hence, if $\angle A_2C_2A_1' = \phi$, we have

$$r\phi = R\theta.$$

Now the equilibrium will be stable if G_2 is to the left of the vertical through A_2; let $A_1'C_2$ cut this vertical in H.

In the triangle A_2HC_2 we have,

$$\frac{C_2H}{r} = \frac{\sin \theta}{\sin (\theta+\phi)} = \frac{\theta}{\theta+\phi} \text{ approximately}$$

since the angles θ and ϕ are small.

Hence $\quad A_1'H = r_1 - C_2H = r\left(1 - \frac{\theta}{\theta+\phi}\right) = \frac{r\phi}{\theta+\phi} = \frac{Rr}{R+r},$

and the equilibrium is stable if

$$h < \frac{Rr}{R+r},$$

or

$$\frac{1}{h} > \frac{1}{r} + \frac{1}{R}.$$

When $1/h = 1/r + 1/R$ it would appear that the equilibrium is neutral. It must be remembered, however, that in obtaining the condition for stability we used the approximations $\sin \theta = \theta$, and $\sin (\theta+\phi) = \theta+\phi$. These approximations are not sufficiently accurate for the critical case in which G_2 falls on the vertical through A_2, and it can be shown that in this case the equilibrium is really unstable.

Another method of determining the stability of a body will be given in Chapter Nine.

7.17 EXAMPLE (i)

A cubical block of edge a rests on a horizontal plane and is gradually undermined by cutting slices away by planes parallel to a horizontal edge, inclined at 45° to the horizontal. Find the centre of mass of the remainder when a length x has been removed from each of the four edges, and show that the block will fall when 9x = 5a, approximately. (I.S.)

Let ABCD (Fig. 7.17) represent a vertical section of the cube through its centre; the cutting plane will then be parallel to DB.

Let EF be the line in which the cutting plane meets ABCD, and AE = x. The centre of gravity of the prism removed will be at the centre of gravity of the triangle AEF.

The weights of the cube and part removed are proportional to the areas of the sections, a^2 and $\frac{1}{2}x^2$.

Tabulating the weights and distances of centres of gravity from BC and AB.

	Weight	Distance of C.G. from BC	Distance of C.G. from AB
Cube	a^2	$\frac{1}{2}a$	$\frac{1}{2}a$
Prism AFE	$\frac{1}{2}x^2$	$a-\frac{1}{3}x$	$\frac{1}{3}x$
Remainder	$a^2-\frac{1}{2}x^2$	X	Y

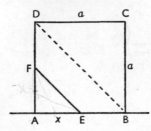

FIG. 7.17

Taking moments about BC, we have

$$(a^2-\tfrac{1}{2}x^2)X = \tfrac{1}{2}a^3-\tfrac{1}{2}x^2(a-\tfrac{1}{3}x)$$

$$\therefore \quad X = \frac{a^3-x^2(a-\tfrac{1}{3}x)}{2a^2-x^2}.$$

Taking moments about AB, we have

$$(a^2-\tfrac{1}{2}x^2)Y = \tfrac{1}{2}a^3-\tfrac{1}{6}x^3$$

$$\therefore \quad Y = \frac{a^3-\tfrac{1}{3}x^3}{2a^2-x^2}.$$

The block will fall over if the centre of gravity comes to the left of E, i.e. if $X>a-x$; hence it will be on the point of falling over when

$$a^3-x^2(a-\tfrac{1}{3}x) = (2a^2-x^2)(a-x)$$

or $\quad a^3-ax^2+\tfrac{1}{3}x^3 = 2a^3-2a^2x-ax^2+x^3$

or $\quad \tfrac{2}{3}x^3-2a^2x+a^3 = 0.$

Putting $x = \tfrac{5}{9}a$, the left side of this gives

$$\frac{250}{2187}a^3- \frac{10}{9}a^3+a^3 = \frac{250}{2187}a^3-\tfrac{1}{9}a^3$$

which is nearly zero.

Alternatively, we can write

$$x = \tfrac{1}{2}a + (x^3/3a^2).$$

Therefore as a first approximation, $x = \tfrac{1}{2}a$.

A second approximation is

$$x = \tfrac{1}{2}a + \frac{(\tfrac{1}{2}a)^3}{3a^2} = \tfrac{13}{24}a = 0.54a.$$

EXAMPLE (ii)

Determine the position of the centre of gravity of a solid uniform prism, whose principal cross-section is shown in Fig. 7.18. Determine

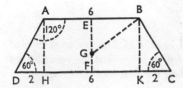

FIG. 7.18

whether such a prism can rest with the face BC *in contact with a horizontal plane.* (I.S.)

The centre of gravity lies in the section (since it was called the principal section), and from symmetry it is clear that it will lie in the line EF, joining the mid-points of AB and DC.

Draw AH, BK perpendicular to DC.

Then DH = 2, and AH = 2 tan 60° = $2\sqrt{3}$.

The prism may now be regarded as made up of three prisms, whose principal sections are the two triangles and the rectangle in the figure, and the weights of the whole and parts will be proportional to the areas of ABCD, and the triangles and rectangle. Tabulating areas and distances of centres of gravity from DC.

	Area	Distance of centre of gravity from DC
ABCD	$16\sqrt{3}$	x
ABKH	$12\sqrt{3}$	$\sqrt{3}$
ADH	$2\sqrt{3}$	$\tfrac{2}{3}\sqrt{3}$
BCK	$2\sqrt{3}$	$\tfrac{2}{3}\sqrt{3}$

Taking moments about DC, we have

$$16\sqrt{3}x = 36 + 4 + 4 = 44$$

$$\therefore \quad x = \frac{44}{16\sqrt{3}} = \frac{11\sqrt{3}}{12}.$$

The centre of gravity is therefore at G in EF, where $FG = \frac{11}{12}\sqrt{3}$

$\therefore \qquad\qquad EG = 2\sqrt{3} - \frac{11}{12}\sqrt{3} = \frac{13}{12}\sqrt{3}.$

Now the prism can rest on the face BC, provided $\angle CBG$ is less than a right angle, i.e. provided $\angle EBG$ is greater than $30°$.

Now $\qquad\qquad\qquad \tan EBG = \dfrac{EG}{EB} = \dfrac{13\sqrt{3}}{36}$

and $\qquad\qquad\qquad \tan 30° = \dfrac{\sqrt{3}}{3} = \dfrac{12\sqrt{3}}{36}$

$\therefore \qquad\qquad\qquad\qquad\qquad \angle EBG > 30°.$

Therefore the prism can rest on the face BC.

EXAMPLE (iii)

A uniform cube rests with a face touching the highest point of a fixed rough sphere. Prove from fundamental principles that the equilibrium is stable if the edge of the cube is less than the diameter of the sphere.

Let AB (Fig. 7.19) represent the vertical section through the centre of gravity G of the cube, C the point of contact, and O the centre of the sphere (radius r).

The sphere is assumed to be rough enough to prevent sliding, and we only consider the stability for tilting.

Let the edge of the cube be $2a$, and let A', C', B', G' represent the positions of A, C, B, G, after tilting through a *small* angle θ, D being the new point of contact.

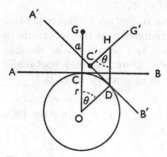

FIG. 7.19

Now the cube will tend to return to its original position if G' is to the left of the vertical through D.

Let $C'G'$ cut this vertical in H.

Since there is no slipping,

$$C'D = \text{arc } CD = r\theta.$$

Also the angle $C'HD = \theta$, and

$$C'H = C'D \cot \theta = r\,\frac{\theta}{\tan \theta}.$$

The equilibrium is therefore stable if

$$r\,\frac{\theta}{\tan \theta} > C'G'.$$

Now when θ is very small the ratio of θ to $\tan \theta$ is approximately unity, and since $C'G' = a$, the equilibrium is stable if

$$r > a,$$

i.e. if the edge of the cube is less than the diameter of the sphere.

EXAMPLE (iv)

The larger and smaller radii of the plane ends of a frustum of a solid right circular cone are R, r respectively, and its vertical height is H. Show that the distance of its centre of gravity from the centre of the larger plane end is

$$\frac{H}{4}\left(\frac{R^2+2Rr+3r^2}{R^2+Rr+r^2}\right).$$

The shape of a solid is a frustum of a right circular cone as above with a hemisphere on its larger plane end. Show that it will be in equilibrium when placed in any position on a horizontal plane if

$$3R^4 = H^2(R+2Rr+3r^2).$$

(The distance of the centre of gravity of a solid hemisphere from its centre is $\frac{3}{8}$ of the radius.)   (Ex.)

Let ABCD (Fig. 7.20) represent a section of the frustum through

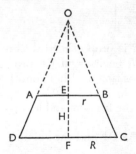

FIG. 7.20

its axis EF; let $EB = r$, $FC = R$, $EF = H$.

Suppose the cone completed by producing DA and CB to meet in O.

By similar triangles OEB, OFC we have

$$\frac{OF}{OE} = \frac{R}{r}$$

or

$$\frac{H+OE}{OE} = \frac{R}{r}$$

giving $OE = \dfrac{Hr}{R-r}$ and $OF = \dfrac{HR}{R-r}$.

The volumes (and weights) of the top cone and the whole are proportional to OE^3 and OF^3 and the volume of the frustum to $OF^3 - OE^3$. The distances of the centres of gravity of these cones from F are $\frac{1}{4}OE + H$ and $\frac{1}{4}OF$.

Hence if x is the distance of the centre of gravity of the frustum from F, taking moments about F, we have

$$(OF^3 - OE^3)x = \tfrac{1}{4}OF^4 - OE^3(\tfrac{1}{4}OE + H)$$

$$\therefore \quad \left(\frac{R^3}{r^3}OE^3 - OE^3\right)x = \tfrac{1}{4}\frac{R^4}{r^4}OE^4 - \tfrac{1}{4}OE^4 - H \times OE^3$$

$$\therefore \quad \frac{R^3 - r^3}{r^3}x = \tfrac{1}{4}\frac{R^4 - r^4}{r^4}OE - H$$

$$= \tfrac{1}{4}\frac{R^4 - r^4}{r^4} \times \frac{Hr}{R-r} - H$$

$$= \frac{H}{4}\left[\frac{(R^2 + r^2)(R+r) - 4r^3}{r^3}\right],$$

$$\therefore \quad x = \frac{H}{4}\left[\frac{R^3 + Rr^2 + R^2r - 3r^3}{R^3 - r^3}\right],$$

$$= \frac{H}{4}\left[\frac{R^2 + 2Rr + 3r^2}{R^2 + Rr + r^2}\right].$$

The weight of the hemisphere of radius R is proportional to its volume, $\frac{2}{3}\pi R^3$, and its centre of gravity is $\frac{3}{8}R$ below F in EF produced. The combined figure will rest in any position on a horizontal plane if its centre of gravity is at F.

In this case the moments of the weights of the frustum and hemisphere about F must be equal and opposite.

Now the volume of the whole cone is $\frac{1}{3}\pi R^2 \times OF$ and the volume of the frustum is $\dfrac{OF^3 - OE^3}{OF^3}$ of this.

Also

$$\frac{OF^3 - OE^3}{OF^3} = \frac{R^3 - r^3}{R^3}.$$

$\therefore$ the volume of frustum $= \frac{1}{3}\pi R^2 \times OF \times (R^3 - r^3)/R^3$

$$= \frac{1}{3}\frac{\pi H(R^3 - r^3)}{R - r}.$$

Hence, taking moments about F, we get

$$\frac{2}{3}\pi R^3 \times \frac{3}{8}R = \frac{1}{3}\pi\frac{H(R^3 - r^3)}{R - r} \times \frac{H}{4} \times \frac{R^2 + 2Rr + 3r^2}{R^2 + Rr + r^2}$$

$$= \frac{1}{12}\pi H^2(R^2 + 2Rr + 3r^2)$$

$\therefore \qquad\qquad 3R^4 = H^2(R^2 + 2Rr + 3r^2).$

EXAMPLES 7.4

1. A triangular portion is cut from a square by a line parallel to a diagonal. Show that the remainder can stand on the remaining part of either of the cut sides, if that part is 0·5 of the side, but cannot stand if it is 0·4 of the side. (I.S.)

2. ABCD is a vertical face of a rectangular block, the horizontal face which has BC for an edge being in contact with the ground; BC is 40 cm, CD is 25 cm; E is a point in BC, 15 cm from B. Find the position of a point F in CD, such that the block will be on the point of toppling over when a prism is cut from it by cutting along EF at right angles to the face ABCD. (H.C.)

3. ABCD is a uniform plane quadrilateral lamina, in which AB is parallel to DC. The length of AB is a and the length of DC is b. Prove that the centre of gravity of the lamina coincides with that of four particles of masses proportional to a, $a+b$, b, $a+b$, placed at A, B, C, D respectively. If AD $=$ BC $= c$, and $b>a$, find the coordinates of the centre of gravity, referred to rectangular axes, of which one is DA and the other is the perpendicular to DA drawn through D, and show that a uniform prism, which has such a quadrilateral for its cross-section, cannot rest on a horizontal plane with the faces corresponding to BC or AD in contact with the plane, if $2c^2(a+2b)$ is less than $b^3 - a^3$. (H.C.)

4. ABC is a vertical sheet of metal, A being a right angle and AC in contact with a horizontal plane; D is the mid-point of AC, and the triangle ABD is cut away. Show that the remaining portion of the sheet is just on the point of falling over.

5. A uniform right prism whose cross-section is an isosceles triangle BAC lies on the table with the face containing BC horizontal and

in contact with the table; the prism is gradually sliced away by cutting slices parallel to the face in which AB lies, beginning from the edge through C. What fraction of the whole prism can be cut away without the remainder toppling over, AB, AC being the equal sides?   (H.S.C.)

6. ABCD is the section of a cube through its centre, and E is a point in AD. If the part cut off by the plane through EC, perpendicular to the section, be removed, find the distance from AB and AE of the centre of gravity of the remainder. If the block be placed on a horizontal plane, find the least length of AE that the block may not upset.

7. The cross-section of a uniform prism is the figure BDEC, where ABC is an equilateral triangle and D and E are the mid-points of the sides AB and AC. Determine in how many possible positions it may rest with a face in contact with a plane, rough enough to prevent sliding, which is inclined to the horizontal at an angle of 30°; all the cross-sections being in vertical planes through lines of greatest slope of the inclined plane.   (H.S.C.)

8. A regular tetrahedron rests with one face in contact with a plane, rough enough to prevent slipping, inclined at an angle α to the horizontal. Of the face in contact with the plane, one edge is horizontal and above the opposite vertex. Show that $\tan \alpha < 2\sqrt{2}$.
                                                                      (H.S.C.)

9. A solid hemisphere and a solid cylinder have the same radii and are made of the same homogeneous material, and one end of the cylinder is cemented to the base of the hemisphere. The height of the cylinder is $\frac{2}{3}$ of its radius. Show that the centre of gravity of the whole solid is $\frac{1}{48}$ of the radius from the common plane. (I.E.)

10. A hollow vessel, formed by uniformly thin material, is in the form of a frustum of a circular cone and has a flat base, the radius of the mouth being twice that of the base. If the semi-vertical angle of the cone is 30°, determine whether the vessel can rest with its curved surface in contact with a horizontal plane.   (H.S.C.)

11. The height of a solid circular cone is three times the radius of its base. Another solid in the form of a hemisphere, of which the base of the cone is the plane boundary, is fastened to the cone. Find the least ratio of the specific gravities of the materials of the hemisphere and cone, if when the combined solid is placed on a horizontal table with the hemispherical surface in contact with the table it cannot be upset.   (I.E.)

12. A hollow vessel, the material of which is of the same thickness throughout, consists of a hollow cone closed by a hollow hemisphere described on its base. Find the greatest possible ratio of the height of the cone to the radius of its base if the vessel can rest in stable equilibrium with the hemispherical surface in contact with a horizontal table. (The distance of the centre of gravity of a hollow hemisphere from its centre is half the radius.) (H.S.C.)

13. A pile in the shape of a flight of stairs is formed of n equal uniform cubical blocks, of edge a, each block being displaced a small distance c horizontally with reference to the block below. Determine the position of the centre of gravity of the pile and deduce that the pile must topple over if $(n-1)c>a$. (H.C.)

14. A body of uniform material consists of a solid right circular cone and a solid hemisphere on opposite sides of the same circular base of radius r. Find the greatest possible height of the cone if the body can rest on a horizontal plane in stable equilibrium with the cone uppermost. (C.S.)

15. A table consists of a 1 cm board, 60 cm square, and having at its corners legs of the same material, 60 cm long, and of 2 cm square section. Find the height of the centre of gravity and determine the greatest angle through which the table can be tilted on two legs without being overturned. (I.A.)

16. ABCD is a heavy uniform square plate, and the portion CBH is removed, where H is a point in AB. The remainder is placed with its plane vertical and AH in contact with a smooth horizontal plane. Show that equilibrium will be impossible unless AH : AB is greater than $\frac{1}{2}(\sqrt{3}-1)$. (C.W.B.)

7.18. Centre of gravity of a uniform circular arc

Let ACB (Fig. 7.21) be a circular arc of radius r, subtending an angle 2α at its centre O, and let C be the mid-point of the arc.

From symmetry it is clear that the centre of gravity will lie on OC, since the arc is uniform.

Take OC as x-axis, and OY perpendicular to OC as y-axis.

Let PQ be an element of the arc such that $\angle POC = \theta$, and $\angle POQ = \delta\theta$. The length of the arc PQ is $r\delta\theta$, and its weight is $wr\delta\theta$, where w is the weight per unit of length.

Since the distance of PQ from OY is $r \cos \theta$, the moment of its weight about OY is $r \cos \theta \times wr\delta\theta$ or $wr^2 \cos \theta \times \delta\theta$.

The sum of the moments of all the elements of the arc is the integral of this between the limits $\theta = -\alpha$ and $\theta = +\alpha$, i.e. $\int_{-\alpha}^{+\alpha} wr^2 \cos \theta \, d\theta$. The weight of the whole arc is $2wr\alpha$, and if the

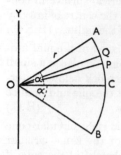

FIG. 7.21

distance of its centre of gravity from OY is $\bar{x}$ we have, taking moments about OY,

$$2wr\alpha\bar{x} = wr^2 \int_{-\alpha}^{+\alpha} \cos \theta \, d\theta = wr^2 \Big[\sin \theta \Big]_{-\alpha}^{+\alpha} = 2wr^2 \sin \alpha$$

$$\therefore \qquad \bar{x} = r\frac{\sin \alpha}{\alpha}.$$

For a semicircular arc, in which $\alpha = \pi/2$, this becomes $2r/\pi$.

7.19. Centre of gravity of a sector of a circle

Let AOB (Fig. 7.22) be a sector of a circle of radius r whose centre is at O, and let $\angle AOB = 2\alpha$. Let C be the mid-point of the arc AB and OC the bisector of the angle AOB.

From symmetry it is clear that the centre of gravity lies in OC.

Take OC as x-axis, and OY perpendicular to OC as y-axis.

Let POQ be an elementary sector, such that $\angle POC = \theta$, and $\angle POQ = d\theta$.

The area of this sector is $\tfrac{1}{2}r^2 \, d\theta$, and its weight is $\tfrac{1}{2}wr^2 \, d\theta$, where w is the weight per unit area, assumed constant.

Since PQ is very small, OPQ is very approximately a triangle, and its centre of gravity is at a distance $\tfrac{2}{3}r$ from O.

The distance of this centre of gravity from OY is therefore

$$\tfrac{2}{3}r \cos \theta.$$

The moment of the weight of POQ about O Y is

$$\tfrac{2}{3}r \cos \theta \times \tfrac{1}{2}wr^2 \,\delta\theta = \tfrac{1}{3}wr^3 \cos \theta \,\delta\theta.$$

The sum of the moments of all the elementary sectors is

$$\int_{-\alpha}^{+\alpha} \tfrac{1}{3}wr^3 \cos \theta \,\mathrm{d}\theta = \tfrac{1}{3}wr^3 \int_{-\alpha}^{+\alpha} \cos \theta \,\mathrm{d}\theta.$$

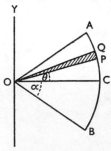

FIG. 7.22

The weight of the whole sector is $wr^2\alpha$, and if $\bar{x}$ is the distance of its centre of gravity from OY, then, taking moments about OY, we have

$$wr^2\alpha \times \bar{x} = \tfrac{1}{3}wr^3 \int_{-\alpha}^{+\alpha} \cos \theta \,\mathrm{d}\theta = \tfrac{2}{3}wr^3 \sin \alpha$$

$$\therefore \qquad \bar{x} = \frac{2r}{3} \times \frac{\sin \alpha}{\alpha}.$$

For a complete semicircle, $\alpha = \pi/2$, and the distance of the centre of gravity from the diameter of the semicircle is

$$4r/3\pi.$$

These formulae can also be obtained by making use of the result of the last paragraph. Suppose the area divided into concentric strips of breadth $\mathrm{d}x$. The weight of a strip of radius x is $2wx\alpha \,\mathrm{d}x$, and the distance of its C.G. from OY is $(x \sin \alpha)/\alpha$.

Hence, taking moments about OY,

$$wr^2\alpha\bar{x} = \int_0^r 2wx\alpha \times x\frac{\sin \alpha}{\alpha}\mathrm{d}x = 2w \sin \alpha \left[\frac{x^3}{3}\right]_0^r$$

$$= \tfrac{2}{3}wr^3 \sin \alpha$$

$$\therefore \qquad \bar{x} = \frac{2r}{3}\frac{\sin \alpha}{\alpha}.$$

7.20. Centre of gravity of a segment of a circle

This is found most easily by considering the segment as the difference between a sector of a circle and a triangle.

Let ACB (Fig. 7.23) be a uniform segment of a circle of radius r, and O the centre of the circle.

The segment is the difference between the sector OAB and the triangle OAB.

Let $\angle AOB = 2\alpha$, and let OC bisect this angle.

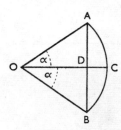

Fig. 7.23

From symmetry we see that the centre of gravity must lie in OC.

If w is the weight per unit area the weight of the sector is $wr^2\alpha$, and that of the triangle $\frac{1}{2}wr^2 \sin 2\alpha$.

Tabulating areas and distances of centres of gravity from O, we get

	Weight	Distance of C.G. from O
Sector	$wr^2\alpha$	$\dfrac{2r}{3}\dfrac{\sin \alpha}{\alpha}$
Triangle	$\frac{1}{2}wr^2 \sin 2\alpha$	$\frac{2}{3}r \cos \alpha$
Segment	$wr^2(\alpha - \frac{1}{2}\sin 2\alpha)$	x

Hence, taking moments about O, we have

$$wr^2(\alpha - \tfrac{1}{2}\sin 2\alpha)x = \tfrac{2}{3}wr^3 \sin \alpha - \tfrac{1}{3}wr^3 \sin 2\alpha \cos \alpha$$

$$\therefore \qquad x = \frac{2r}{3}\frac{\sin \alpha - \cos^2 \alpha \sin \alpha}{\alpha - \frac{1}{2}\sin 2\alpha}$$

$$= \frac{4r}{3}\frac{\sin^3\alpha}{2\alpha - \sin 2\alpha}.$$

If we are given the radius of the circle and height of the segment CD it is best to proceed as above and then express α, sin α, etc., in terms of CD and r.

When the segment is a semicircle and $α = π/2$ the above result reduces to $4r/3π$, as obtained in the last paragraph.

7.21. Centre of gravity of a uniform solid hemisphere

Let ACB (Fig. 7.24) represent a section of the hemisphere perpendicular to the plane base, of which AB is a diameter and O the centre.

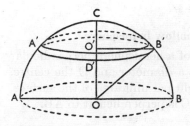

FIG. 7.24

Let r be the radius of the hemisphere, w the weight per unit volume, and C the highest point, so that OC is perpendicular to AB.

Suppose the hemisphere divided into infinitely thin slices parallel to the plane base.

From symmetry it is clear that the centres of gravity of all these slices, and therefore the centre of gravity of the hemisphere, will lie in OC.

If A′D′B′ represents one of the slices, of thickness $δx$, and O′ its centre then, if $OO' = x$,

$$O'B' = \sqrt{(r^2-x^2)}.$$

The volume of the slice is therefore

$$π(r^2-x^2)\,δx$$

and its weight is

$$wπ(r^2-x^2)\,δx.$$

This weight may be supposed concentrated at O′, and its moment about O is

$$wπx(r^2-x^2)\,δx.$$

Considering all the slices in the same way the sum of their moments about O is

$$\int_0^r w\pi(r^2x - x^3)\,\mathrm{d}x = w\pi\left[\tfrac{1}{2}r^2x^2 - \tfrac{1}{4}x^4\right]_0^r$$
$$= \tfrac{1}{4}w\pi r^4.$$

The weight of the whole hemisphere is $\tfrac{2}{3}\pi w r^3$, and if $\bar{x}$ is the distance of its centre of gravity from O,

$$\tfrac{2}{3}\pi w r^3 \bar{x} = \tfrac{1}{4}w\pi r^4$$
$$\therefore \qquad\qquad \bar{x} = \tfrac{3}{8}r.$$

7.22. Centre of gravity of a thin hollow hemisphere

Let ACB (Fig. 7.25) be a section of a hemisphere perpendicular to its plane base, of which AB is a diameter and O the centre.

Let r be the radius, w the weight per unit area of the surface, and C the highest point so that OC is perpendicular to AB.

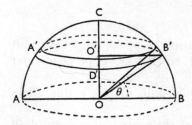

FIG. 7.25

Suppose the surface divided into infinitely narrow bands, such as A′D′B′, by planes parallel to the base.

If O′ is the centre of A′D′B′ and $\angle BOB' = \theta$ the radius of the band O′B′ is equal to $r\cos\theta$.

It is clear from symmetry that the centres of gravity of all the bands lie in OC.

The angle subtended by the arc of the band at O is $\delta\theta$, and its breadth is therefore $r\,\delta\theta$.

The whole surface of the band is therefore $2\pi r \cos\theta\, r\,\delta\theta$, and its weight is $2\pi w r^2 \cos\theta\,\delta\theta$.

This weight may be supposed to act at O′, and its moment about O is therefore

$$r\sin\theta\, 2\pi w r^2 \cos\theta\,\delta\theta.$$

The sum of the moments of all the bands about O is

$$2\pi wr^3 \int_0^{\frac{\pi}{2}} \sin\theta \cos\theta \, d\theta$$

$$= 2\pi wr^3 \left[\frac{\sin^2\theta}{2}\right]_0^{\frac{\pi}{2}} = \pi wr^3.$$

The weight of the whole surface is $2\pi r^2 w$, and if $\bar{x}$ is the distance of its centre of gravity from O,

$$2\pi r^2 w \times \bar{x} = \pi wr^3$$

$$\therefore \qquad \bar{x} = \frac{r}{2}.$$

7.23. The centre of gravity of a portion or zone of the surface between two planes parallel to the base in positions such that $\theta = \alpha$, $\theta = \beta$ can be obtained by evaluating the weight and moment integrals of the last paragraph between the limits α and β instead of 0 and $\pi/2$.

The weight of an elementary zone is $2\pi wr^2 \cos\theta \, \delta\theta$.

The weight of the zone between $\theta = \alpha$ and $\theta = \beta$ is

$$2\pi wr^2 \int_\alpha^\beta \cos\theta \, d\theta = 2\pi wr^2(\sin\beta - \sin\alpha).$$

But $r(\sin\beta - \sin\alpha)$ is the distance between the cutting planes, i.e. h, the height of the zone.

$$\therefore \qquad \text{the weight is } 2\pi wrh.$$

The moment about O of the weight of an elementary zone is $2\pi wr^3 \sin\theta \cos\theta \, \delta\theta$.

The sum of these moments is

$$2\pi wr^3 \int_\alpha^\beta \sin\theta \cos\theta \, d\theta = \pi wr^3 (\sin^2\beta - \sin^2\alpha),$$

$$= \pi wr^2 h(\sin\beta + \sin\alpha).$$

Hence, if x is the distance of the centre of gravity of the zone from O

$$x = \frac{\pi wr^2 h(\sin\beta + \sin\alpha)}{2\pi wrh},$$

$$= \tfrac{1}{2}r(\sin\beta + \sin\alpha).$$

Now $r \sin \beta + r \sin \alpha$ is the sum of the distances of the cutting planes from the base, so that $\frac{1}{2}r(\sin \beta + \sin \alpha)$ is the distance of the plane half-way between them.

The centre of gravity of the zone is therefore half-way between the planes which cut it off from the sphere.

7.24. The position of the centre of gravity of a complete hemispherical surface, and of a zone bounded by planes parallel to the base, can both be deduced from the known geometrical fact that the area of each elementary band or zone cut off by planes parallel to the base is equal to the corresponding band which they cut off from the circumscribing cylinder. The centre of gravity of the spherical surfaces is therefore at the same height as that of the circumscribing cylinder, i.e. half-way between the base and vertex in the case of the complete hemisphere, and half-way between the bounding planes in the case of a zone.

7.25. EXAMPLE (i)

Find the centre of gravity of a uniform parabolic plate whose boundary consists of the curve $y^2 = ax$ and the line $x = b$. Find also the position of the centre of gravity of the solid formed by revolution of the above area about the axis of x.

Let ABMC (Fig. 7.26) represent the plate, AX the axis of the parabola (also the axis of x), and AM $= b$.

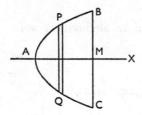

FIG. 7.26

It is clear from symmetry that the centre of gravity lies in AM. The area of a strip PQ parallel to BC, of breadth δx and distant x from A, is $2y \, \delta x$ or $2a^{\frac{1}{2}}x^{\frac{1}{2}} \, \delta x$, and its weight is $2wa^{\frac{1}{2}}x^{\frac{1}{2}} \, \delta x$, where w is the weight per unit area.

In this case we cannot write down the weight of the whole plate, but the value is obtained by integrating $2wa^{\frac{1}{2}}x^{\frac{1}{2}}\,\delta x$ between $x = 0$ and $x = b$. If W is the weight of the plate,

$$W = 2wa^{\frac{1}{2}}\int_0^b x^{\frac{1}{2}}\,\mathrm{d}x = \tfrac{4}{3}wa^{\frac{1}{2}}\left[x^{\frac{3}{2}}\right]_0^b = \tfrac{4}{3}wa^{\frac{1}{2}}b^{\frac{3}{2}}.$$

The weight of each strip acts at its mid-point, and the moment of the weight of PQ about A is

$$2wa^{\frac{1}{2}}x^{\frac{3}{2}}\,\delta x.$$

The sum of the moments about A is

$$2wa^{\frac{1}{2}}\int_0^b x^{\frac{3}{2}}\,\mathrm{d}x = \tfrac{4}{5}wa^{\frac{1}{2}}\left[x^{\frac{5}{2}}\right]_0^b = \tfrac{4}{5}wa^{\frac{1}{2}}b^{\frac{5}{2}}.$$

If $\bar{x}$ is the distance of the centre of gravity from A,

$$\tfrac{4}{3}wa^{\frac{1}{2}}b^{\frac{3}{2}}\times\bar{x} = \tfrac{4}{5}wa^{\frac{1}{2}}b^{\frac{5}{2}},$$

$$\therefore \qquad \bar{x} = \tfrac{3}{5}b.$$

When the area is rotated about AX, the section of the solid so formed by a plane perpendicular to AX will be circular. Dividing the solid into circular slices by planes perpendicular to AX, the volume of a slice distant x from A will be

$$\pi y^2\,\delta x = \pi a x\,\delta x$$

and its weight $= \pi wax\,\delta x$.

The weight of the whole solid, W, is given by

$$W = \pi wa\int_0^b x\,\mathrm{d}x = \tfrac{1}{2}\pi wab^2.$$

The weight of each slice acts at its centre, and the moment about A of the weight of the slice distant x from A is

$$\pi wax^2\,\delta x.$$

Hence, the sum of the moments about A is

$$\pi wa\int_0^b x^2\,\mathrm{d}x = \tfrac{1}{3}\pi wab^3.$$

If $\bar{x}$ is the distance of the centre of gravity from A,

$$\tfrac{1}{2}\pi wab^2\times\bar{x} = \tfrac{1}{3}\pi wab^3$$

$$\therefore \qquad \bar{x} = \tfrac{2}{3}b.$$

EXAMPLE (ii)

A solid homogeneous circular cylinder of radius r is bisected by a plane passing through its axis, and on one-half as base is constructed a triangular prism of isosceles section and of the same substance; the whole is placed in equilibrium on the top of a fixed circular cylinder of radius 2r with axis horizontal, the axes of the cylinders being parallel and the curved surfaces in contact. Show that the greatest height of the prism consistent with stability for a small rolling displacement is

$$\tfrac{1}{2}r[(9-2\pi)^{\frac{1}{2}}-1].$$ (C.S.)

Let ABCD (Fig. 7.27) represent a section of the half cylinder and

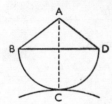

FIG. 7.27

prism, C being the point of contact with the fixed cylinder, and let H be the height of the prism.

The weights of the cylinder and prism are proportional to the areas of BCD and ABD, i.e. to $\tfrac{1}{2}\pi r^2$ and rH.

The distance of the centre of gravity of BCD from BD is $4r/3\pi$, and from C it is $r-4r/3\pi$.

The distance of the centre of gravity of the prism from C is $r+\tfrac{1}{3}H$.

If h is the height of the centre of gravity of the combined solid above C, then, taking moments about C, we have

$$h[\tfrac{1}{2}\pi r^2+rH] = \tfrac{1}{2}\pi r^2\left(r-\frac{4r}{3\pi}\right)+rH(r+\tfrac{1}{3}H).$$

The equilibrium is stable if

$$\frac{1}{h}>\frac{1}{r}+\frac{1}{2r}\ \text{or}\ h<\frac{2}{3}r.$$

Hence for stability, we get

$$\tfrac{1}{2}\pi r^2\left(r-\frac{4r}{3\pi}\right)+rH(r+\tfrac{1}{3}H)<\frac{2}{3}r(\tfrac{1}{2}\pi r^2+rH)$$

$$\therefore\qquad \tfrac{1}{2}\pi r^3-\frac{2}{3}r^3+Hr^2+\tfrac{1}{3}H^2r<\frac{1}{3}\pi r^3+\frac{2}{3}Hr^2$$

$$\therefore \qquad \frac{H^2r}{3}+\frac{Hr^2}{3}<\frac{2r^3}{3}-\frac{\pi r^3}{6}$$

$$\therefore \qquad H^2+rH<2r^2-\tfrac{1}{2}\pi r^2$$

$$\therefore \qquad (H+\tfrac{1}{2}r)^2<r^2\left(\frac{9-2\pi}{4}\right)$$

$$\therefore \qquad H+\tfrac{1}{2}r<\tfrac{1}{2}r(9-2\pi)^{\frac{1}{2}}$$

$$\therefore \qquad H<\tfrac{1}{2}r[(9-2\pi)^{\frac{1}{2}}-1].$$

EXAMPLES 7.5

1. Find the centre of gravity of a uniform semicircular disc.

2. A piece of metal, of uniform thickness, consists of a semicircular portion ABC with diameter AC, and a triangular portion ACD having AD = CD. Find the ratio of the height of the triangle to the radius of the semicircle in order that, when placed on a smooth horizontal plane with the plane ABCD vertical, the metal may rest in equilibrium whatever point of the circular arc may be in contact with the horizontal plane. (Ex.)

3. Find the centre of gravity of a thin wire in the form of a circular arc. A thin strip of metal of uniform width and thickness has part of it bent into the form of a semi-cylinder of radius r, leaving a plane piece of length l tangential to the semi-cylinder along one of its edges. Show that the body can rest with the plane piece in contact with a horizontal plane provided that l is greater than $2r$. (I.E.)

4. A solid body consists of a solid hemisphere surmounting a cylinder of the same radius whose height is three times the radius of its base. It is placed with its plane face in contact with a rough plane, and the inclination of the plane to the horizontal is gradually increased. Prove that it will slide down the plane without toppling over, provided that the coefficient of friction between the body and the plane is less than $\frac{44}{81}$. (I.E.)

5. A uniform solid circular cylinder is cut in two by a plane through its axis. One half is placed with its curved surface in contact with an inclined plane rough enough to prevent slipping. The line of contact is perpendicular to the line of greatest slope. Show that there are two positions of equilibrium if the inclination of the plane to the horizontal lies between $\sin^{-1}(4/3\pi)$ and $\tan^{-1}(4/3\pi)$, and one position if it is less than the latter angle. (H.C.)

6. A chimney of brickwork 45 cm thick has an external diameter of 3·9 m at the base, and 2·7 m at the top, which is 30 m above the

base. Show that the centre of gravity of the chimney is about 1·05 m below the mid-point of the axis.   (C.S.)

7. Show that the centre of gravity of a uniform semicircular rod is at a distance from the centre equal to $2/\pi$ times the radius. A circular disc, whose weight per unit area is σr, where σ is a constant and r the distance from the centre, is divided into two by a diameter. Find the centre of gravity of either half.   (C.S.)

8. Find by integration the position of the centre of gravity of a uniform solid right circular cone.

9. Prove that the centre of gravity of a uniform thin hemispherical cup of radius r is at a distance $r/2$ from the centre. Such a cup stands on a circular base of the same material, thickness, and radius as the cup, while the intervening stem is of length equal to the radius of the cup, and its weight is one-quarter of that of the hemisphere. Find the height of the centre of gravity above the base.   (I.S.)

10. Show that the centre of mass of a semicircular lamina of radius r is at a distance $4r/3\pi$ from the centre. A solid right circular cylinder is cut into two equal parts by a plane through its axis. One of these parts is placed with its curved surface on a rough plane, which is inclined to the horizontal at an angle θ, with its generators horizontal. What will be the inclination of the rectangular plane face of the solid to the horizontal when it rests in equilibrium, assuming the plane to be rough enough to prevent slipping?   (H.S.D.)

11. Assuming that the centre of gravity of a uniform circular arc subtending an angle 2θ at the centre of a circle of radius a is at a distance $(a \sin \theta)/\theta$ from the centre, find the centre of gravity of the uniform sector bounded by that arc and the radii to its extremities. Find the centre of gravity of the segment cut off by a chord equal to the radius.   (I.E.)

12. Find the centre of gravity of a hemispherical surface. A hemispherical bowl of weight W is placed with its spherical surface on a smooth horizontal plane; what weight must be placed on the rim in order that the bowl may rest with its plane surface inclined at $\alpha°$ to the horizontal?   (H.S.D.)

13. A uniformly tapering circular spar is 15 m long. The butt diameter is 35 cm and the top diameter is 25 cm. Find the mass of the spar and the position of the centre of mass, assuming the wood has a mass of 700 kgm^{-3}.

14. A uniform wire bent in the form of a semicircle hangs in a vertical plane over a rough horizontal peg and is just on the

point of slipping when the line joining its ends makes an angle of 30° with the horizontal. Find the coefficient of friction for the wire and the peg.

15. A solid uniform hemisphere rests with its convex surface in contact with a rough inclined plane; show that the greatest possible inclination of the plane to the horizontal is $\sin^{-1} \frac{3}{8}$.

16. If a solid uniform hemisphere rest in equilibrium with its curved surface in contact with a rough plane inclined to the horizontal at an angle $\sin^{-1} \frac{1}{8}$, find the inclination of the plane base of the hemisphere to the horizontal.

17. Weights 5, 4, 3, 2, 1 are placed at five of the vertices of a cube, the weights 4, 3, 2 being at the other ends of the edges through the weight 5, and the weight 1 at the other end of the diagonal through the weight 5. What weights must be placed at the remaining three vertices in order that the centre of gravity of the whole system may be at the centre of the cube?   (Ex.)

18. A triangle of uniform rods of different densities has its centre of gravity at the centre of its circum-circle. Prove that the weights of the rods are proportional to the tangents of the opposite angles.   (H.S.D.)

19. A letter D is made from a uniform semicircular plate of radius $(a+x)$ by removing the concentric semicircular part of radius a from it and joining across the two ends of the curved strip a rectangular plate of the same material of length $2(a+x)$ and breadth x, the complete letter being similar to that illustrated in Fig. 7.28. Find the position of the centroid of this plate, and

Fɪɢ. 7.28

show that when x is small it is approximately at a distance $2a/(\pi+2)$ from the centre of the circles.

20. Find the centre of gravity of the portion of a uniform spherical shell bounded by two parallel planes. From a spherical shell two segments are cut off by planes meeting at one point on the sphere. The angle of the sectors is in each case 120°. Find the position of the centre of gravity of the remaining portion of the shell.
                                                                    (I.E.)

21. (i) A line is drawn parallel to BC through the centroid of the triangular area ABC, to intersect AB and AC at E and F. Show that the centroid of the area BEFC lies on the median of the triangle drawn from A and find in what ratio it divides this median.

(ii) Prove that the distance of the centroid of a sector of a circle, of radius a, from the centre of the circle is $(2a \sin \alpha)/3\alpha$ where $2\alpha(<\pi)$ is the angle of the sector.

If PQ is a quadrant arc of the circle and the tangents at P and Q meet at T, find the distance from the centre of the circle of the centroid of the area enclosed between PT, QT and the arc PQ.   (L.A.)

22. Prove that the centre of gravity of a uniform solid hemisphere is at a distance $3r/8$ from the plane face, r being the radius.

The hemisphere is placed with its curved surface in contact with a fixed inclined plane making an angle of 10° with the horizontal. A particle of mass one-quarter that of the hemisphere is attached to the lowest point of the plane face of the hemisphere. Calculate the inclination of the plane face of the hemisphere to the horizontal. It may be assumed that the contact between the hemisphere and the plane is sufficiently rough to prevent slipping.   (L.A.)

23. Prove by integration that the centre of gravity of a uniform semicircular lamina of radius a is at a distance $4a/3\pi$ from the centre.

ACB is such a lamina with diameter AOB, and OC is the radius perpendicular to AB. A square portion OPQR is cut out of the lamina, P being on OB and OP having length $\frac{1}{2}a$. Find the distances from OA and OC of the centre of gravity of the remaining portion.

Hence show that, if the remaining portion is suspended from A and hangs in equilibrium, the tangent of the angle made by AB with the vertical is just less than $\frac{1}{2}$.   (C.A.)

24. A smooth hemispherical bowl is fixed with its axis vertical. A rod AB, of length l, rests partly inside and partly outside the bowl, with A at the lowest point of the bowl, and with one-third of its length outside the bowl. The rod is of weight W and its centre of gravity is at a distance c from its end A. Equilibrium is maintained by means of a horizontal force applied to the rod at B. Find this force and the reactions on the rod due to the bowl.

Show that this position of equilibrium is impossible if $c > \frac{1}{3}l$.   (C.A.)

25. One end of a light inextensible cord is attached to the rim of a uniform solid hemisphere of radius a and of weight W. The other end of the cord is fixed to a smooth vertical wall. The hemisphere rests in equilibrium with its curved surface against the wall.

If the reaction of the wall on the hemisphere is $3W/8$, find the inclination of the plane face to the vertical and show that the length of the cord is $\frac{2}{15}a\sqrt{73}$. (L.A.)

REVISION EXAMPLES B

1. ABCDE is a chain formed of four uniform equal heavy rods freely jointed at B, C, and D. The ends A and E of the chain are hinged to two fixed points in the same horizontal line so that the chain hangs symmetrically with C as its lowest point. If θ is the angle of inclination of the upper pair of rods to the horizontal in the equilibrium position and ϕ the corresponding angle for the lower pair, prove that $\tan \theta = 3 \tan \phi$.

   If, instead of being hinged, A and E are attached to small rings able to slide along a fixed rough horizontal rod. determine the coefficient of friction given that in limiting equilibrium $\theta = 60°$. (L.A.)

2. Two equally rough parallel bars are each fixed horizontally, the distance between them being l and their plane being inclined to the horizontal at an angle θ. A uniform heavy rod rests in contact with both bars passing over the upper and under the lower one. If the rod is at right angles to the bars and in limiting equilibrium, show by consideration of force moments, or otherwise, that the centre of gravity of the rod must be higher than the upper bar. Deduce that the length of the rod is at least

$$l(\mu + \tan \theta)/\mu$$

where μ is the coefficient of friction. (L.A.)

3. Two small rings, each of mass m, are threaded on a rough straight rod which is fixed at an inclination α to the horizontal. The rings are connected by a light inelastic string to the midpoint of which is attached a particle of mass m. The system is in equilibrium, the angle between the two parts of the string being $2\theta(>2\alpha)$. By considering the equilibrium of the higher ring and of the particle show that, if μ is the coefficient of friction between either ring and the rod,

$$\mu \geqslant \left(\frac{3 \tan \alpha + \tan \theta}{\tan \alpha + 3 \tan \theta}\right) \tan \theta. \qquad \text{(L.A.)}$$

4. A uniform rod of length $2l$ and weight W is smoothly hinged at

one end to a horizontal plane. It is supported in an inclined position by a cylinder of radius r which has one generator on the plane and at right angles to the vertical plane through the rod, the point of contact being the mid-point of the rod. The coefficient of friction between the rod and the cylinder is μ and the horizontal plane is sufficiently rough to prevent the cylinder sliding on it. Prove that if the cylinder is about to roll, $\mu = r/l$. Find the reaction at the hinge.   (L.A.)

5. A uniform ladder AB, of weight W, leans at an angle α to the vertical against a rough wall BC with its other end on a rough horizontal floor AC, the coefficient of friction at each point of contact being $\mu(<\tan \frac{1}{2}\alpha)$. The mid-point of the ladder is attached to the point C by means of a taut string. If the ladder is on the point of slipping, prove that the tension in the string is

$$\frac{W}{2\mu}\left\{ (1-\mu^2)\sin\alpha - 2\mu\cos\alpha \right\}$$

and find the normal components of the reactions at A and B.
                                                                        (C.W.B.)

6. Explain the term 'angle of friction'. A rectangular block rests in limiting equilibrium with one edge on the floor and another edge in contact with a vertical wall, the angle of friction λ being the same at both contacts. Prove that, if the cross-section at right angles to the wall is a rectangle, ABCD, with the corner A on the floor and B on the wall, then the inclination θ of AB to the vertical is given by

$$\tan\theta = \tan 2\lambda + \frac{BC}{AB}\sec 2\lambda. \qquad \text{(H.C.)}$$

7. Explain the terms coefficient of limiting friction, angle of limiting friction, and the relation between them. Two equal light rods AB, BC, of length $2a$, rigidly joined at right angles to each other at B, are placed astride a fixed rough horizontal circular cylinder of radius a, so that they are equally inclined to the vertical. Weights w, $W(W>w)$ are hung from A and C respectively, W being adjusted so that slipping is about to take place. Find the normal reactions of the cylinder on the rods and prove that

$$W/w = (1+\mu+\mu^2)/(1-\mu+\mu^2),$$

where $\mu(<1)$ is the coefficient of limiting friction.   (H.S.D.)

8. The diagram (Fig. 7.29) shows a smoothly jointed framework of light rods AB, BC, AC, CD smoothly pivoted at A and D to a vertical wall. A weight of mass 100 kg is suspended at B and the system is at rest in a vertical plane. Find the stresses in all the

rods, distinguishing tension from compression, and find also the
forces exerted on the wall at A and D.   (I.S.)

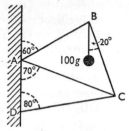

FIG. 7.29

9. The framework shown in the diagram (Fig. 7.30) consists of
   five light rods freely jointed at their ends. A load of 2 units is
   hung from D and the framework is kept in equilibrium, with AD
   horizontal and AB vertical, by a horizontal force at A and a
   force at B.

   Calculate these forces and determine, graphically or otherwise, the forces in the rods, stating which rods are in tension and
   which in compression.   (L.A.)

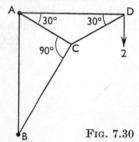

FIG. 7.30

10. The diagram (Fig. 7.31) shows a framework of five freely
    jointed light rods AB, BC, CD, DA, AC in the form of a parallelo-

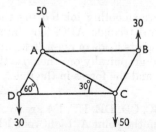

FIG. 7.31

gram ABCD in a vertical plane with AB horizontal, and acted
on by vertical forces, as shown, at the four corners. Prove that

the frame is in equilibrium, and find graphically the stresses in all the rods, distinguishing tension from compression.   (I.S.)

11. Six light rods of equal length are smoothly jointed together to form a regular hexagon OABCDE, and three other light rods join OB, OC, OD. Particles of weight w are attached to A and E. The system rests in a vertical plane with OC horizontal (A vertically above E), the point O being fastened to a fixed peg. The framework is kept in position by means of a string attached to B, whose line of action is along CB produced. Find the tension of the string. Find the stresses in AO, OB, OC, OD, OE, and state whether these rods are struts or ties.   (H.C.)

12. ABCD is a rhombus of smoothly jointed light rods suspended from a point O by a light vertical rod OA smoothly jointed at A and equal strings OB, OD, the diagonal AC being vertical, and ABC = 120°, BOD = 30°. Find graphically, or calculate, the stresses in the rods and strings when a weight W is suspended from C, and indicate which of them are tensions.   (H.C.)

13. The diagram (Fig. 7.32) represents a framework of seven smoothly jointed light rods hinged to a fixed point at A. The outer rods form a square, and A, C, B are collinear, with AC = 3CB. Determine, graphically or otherwise, the stress in each rod when a force P is applied at B as shown, distinguishing between tensions and thrusts.   (H.C.)

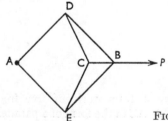

Fig. 7.32

14. Three light rods equal in length are smoothly jointed together at their ends to form an equilateral triangle ABC. The frame hangs freely at rest from A and supports loads of mass 30 kg and 10 kg at B and C respectively. Find graphically or otherwise the inclination of BC to the horizontal and the forces in the rods.   (L.A.)

15. Six equal light uniform rods AB, BC, CD, DE, EF, FA are freely hinged together and suspended from the point A. Light rods FB, BE, EC, of such lengths that ABCDEF is a regular hexagon, are inserted and are freely hinged to the hexagon at their ends. By

drawing, find the force in each of the nine rods when weights of mass 10 kg are suspended from each of C, D, E.   (L.A.)

16. The framework shown in the diagram (Fig. 7.33) consists of seven equal light rods freely jointed together. It is in equilibrium in a vertical plane with AB horizontal, there being vertical supports at A and B and vertical loads, as indicated, at D and E.

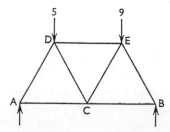

FIG. 7.33

Find the reactions at the supports and the forces in the rods CD, CE, and DE, stating which of these rods are in tension and which in compression.   (L.A.)

17. Find the centres of gravity of: (i) a uniform triangular lamina; (ii) a uniform thin hollow right circular cone. The cone has a base made of the same material. When it is suspended from a point O on the rim the diameter of the base and the generator through O make equal angles with the vertical. Find the vertical angle of the cone.   (H.C.)

18. Show that the centre of gravity of three equal particles placed one at each vertex of a triangle is at the intersection of the medians of the triangle. A uniform square plate ABCD of side $2a$ and centre O is pierced by three small circular holes, of radius b, whose centres are at the mid-points of OA, OB, and OD. Find the centre of gravity of the plate.   (H.C.)

19. A uniform lamina is in the form of the quadrilateral whose vertices are the points (4, 0), (5, 0), (0, 12), (0, 3) referred to rectangular axes. Find the coordinates of the centre of mass of the lamina.

Find also the coordinates of the centre of mass of a uniform wire bent into the form of the perimeter of the quadrilateral.
          (L.A.)

20. Prove that the centre of gravity of a uniform semicircular lamina of radius a is at a distance $4a/3\pi$ from its centre.

AOB is the base of a uniform semicircular lamina of radius $2a$, O being its centre. A semicircular lamina of radius a and base AO is cut away and the remainder suspended freely from A.

Find the inclination of AOB to the vertical in the equilibrium
position.   (L.A.)

21. Show that the centre of gravity of a thin uniform hemispherical
    shell is at the mid-point of the radius of symmetry. A thin hemi-
    spherical shell of weight W rests with its curved surface on a
    horizontal plane. To a point on the rim a particle of weight $\frac{1}{2}W$ is
    attached. Show that, in the position of equilibrium the plane of
    the rim makes an angle of $45°$ with the horizontal.   (I.S.)

22. (i) Show that the centre of mass of a triangle formed of uniform
       rods of the same mass per unit length is at the incentre of the
       triangle whose corners are the mid-points of the rods.

    (ii) If a uniform lamina in the shape of a symmetrical trapezium
       has one of its parallel sides double the other, prove that its
       centre of mass is distant from the centre of mass of four
       equal particles situated at the corners of the trapezium, by
       a distance equal to one-eighteenth of that between the two
       parallel sides.   (H.S.D.)

23. Show that the centre of gravity of a uniform solid hemisphere of
    radius a is at a distance $3a/8$ from the centre of the base. A uni-
    form cube is attached by one of its faces to the base of a uniform
    hemisphere, a diagonal of this face of the cube being a diameter
    of the base of the hemisphere. If ρ_1 be the density of the material
    of the hemisphere and ρ_2 of the material of the cube, show that
    the combined solid will be in equilibrium with any point of the
    curved surface of the hemisphere in contact with a horizontal
    plane if $\pi\rho_1 = 8\rho_2$.   (I.S.)

24. The central cross-section of a solid right prism is a trapezium
    ABCD in which A and D are right angles, $AD = CD = a$ and
    $AB = b$. Find the distances of the centre of mass of the solid
    from AB and AD.

    Show that if the solid can stand at rest with the face through
    AB on a horizontal plane, then the least possible value of b/a
    is $\frac{1}{2}(\sqrt{3}-1)$.   (L.A.)

25. A hollow vessel made of uniform material of negligible thickness
    is in the form of a right circular cone of surface density ρ mounted
    on a hemisphere of surface density σ whose radius is equal to that
    of the circular rim of the cone. If the vessel can just rest with a
    generator of the cone in contact with a smooth horizontal plane,
    prove that the semi-vertical angle α of the cone is given by the
    equation
    $$\rho(\cot^2\alpha + 3) = 3\sigma(\cos\alpha - 2\sin\alpha).$$ (L.A.)

26. Show that the centroid of a thin hollow hemispherical shell is at the mid-point of the radius of symmetry.

If this hemisphere rests in equilibrium with its curved surface in contact with a plane inclined to the horizontal at an angle α and sufficiently rough to prevent sliding, prove that α must be less than or equal to 30°. (L.A.)

27. A solid consists of a uniform right circular cone of density ρ, radius r, and height $4r$, mounted on a uniform hemisphere of density σ and radius r, so that the plane faces coincide. Show that the distance of the centre of mass of the whole solid from the common plane face is

$$\frac{r}{8}\left[\frac{16\rho-3\sigma}{2\rho+\sigma}\right].$$

If $\rho = \sigma$ and the solid is suspended freely by a string attached to a point on the rim of the common plane face, find the inclination of the axis of the cone to the vertical. (L.A.)

28. A uniform solid consists of a cylinder of radius r and height r surmounted by a hemisphere of radius r, the centre of the plane face of the hemisphere being on the axis of the cylinder. The solid rests with its plane face on a rough plane which is slowly tilted from a horizontal position until equilibrium is broken. Show that the body will slide and not topple if the coefficient of friction is less than 20/17. (L.A.)

29. Prove that if a rigid body is in equilibrium under the action of three coplanar forces, then the lines of action of these forces are either parallel or concurrent.

A hollow conical vessel, of internal height h and vertical angle 90°, is fixed with its axis vertical and vertex downwards. A smooth uniform rod rests in equilibrium with one end inside and the other end outside the vessel. If the rod is inclined at an angle θ ($<45°$) to the horizontal, show that its length is

$$\frac{4h}{\cos\theta(\cos\theta+\sin\theta)^2}.$$ (L.A.)

30. Find the position of the centre of mass of a uniform solid hemisphere of radius a.

Prove that the centre of mass of a uniform hemispherical shell, whose inner and outer radii are a and b, is at a distance

$$\frac{3}{8}\frac{(a+b)(a^2+b^2)}{a^2+ab+b^2}$$

from the centre and deduce the position of the centre of mass of a thin hemispherical shell. (L.A.)

31. A uniform solid sphere is divided into two parts by a spherical surface of the same radius, the centre of the spherical surface being on the surface of the solid sphere. Find the position of the centre of mass of the larger portion into which the solid sphere is divided.

   Show that, when this portion hangs at rest from a point on its circular rim, its axis of symmetry is inclined at an angle θ to the horizontal, where $\tan \theta = 16\sqrt{3}/33$.   (L.A.)

32. There are three pulleys in each block of a 'block and tackle'. The rope is attached to the fixed block, passes round each pulley, and the free end comes off a pulley of the fixed block. Find the effort needed to raise a load of 60 N if the weight of the movable block is negligible and the efficiency is 63 per cent.   (I.S.)

33. In a machine of velocity ratio r, the applied force P is connected with the load W by the equation $P = a+bW$, where a and b are positive constants. Prove that the efficiency increases with the load from 0 to $1/br$ as a limit. If the efficiency for an applied force P_2 is double that for an applied force P_1, prove that

$$\frac{2}{P_1} - \frac{1}{P_2} = \frac{1}{a}.$$ (H.S.C.)

34. The diagram (Fig. 7.34) indicates a system of frictionless pulleys by means of which an effort P raises a load W; pulleys A, B are movable and of mass 2 and 3 kg respectively, while C is fixed. What is the *velocity ratio* of this system? Find the *mechanical advantage* and the *efficiency* when a load of mass 50 kg is being raised. If there were n movable pulleys, each of weight w, what effort would raise a load W?   (I.S)

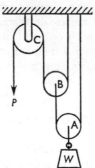

Fig. 7.34

35. When lifting a load the mechanical advantage of a machine is given by the formula $4-4w/(W+w)$, where w is a constant and W the load. If the velocity ratio of the machine is 5, find expressions in terms of w and W for the effort and the efficiency

and show that the latter approaches the value $\frac{4}{5}$ as W increases. If the machine consists of two suitable pulley blocks, draw a clear diagram to show their arrangement and the run of the connecting rope. Suggest an interpretation of w. (H.S.C.)

36. The screw of a screw-jack is rotated by an arm, the end of which describes a circle of radius 90 cm, and the load is raised 0·6 cm by one complete turn of the screw. It is found that to lift a load of mass 500 kg a force of 20 N must be applied to the end of the arm, while to lift 1000 kg a force of 36 N must be applied. Assuming the law of the machine is of the form $P = aW + b$, where a and b are constants, find the force, P, required to raise a load, W, of mass 2000 kg. Calculate the efficiency of the jack when W equals the weight of 1, 2, 3 Mg, and show that it is always less than $49/48\pi$. Sketch the graph of the efficiency against the load.

37. Two uniform rods AB and BC, each of weight W and length $2a$, are hinged at B. The joint at B is stiff and requires the application of a couple of moment M to bend it. The system rests in equilibrium in a vertical plane with the ends A and C in contact with a rough horizontal plane; the coefficient of friction at A and C is μ. Show that

$$M \geqslant Wa(\sin \alpha - 2\mu \cos \alpha),$$

where α is the angle of inclination of the rods to the vertical.

Find the condition that equilibrium should be possible for all values of α. (C.A.)

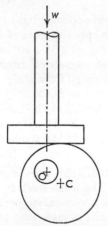

FIG. 7.35

38. A rod is moved in a smooth vertical guide by means of a cam mounted on a slowly rotating horizontal shaft (Fig. 7.35). The

cam has the form of a circle of radius $2a$ with its centre C at distance a from the centre of rotation O. The rod presses down on the cam with a vertical force W, and the coefficient of friction between the cam and the rod is μ. Ignoring the weight of the cam and the friction in the bearings of the shaft, show that, if $\mu > 1/\sqrt{3}$, the torque required to turn the shaft is always positive and varies between

$$Wa[2\mu + \sqrt{(1+\mu^2)}] \text{ and } Wa[2\mu - \sqrt{(1+\mu^2)}]. \qquad \text{(C.A.)}$$

39. The figure (Fig. 7.36) represents a framework of freely jointed light rods, AB, BC, CD, DE, CE, CF, BF, EF, attached to two

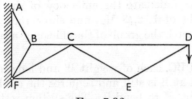

Fig. 7.36

pins fixed to a vertical wall at A and F. The angles in the triangles are either 30° or 120°. By means of a stress diagram, or by calculation, find the stresses in the rods when a weight of mass 1 Mg is hung from D, stating which are tensions and which are thrusts.

Find also the magnitude and direction of the reaction on the pin at F. (O.C.)

40. A framework is formed of five freely jointed coplanar light rods PQ, QR, RS, SP, and QS, with P and R on opposite sides of QS, the angles PQS, RQS, PSQ, and RSQ being 60°, 75°, 90°, and 30° respectively. The framework is at rest suspended from Q with loads of mass 1 kg, 1 kg, and w kg suspended from P, S, and R respectively. Show that, if QS is vertical, $w = 2\sqrt{3}$.

Draw the force diagram for this case. Show that the tension in the rod QR equals the weight of a mass of $(\sqrt{3}-1)\sqrt{6}$ kg and find the magnitude and nature of the stresses in the remaining rods. (L.A.)

41. In the framework of seven freely jointed light rods shown (Fig. 7.37) all the triangles are equilateral. The framework is in a vertical plane with AD horizontal. There is a vertical support at E and a smooth hinge at A. A load of 10 units is suspended at D and a horizontal force of 4 units acts at C in the sense indicated. Find the reaction at A and the supporting force at E.

Also, by scale drawing or otherwise, determine the stresses in the rods, distinguishing between rods in tension and under compression.   (L.A.)

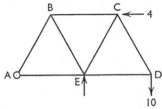

FIG. 7.37

42. The diagram (Fig. 7.38) illustrates a framework of seven light rods, freely jointed at P, Q, R, S and T. The framework is free to rotate in a vertical plane about a smooth fixed pivot at P. A horizontal wire joins Q to a fixed point N in the plane of the

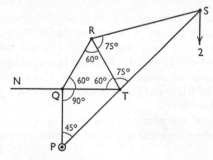

FIG. 7.38

framework. PQ is vertical. A load of 2 units is suspended at S. Find *graphically* the tension in the wire, the reaction at the pivot, and the thrusts in the rods. Distinguish between rods which are in tension and those which are under compression.   (L.A.)

43. A framework consists of five freely jointed rods, AB, BC, CD, DA, and DB. The joints A, B, and C form a triangle, right-angled at B, with the joint D as the mid-point of AC. The framework rests in a vertical plane, freely hinged at B, with equal weights W suspended from D and from C and is kept in equilibrium, with BC horizontal and A above it, by a horizontal force applied at A.

By means of a force diagram, or otherwise, determine which of the members are in compression and show that the forces in BD, CD, and AD are in the proportion $1 : 2 : 3$. Show further that the stress in the horizontal rod is two-thirds of the hori-

zontal force applied at A and that the stress in the vertical rod
is $\frac{3}{2}W$.   (L.A.)

44. Prove that the centre of mass of a uniform solid hemisphere of
    radius a is at a distance $3a/8$ from the centre of the plane base
    of the hemisphere.

    The plane base of the hemisphere coincides with the base of a
    uniform solid right circular cone, of base-radius a and height a.
    The two solids are made out of the same material and together
    form one uniform composite solid. Show that the centre of mass
    of this composite solid is at a distance $7a/6$ from the vertex of
    the cone.

    This solid, of weight W, rests with a point of the curved
    surface of the hemisphere on a horizontal plane and with its
    axis inclined at an angle θ to the horizontal, equilibrium being
    maintained by a couple acting in a vertical plane. Calculate the
    moment of this couple and indicate its sense in a diagram.

                                                                      (N.U.)

45. A heavy uniform circular hoop hangs on a small rough horizontal
    peg. The hoop is pulled by a gradually increasing horizontal force
    which acts in the plane of the hoop and is applied at the other
    end of the diameter through the peg. If the hoop has not slipped
    when this diameter makes an acute angle θ with the vertical,
    show that the ratio of the frictional force to the normal reaction
    at the peg is $\tan\theta/(2+\tan^2\theta)$.

    Find the least value of the coefficient of friction at the peg
    which will ensure that the hoop will not slip however hard it is
    pulled.   (N.U.)

46. A lamina in the form of a triangle ABC with an obtuse angle at
    A stands in a vertical plane with the side AB resting on a rough
    horizontal plank. Prove that the lamina will not tilt about A
    unless (with the usual notation for the sides of a triangle)
    $a^2 > b^2 + 3c^2$.

    If the plank is now gradually tilted in the plane of the triangle
    so that B rises above the level of A, prove that the lamina will
    tilt at A before it will slide down the plank if

    $$\mu > \frac{b^2 + 3c^2 - a^2}{4\Delta},$$

    where Δ stands for the area of the triangle and μ is the coefficient
    of friction between the lamina and the plank.   (L.A.)

47. Two equal uniform rods AB and BC, each of weight W, are
    smoothly hinged together at B. To the ends A and C are attached
    small light rings which are threaded on a rough bar fixed at 45°

to the horizontal, the coefficient of friction at A and C being μ. The rods are below the bar, with AB horizontal and BC vertical. Find the least value of μ necessary for equilibrium.

A horizontal force P, parallel to AB and acting in the sense from B to A, is now applied at B. If $\mu = 1\cdot5$, show that slipping will take place if $\dfrac{W}{10}<P<\dfrac{5W}{2}$. If the force P acts as before but in the sense from A to B, find for what values of P, if any, slipping can take place. (N.U.)

48. Two uniform spheres of equal weight and of equal radius rest in contact with each other, the lower one resting on level ground and the upper one resting against a vertical wall. The line of centres of the spheres makes an angle of 45° with the wall and with the ground. If there is limiting equilibrium at each point of contact, show that the coefficient of friction between wall and sphere is unity and find the coefficients of friction between sphere and sphere and between sphere and ground. (L.A.)

49. Two uniform circular cylinders A and B, of radii $4a$ and a respectively but of equal weight W, rest in contact on a horizontal plane with their axes horizontal. Their central circular cross-sections are in the same plane p, and the coefficient of friction is the same at all three contacts. The cylinder A is pushed against the cylinder B by a horizontal force P acting in the plane p and through the centre of gravity of A. Indicate in a diagram the directions of the frictional forces and normal reactions on each cylinder, and show that, so long as equilibrium is maintained, the frictional force is the same at each contact. Show also that in equilibrium the normal reaction between the cylinders is equal to P, and determine the normal reactions at the other two contacts.

If the coefficient of friction is greater than $\frac{1}{2}$, determine how equilibrium will be broken as the force P is gradually increased. (N.U.)

50. Show that four smooth equal spheres of radius a can stand in equilibrium in a fixed hollow sphere of radius b $(b>3a)$ in such a way that three of them touch the hollow sphere and each other and together support the fourth, provided that $b<a(2\sqrt{11}+1)$. (O.C.)

CHAPTER EIGHT

SHEARING FORCE AND BENDING MOMENT: SUSPENSION BRIDGES AND CATENARY

8.1. In the preceding chapters we have considered the equilibrium of the external forces acting on a rigid body without taking into account the stresses caused by these forces in the material of the body itself, except in the case of a light rod, when the external forces act at the ends and the stress is therefore either a thrust or a tension along the rod.

In other cases, where the external forces acting on a rod are not in the direction of the length, and may be applied at points other than the ends, the stresses are more complicated. We know, however, that if we consider any portion of the rod it is in equilibrium under the stresses exerted on its ends by the adjoining portions of the rod and any external forces (including its weight) which may act on it.

8.2. Consider any plane section ABCD of a rectangular bar (Fig. 8.1) perpendicular to its length. It is clear that at any

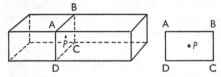

FIG. 8.1

element P of the section there will be equal and opposite forces acting on the portions of the bar on each side of the section.

If we confine our attention to the forces acting on one side of the section the force at P may be resolved into two components, one perpendicular to the section and the other in its plane. The forces perpendicular to the section on the different elements such as P can, in general, be replaced by a force T at the centre of the section, together with a couple M whose plane is perpendicular to the section.

The force T which is in the direction of the length of the bar is the tension (or thrust); it may vary from section to section.

The couple M is called the *Bending Moment* at the section. Either T or M may be zero.

There will, of course, be an equal and opposite force T, and a couple M in the opposite sense, acting on the portion of the bar on the other side of the section.

Similarly, if we consider the component forces acting in the plane of the section these can be replaced by a single force S acting at the centre of the section and a couple N tending to twist the bar about its axis. S is called the *Shearing force* at the section.

N is called the *Torsional couple* at the section.

All these stresses tend to distort the bar; T tends to stretch or compress it along its length, M tends to bend it, S to break it across, and N to twist it about its axis.

We shall suppose that the bar is rigid enough to resist these distortions, but it is important to be able to calculate the magnitude of the stresses so that we know what tendency there is for the bar to break or bend at any point.

In the present book we shall only consider cases where all the forces acting on the bar are in the vertical plane containing the length of the bar. The shearing force will therefore be in this vertical plane and perpendicular to the rod, and there will be no torsional couple.

8.3. Let ABCD (Fig. 8.2) represent a uniform bar of weight W fixed horizontally into a wall at A, and let G be its centre of gravity.

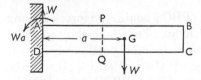

Fig. 8.2

It is clear that the weight W is supported by the wall, and there must therefore be an upward force W acting at A.

This forms a couple with the weight W, and to balance it there must be an equal and opposite couple applied across the section AD where the bar meets the wall. The moment of this couple is Wa, where a is the distance of G from the wall.

If PQ is any vertical section of the bar, Fig. 8.3 shows all the forces acting on the two portions PBCQ and PQDA.

The portion PBCQ is acted on by its weight w at the mid-point G', and is supported by the part of the bar to the left of PQ. This must therefore exert an upward force along QP, equal to w, and a couple.

The upward force is the *shearing force* at the section PQ.

If there are other vertical forces applied to the right of PQ the shearing force must also support these, and its magnitude is

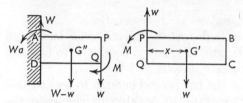

FIG. 8.3

clearly equal to the sum of all the vertical forces to the right of the section PQ, measured downwards.

By considering the left-hand portion of the bar we see that the shearing force at PQ is also equal to the sum of all the vertical forces to the left of PQ, measured upwards. (These include the weight $(W-w)$ of the left-hand portion and the vertical force W exerted by the wall.) *We can therefore find the magnitude of the shearing force at any section by finding the vector sum of all the vertical forces acting on either side of that section.* The direction of the shearing force is opposite to that of the resultant of these forces.

Now, considering the portion PBCQ again, the forces w along QP and at G' form a couple which must be balanced by an equal and opposite couple M due to the left-hand portion of the bar. This is the *bending moment* couple at the section PQ. If x be the distance of the centre of gravity, G', of PBCQ from PQ, the moment of this couple is wx.

Hence the bending moment at PQ is wx.

If there are other forces acting on this portion of the bar in addition to its weight and in the same plane with it, then the moment of the couple due to the left-hand portion APQD of the bar must, for equilibrium, be equal and opposite to the sum of the moments of all these forces about the axis through the centre of PQ perpendicular to the plane of the forces. This follows by taking moments about that axis.

We shall disregard the thickness of the bar and can then say that the bending moment at any point of the bar is equal to the sum of the moments about that point of all the forces acting on the bar on either side of the point.

The bending moment at any point is therefore obtained by finding the sum of the moments of all the forces acting on the bar on either side of that point. These forces include the weight of that portion of the bar, any attached weights or external forces, and any reactions due to contact with supports, etc.

In short, the shearing force and bending moment at any point of a bar or rod can be found by writing down the conditions of equilibrium of the portion of the rod to the right (or the portion to the left) of the point. In the following examples we shall consider the portion to the right, and use the convention that the shearing force is positive if vertically upwards and the bending moment positive if in a counterclockwise direction.

8.4 EXAMPLE (i)

A horizontal rod AB, *of length* 2 m *and of negligible weight, is built into a wall at* A. *What are the magnitudes of the shearing force and bending couple at x m from* A *when a weight of mass* 6 kg *is suspended from* B?

Let P (Fig. 8.4) be the point such that AP = x m.

There must be an upward vertical force of $6g$ N at A to balance

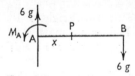

FIG. 8.4

the weight, and the wall must also exert a couple M_A of moment $2 \times 6g$ Nm to balance the couple formed by these two forces acting at A and B. If we take moments of the forces to the left of P we must include this couple.

The sum of the moments about P of all forces acting to the right of P is therefore $6g(2-x)$ in a clockwise direction.

Therefore $M = (12-6x)g$ Nm.

The same result (with opposite sign) is obtained by taking moments about P of the forces to the left of P. We get

$$M = M_A - 6gx = (12-6x)g \text{ Nm.}$$

The shearing force is obviously equal to $6g$ N at every point of the rod.

EXAMPLE (ii)

A light rod AB *of length* 6 m *is supported at* A *and* B *and carries a weight* W *at a point* C, *where* AC = 2 m.

Find the shearing force and bending moment at any point of the rod.

Let P (Fig. 8.5) be any point in the rod such that AP $= x$ m.

The shearing force at P is equal to the sum of the forces acting on PB, so that we shall have to find the reaction at B.

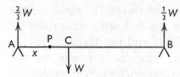

FIG. 8.5

Let R be the reaction at B: then taking moments about A,

$$6R = 2W$$

∴
$$R = \tfrac{1}{3}W.$$

The reaction at A is therefore $\tfrac{2}{3}W$.

First consider P between A and C, that is, $x<2$.

If we sum the forces on PB we get $\tfrac{2}{3}W$ downwards, i.e. the shearing force at P is $\tfrac{2}{3}W$ upwards.

Now as P moves along towards B, and we sum to the right of P, the sum is $\tfrac{2}{3}W$ downwards until we reach C. (It does not matter on which side we add the forces, but *we must keep to the same side throughout the length of the rod.*)

Directly P gets to the right of C the sum is $\tfrac{1}{3}W$ *upwards* and the shear is now $\tfrac{1}{3}W$ downwards, i.e. there is a sudden change in the magnitude and direction of the shearing force at C. This is due to the weight W hanging at C. A similar discontinuity in the value of the shear always occurs at a point where a weight is applied, the difference between the values on the two sides being equal to this weight.

Consider now the bending moment at P.

Taking the sum of the moments of the forces to the right of P, when P is between A and C, we get
$$W(2-x)-\tfrac{1}{3}W(6-x) = -\tfrac{2}{3}Wx$$
in a clockwise direction.

The bending moment M is therefore $-\tfrac{2}{3}Wx$ in a counterclockwise direction, when $x<2$.

When P is to the right of C, that is, $x>2$, the sum of the moments of the forces on PB about P is $\tfrac{1}{3}W(6-x)$ in a counterclockwise direction. The bending moment M is therefore $-\tfrac{1}{3}W(6-x)$ in a counterclockwise direction, when x lies between 2 and 6.

When $x = 2$, these two expressions for M give the same value $-\tfrac{4}{3}W$, and this is the value of the bending moment at C.

There is no discontinuity in M at C as there is in the case of the shearing force.

The variation of shearing force and bending moment at different points is very clearly shown graphically.

Draw a horizontal line ACB (Fig. 8.6) to represent the rod and

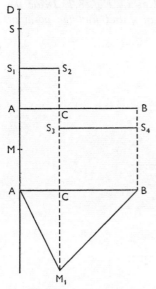

Fig. 8.6

draw AD perpendicular to it.

Take AB, AD as axes and plot distances along AB, and values of the shearing force and bending moment at these distances from A, parallel to AD and to any convenient scale.

We have seen that the shearing force to the left of C is constant and equal to $\frac{2}{3}W$ upwards. This will give a horizontal line S_1S_2 above AB, S_1 and S_2 being the points where this line cuts AD and the perpendicular to AB at C.

To the right of C the shear is $\frac{1}{3}W$ downwards. This gives a horizontal line S_3S_4 below AB, cutting the verticals at C and B in S_3 and S_4. Also $CS_3 = \frac{1}{2}CS_2$.

The bending moment to the left of C is $M = -\frac{2}{3}Wx$.

This is represented by a straight line AM_1 passing through A such that CM_1 represents $-\frac{4}{3}W$ to scale.

To the right of C $\qquad M = -W(2-\frac{1}{3}x)$.

This is represented by the line M_1B.

The curve of shearing force therefore consists of two separate straight lines S_1S_2 and S_3S_4 parallel to AB.

The bending moment curve consists of the two straight lines AM_1 and M_1B meeting at M_1.

EXAMPLE (iii)

A light beam AB *is supported and loaded as in Fig. 8.7. Draw the shearing force and bending moment diagrams, and find the point at which the bending moment is a maximum.*

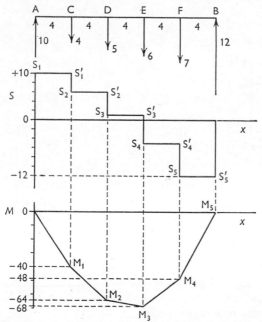

FIG. 8.7

We must first calculate the reaction at A and B by taking moments for the whole beam. They are found to be 10 units at A, and 12 units at B.

The figure should be drawn to scale and verticals drawn at the points where the weights are attached and at A and B.

From A to C the shearing force is 10 units upwards (taking forces on the right), and this is represented in the diagram by a horizontal line S_1S_1' at a distance above AB representing 10 to a convenient scale. From C to D the shear is 6 units upwards and is represented by S_2S_2'. From D to E it is 1 unit upwards, and is represented by S_3S_3'. From E to F it is 5 units downwards, and is represented by S_4S_4'. From F to B it is 12 units downwards, and is represented by S_5S_5'. The shearing force curve is therefore composed of five distinct lines.

If x is the distance of a point P from A the bending moments between A and C, C and D, . . . are as follows:

$$0 < x < 4, \ M = 4(4-x)+5(8-x)+6(12-x)+7(16-x)-12(20-x)$$
$$= -10x$$
$$4 < x < 8, \ M = 5(8-x)+6(12-x)+7(16-x)-12(20-x)$$
$$= -16-6x$$
$$8 < x < 12, \ M = 6(12-x)+7(16-x)-12(20-x)$$
$$= -56-x.$$
$$12 < x < 16, \ M = 7(16-x)-12(20-x)$$
$$= -128+5x$$
$$16 < x < 20, \ M = -12(20-x)$$

The bending moment curve is therefore the series of straight lines OM_1, M_1M_2, M_2M_3, M_3M_4, M_4M_5.

The point at which the bending moment is a maximum is easily seen from the diagram to be at E where its value is 68 units.

In cases where the beam is light and loaded only at certain points the shearing force and bending moment curves are always of the type illustrated in the preceding example.

The shearing force curve is a series of horizontal straight lines, not continuous, while the bending moment curve is a series of inclined lines which join one on to another at the points where the loads are placed.

Cases where the beam is heavy or the load is distributed along the whole or part of the length of the beam will be considered in § 8.6.

EXAMPLE (iv)

A rod of length l, whose weight is negligible, rests horizontally with its ends supported and carries a movable weight w; the rod will break if the

*bending moment at any point is L. Prove that the least value of w which
can break the rod is 4L/l.*   (I.A.)

Let AB (Fig. 8.8) represent the rod, C the position of w, where

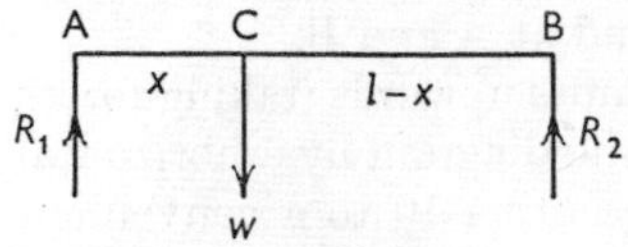

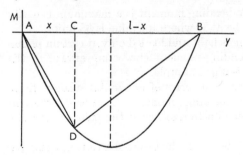

FIG. 8.8

$AC = x$. If R_2 is the supporting force at B we have, taking moments
about A,

$$lR_2 = wx$$

$$\therefore \qquad R_2 = w\frac{x}{l}.$$

The reaction R_1 at A is therefore

$$w(l-x)/l.$$

The bending moment M at a point P, where $AP = y$, may be
obtained by considering the equilibrium of the portion PB of the rod.
We get:

when $\qquad y<x, \; M = w(x-y)-\dfrac{wx}{l}(l-y)$

$$= -wy+\frac{wxy}{l} = -wy(l-x)/l,$$

and when $\qquad y>x, \; M = -wx(l-y)/l.$

The variation of these expressions with y is shown in Fig. 8.9,
by the straight lines AD and DB.

FIG. 8.9

The expressions have the same numerical value when $y = x$, viz.
$wx\left(\dfrac{l-x}{l}\right)$, and this is the maximum moment and occurs at C.

We have now to find the position of C for which this is a maximum, as at this point the value of w required to produce a given moment will be a minimum.

Differentiating and equating to zero, we get

$$w\left(l - \frac{2x}{l}\right) = 0$$

$$\therefore \qquad x = \tfrac{1}{2}l.$$

Hence the bending moment is numerically a maximum when C is at the mid-point of the rod.

When $\qquad x = \dfrac{l}{2}, \ M = w\dfrac{1}{2}\left(l - \dfrac{l}{2}\right) = \tfrac{1}{4}wl$

$$\therefore \qquad \tfrac{1}{4}wl = L$$

$$\therefore \qquad w = 4L/l.$$

8.5. Graphical construction for bending moment

In the case of a beam carrying concentrated loads the funicular polygon for the loads may be used as a bending moment diagram.

Let QQ′ (Fig. 8.10) represent a beam, supported at Q and Q′

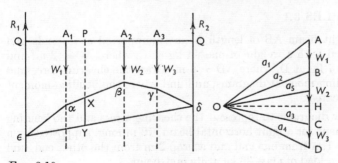

FIG. 8.10

and loaded with weights W_1, W_2, W_3 at A_1, A_2, A_3.

Draw the force polygon as in the right-hand figure, and the corresponding funicular of which $\epsilon\delta$ is the closing side.

Let P be any point in QQ′ between A_1 and A_2, and draw PXY vertically to meet the funicular in X and Y.

We shall show that the bending moment at P is proportional to XY.

From the force diagram we see that

W_1 is equivalent to a_1 along $a\epsilon$ and a_2 along αX.

R_1 is equivalent to a_1 along $\epsilon\alpha$ and a_5 along Yϵ.

Hence W_1 and R_1 are equivalent to a_2 along αX and a_5 along Yϵ.

Hence the sum of the moments of W_1 and R_1 about P is equal to the sum of the moments about P of these forces a_2 and a_5.

Now a_2 along αX is equivalent to a vertical component along XP which has no moment about P, and a horizontal component h (where h is the length of OH in the force diagram), whose moment about P is h . PX. Similarly, the moment of a_5 along Yϵ is equal to h . PY.

Hence the total moment about P of the forces to the left of P is

$$h\text{PY} - h\text{PX} = h\text{XY},$$

i.e. the bending moment at P is proportional to XY, and is equal to this length multiplied by the force represented by the distance of the pole O from the force polygon.

Note. The reader is recommended to apply this method to some of the following Examples 8.1, and so check his calculation.

EXAMPLES 8.1

1. A light beam AB of length 10 m is supported at its ends and loaded with a weight of mass 4 kg at C, where AC = 2 m, and with 7 kg at D, where AD = 6 m. Draw the shearing force and bending moment diagrams, and find where the bending moment is greatest.

2. Draw diagrams to represent the shearing stress and the bending moment for a light horizontal beam, 12 m long, supported by a prop 1 m from one end and a prop 2 m from the other end, and with a load of mass 36 kg at its mid-point.

3. Draw the graphs of bending moment and shearing force for a light horizontal cantilever 10 m long carrying concentrated loads each of $\frac{1}{2}$ unit at its centre and at its free end.   (C.S.)

4. A horizontal bar of length 10 m and of negligible weight rests on two supports at its ends. A load of 1 unit hangs from a point distant x m from one end. Find expressions for : (i) the reactions on the supports; (ii) the shearing force at a distance 2 m from one end; (iii) the bending moment at this point. Consider both the case of $x < 2$ and that of $x > 2$. Draw a graph showing the

variation of bending moment at this point when x increases
from 0 to 10 and find the position of the load for which the
bending moment is greatest.   (I.E.)

5. A horizontal bar AB of length 10 m and of negligible mass
   rests on two supports at its ends. A load of 1 unit hangs at C,
   3 m from A, and one of 2 units at D, 7 m from A. Find the shearing stress and bending moment at P, x m from A, and draw a
   graph showing the variation in this values of the shearing stress
   and of the bending moment as x increases from 0 to 10. Determine also the maximum bending moment.   (I.E.)

6. A beam is supported at the ends and loaded as in Fig. 8.11.

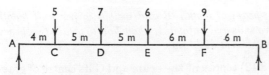

FIG. 8.11

   Sketch the shearing-force and bending-moment diagrams. Calculate the reactions at A, B and the shearing force and bending
   moment at the mid-point of EF.   (I.C.)

7. A horizontal beam of negligible mass rests upon two supports at
   the same level and distant apart 20 m. Vertical loads of 5, 2, 3,
   and 7 units rest on the beam at points 3, 7, 9, and 15 m from the
   left-hand end. Sketch the shearing-force and bending-moment
   diagrams. Also calculate the shearing force and bending moment
   at the point 12 m from the left-hand end.   (I.C.)

8. A beam is freely supported at its ends at the same level. Vertical
   loads of 5, 4, and 8 units act at points distant 3, 6, and 10 m
   respectively from the left-hand end. The length of the beam is
   12 m. Sketch the shearing-force and bending-moment diagrams.
   Also calculate the shearing force and bending moment at a point
   7 m from the left-hand end.   (I.C.)

9. A, B, C, D, E are points taken in order along a horizontal beam,
   and AB = 6 m, BC = 4 m, CD = 4 m, DE = 5 m. The beam
   is freely supported at A and D, and carries vertical loads of 3, 7,
   and 5 units at B, C, E respectively. Show, with proof, how to
   draw the shearing-force and bending-moment diagrams for this
   beam. Find the distance from D of the point where the bending
   moment is zero.   (I.C.)

10. A light girder is supported horizontally at its two ends and carries
    loads at various points of its length. Show how the reactions

at the ends and the bending moment at any point can be obtained by means of a link (or funicular) polygon. The girder is 16 m long and has loads 9, 11, and 8 units at points distant 5 m, 9 m, and 12 m from an end. Find the reactions at the ends, and show that the greatest bending moment is 81 units.

8.6. Heavy beams. Distributed loads

The following examples illustrate cases where the beam is heavy, or the load is distributed continuously along the whole or part of the beam.

EXAMPLE (i)

A heavy uniform beam of length $2l$ and weight w per unit length is supported at its two ends and unloaded. Show how to obtain the shearing-force and bending-moment diagrams.

Let AB (Fig. 8.12) represent the beam and G its centre of gravity. The total weight of the beam is $2lw$ acting at G.

The supporting force at each end is obviously lw.

Let P be any point in the beam, and $AP = x$.

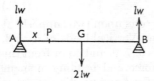 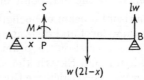

FIG. 8.12   FIG. 8.13

All the forces acting on the portion PB of the beam are shown in Fig. 8.13, $w(2l-x)$ being the weight of PB acting at the mid-point of PB.

The sum of the moments of the forces to the right of P is therefore $\frac{1}{2}w(2l-x)^2 - lw(2l-x) = -\frac{1}{2}w(2l-x)x$ clockwise.

The bending moment M measured counterclockwise is therefore of magnitude

$$M = -\tfrac{1}{2}w(2l-x)x \tag{i}$$

This result holds for all points from A to B, as there is no load suspended at any point.

The shearing force, S, at P measured upwards is given by

$$S = w(2l-x) - lw = w(l-x) \tag{ii}$$

The shearing-force curve is a straight line and the bending-moment curve is a parabola, as shown in Fig. 8.14.

At A the value of M is zero, and it is also zero at B, where $x = 2l$.

At the mid-point where, $x = l$, the value is $\frac{1}{2}wl^2$, and the curve is symmetrical about the perpendicular to AB at the mid-point, the bending moment being a maximum there.

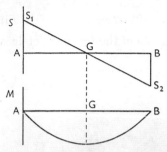

FIG. 8.14

The point of maximum bending moment can be calculated by equating dM/dx to zero.

Now
$$\frac{dM}{dx} = -w(l-x),$$

and this is zero when $x = l$.

It should be noticed that $dM/dx = -S$.

This will be found to be so in other cases; the slope of the bending-moment curve at any point is equal to the shearing force at that point but of opposite sign.

EXAMPLE (ii)

A uniform beam 10 m long, and of mass 2 kg m⁻¹ run, is supported in a horizontal position at its two ends, and carries a load of mass 30 kg at a point distant 4 m from the end. Draw the curves of shearing force and bending moment for any point of the beam and determine where the bending moment is greatest. (I.E.)

Let AB (Fig. 8.15) represent the beam, and C the point where the 30g load is attached.

Draw the perpendiculars to the beam at A, C, and B.

We must calculate first the reactions at A and B.

If R_1 N is the reaction at B, taking moments about A, we get

$$10R_1 = 30g \times 4 + 20g \times 5 = 220g$$

$$\therefore \qquad R_1 = 22g$$

$$\therefore \qquad \text{Reaction at A} = 28g \text{ N}.$$

Let P be any point on the beam, $AP = x$.

When $0 < x < 4$, the shearing force S is given by

$$S = 2g(10-x) - 22g + 30g = (28-2x)g \tag{i}$$

but when $4 < x < 10$, we get

$$S = 2g(10-x) - 22g = -(2+2x)g. \tag{ii}$$

Hence the shearing-force curve consists of the two parallel lines

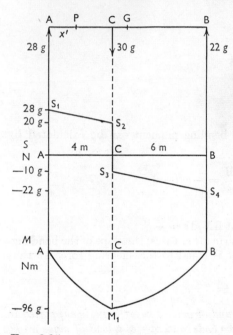

Fig. 8.15

$S_1 S_2$ and $S_3 S_4$ shown in Fig. 8.15. There is a discontinuity of amount $30g$ when $x = 4$.

When $0 < x < 4$, the bending moment, M, is given by

$$M = 30g(4-x) + (20-2x)g(5-\tfrac{1}{2}x) - 22g(10-x)$$
$$= (-28x + x^2)g \tag{iii}$$

and when $4 < x < 10$,

$$M = (20-2x)g(5-\tfrac{1}{2}x) - 22g(10-x)$$
$$= (-120 + 2x + x^2)g. \tag{iv}$$

From (iii) $dM/dx = (-28+2x)g$ and for a maximum $x = 14$, an impossible value, as this expression holds only up to $x = 4$. At this point $M = 96g$ Nm.

Notice that dM/dx is the same as the value of $-S$ given by (i).

From (iv) $dM/dx = (2+2x)g$ and for a maximum or minimum $x = -1$, another impossible value. There is no real maximum or minimum point on either portion of the curve, as neither curve is complete up to its vertex, the greatest numerical value being at C, where $M = -96g$; this value is given for $x = 4$ by either (iii) or (iv).

Notice again that dM/dx is the same as the value of $-S$ given by (ii).

The bending-moment curve consists of the curve AM_1 and the curve M_1B, shown in Fig. 8.15. They are parts of different parabolas.

EXAMPLE (iii)

A train of weight W and length l is in the centre of a bridge of twice its own length. Assuming that the weight of the train is uniformly distributed throughout its length, calculate the bending moment at the centre of the bridge, and compare it with the value when one end of the train just reaches one of the piers of the bridge. (I.S.)

Let AB (Fig. 8.16) represent the bridge, CD the train.
The reactions at A and B are each $\frac{1}{2}W$ from symmetry.

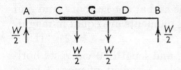

FIG. 8.16

The weight of the train to the right of G is $\frac{1}{2}W$, and acts at the mid-point of GD. The bending moment M at G is therefore given by

$$M = -\frac{W}{2}l + \frac{W}{2} \times \frac{l}{4} = -\frac{3}{8}Wl.$$

If the train is in position AG the reactions at A and B are altered; that at A is $\frac{3}{4}W$ and that at B is $\frac{1}{4}W$.

The whole weight of the train now acts at C, and the bending moment, M, at G is given by

$$M = -\frac{1}{4}Wl = -\frac{1}{4}Wl.$$

The ratio of the two bending moments is therefore $3 : 2$.

EXAMPLES 8.2

1. A uniform rigid rod LM, 2 m long and of mass 10 kg, rests in a horizontal position supported at the ends L and M. A weight of mass 12 kg is suspended from the point C of the rod at a distance of 0·8 m from L. Calculate the shearing force and bending couple at a point 0·7 m from M.   (I.E.)

2. A uniform beam AC 3·2 m long, whose mid-point is B, rests on supports at the same level at its ends, and carries a load of 4 units uniformly distributed along the half AB. Assuming that the weight of the beam itself is negligible, draw a bending-moment diagram, and indicate clearly in the diagram the magnitude of the maximum bending moment and the point at which it acts.   (I.E.)

3. A light 12-m beam AB is supported at its ends, and carries weights of 2 units at C and 4 units at D, where AC = 4 m, DB = 2 m. Draw the shearing-stress and bending-moment diagrams. Draw the corresponding diagrams when the load is not concentrated at C and D but is uniformly distributed over the beam.   (I.E.)

4. A uniform beam of weight W and length l rests on two supports, each at a distance a from the middle section of the beam. Find the bending moments at the middle and at one support, and find a in terms of l if these two bending moments are equal in magnitude but opposite in sign.   (I.E.)

5. A uniform steel bar of mass 16 kg and length 80 cm rests horizontally on two supports at its ends. A weight of mass 4 kg is suspended from the bar at a point 20 cm from the end. Find by calculation at what point of the bar the shearing force is zero and evaluate the bending moment at this point. Draw a diagram showing the distribution of shearing force along the bar.   (I.E.)

6. A uniform beam AB, of weight $3W$, loaded with equal weights W at A and a point of trisection D, rests in a horizontal position on two supports at B and a point of trisection, C, where AC = CD = DB. Sketch diagrams showing the distribution of shearing stress and bending moment along the beam.   (C.S.)

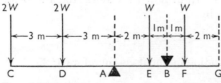

FIG. 8.17

7. Draw the graphs of shearing stress and bending moment for the beam of weight W per metre loaded as shown in Fig. 8.17 and supported at A and B.   (C.S.)

8. A uniform plank 16 m long is supported horizontally at two points distant 4 m from the ends. Draw two sets of diagrams to represent the shearing force and bending moment in the plank when a heavy particle, whose weight is one-half that of the plank, is placed on it: (a) at one end; (b) at the middle.   (C.S.)

9. A uniform beam AB, of weight W, rests horizontally on two supports C, D, as shown in Fig. 8.18. Weights of $3W$, $2W$ are

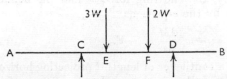

FIG. 8.18

placed at E, F respectively, where AC = 3 m, EF = DB = 2 m, CE = FD = 1 m. Find the bending moment at H the point mid-way between C and D. If the point of support D is raised 0·5 m above the level of C, without the beam or the weights slipping, determine whether the bending moment at H is increased or decreased.   (C.S.)

10. Two equal uniform beams AB, BC of length a, of the same weight per unit of length w, are smoothly hinged at B and supported in a horizontal line by props at A and D, where BD = $\frac{1}{2}$DC. Find expressions for the shearing force and bending moment at any point of each beam, and draw graphs to represent the variations in their values.

11. A beam 18 m long is supported at points 12 m apart, one support being 3 m from the left-hand end of the beam. There is a load of 3 units acting at each end of the beam, and a uniformly distributed load of 1 unit per metre run on the span of 12 m. Determine: (i) the reactions at the supports; (ii) the shearing-force diagram; (iii) the bending moment diagram for the beam.

12. A uniform beam ABC, whose weight per unit length is w, rests on two supports at A and B, and carries a weight equal to one-fourth of its own weight at C. If ABC is horizontal and AB = 10 m, BC = 6 m, draw carefully the graphs of shearing force and bending moment.   (C.W.B.)

13. A uniform beam, 12 m long and of mass 108 kg, is supported at its ends and is horizontal. A weight of mass 36 kg is placed at a

distance of 3 m from one end. Draw the shearing-stress and bending-moment diagrams for the beam. Prove that the shearing stress is zero at a point 2 m from the load and 1 m from the mid-point of the beam. (I.E.)

14. A beam AB 12 m long can turn freely about a pin at A and is supported horizontally by means of a strut at a point C, 8 m from A. The beam is uniform and has a mass of 36 kg. Draw diagrams showing the shearing force and bending moment at any point of the beam, and find the maximum bending moment. (I.E.)

15. Prove that, when a beam whose weight per unit length is w is supported horizontally, the shearing force F and the bending moment M are given by the equations

$$\frac{\mathrm{d}F}{\mathrm{d}x} = -w, \frac{\mathrm{d}M}{\mathrm{d}x} = -F.$$

The sectional area of a cantilever of length l projecting horizontally from a wall varies as $l^2 - x^2$, where x denotes distance from the wall. Prove that the bending moment is given by the equation

$$M = \frac{w_0}{12l^2}(l-x)^3 (3l+x),$$

where w_0 is the weight per unit length at the wall. (C.W.B.)

16. A simply supported beam has a span of 12 m and supports a uniform dead load of mass 1.5 Mg m^{-1}. A uniform live load of mass 3 Mg m^{-1} and of length greater than 12 m moves across the beam. Find the maximum and minimum values of the shearing force due to the combined loads at: (i) the ends of the beam, and (ii) the mid-point of the beam.

17. Prove that, if w is the load per unit length, S the shearing force, and M the bending moment at a distance x along a horizontal beam

$$w = -\frac{\mathrm{d}S}{\mathrm{d}x} = \frac{\mathrm{d}^2M}{\mathrm{d}x^2}.$$

A train of length l and weight W, which may be considered as uniformly distributed over its length, crosses a bridge of span $2l$. Find the numerically greatest shearing force and the numerically greatest bending moment due to the train when it is on the bridge with one end at one of the supports.

Sketch the shearing-force and bending-moment diagrams for this position of the train. (O.C.)

18. A beam AD, 24 m long, is supported at B and C, where AB = 6m, BC = 15 m, and CD = 3 m, AB and CD being overhanging ends. It carries loads of 1 unit at A, 2 units at D, and 2 units at E,

the mid-point of BC, and in addition a uniformly distributed load of 1 unit per metre from B to C.

Draw the shearing-force and bending-moment diagrams.

19. The shearing-force diagram shown in Fig. 8.19 is for a beam which rests on two supports, one being at the left-hand end.

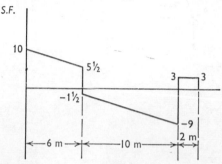

FIG. 8.19

Deduce directly from the shearing-force diagram: (i) the bending moment at 2 m intervals along the beam; (ii) the loading on the beam.

Draw the bending-moment diagram.

20. A portion PQ of a horizontal beam, whose weight per unit length is w, is of length b; it carries an isolated load W at the point of trisection of PQ nearer to Q. Write down equations connecting the values of the shearing forces and bending moments at P and Q.

A rod AB of length h has variable density, the density at distance x from A being $2kx$, where k is a constant. The rod rests horizontally on rigid props at A and B. Prove that the magnitude of the bending moment at the point distance x from A is $\frac{1}{3}kx(h^2-x^2)$.

Find the value of the greatest bending moment. (C.W.B.)

8.7. Particles connected by a string

If the ends of a light string are attached to two fixed points, and weights are attached to the string at different points, the figure formed by the string is called a funicular or rope polygon. This is the origin of the name given to the corresponding polygon in graphical statics (see 4.2).

Let A, B (Fig. 8.20) be the points to which the ends of the string are attached, and let $A_1, A_2, \ldots A_n$ be the points at which are attached weights $W_1, W_2, \ldots W_n$ respectively.

Let the lengths of the portions of the string AA_1, A_1A_2, ...
A_nB be a_1, a_2, ... a_{n+1} respectively, let their inclinations to the
horizontal be θ_1, θ_2, ... θ_{n+1}, and let T_1, T_2, ... T_{n+1} respec-

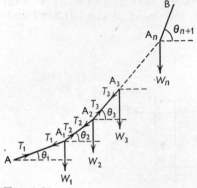

FIG. 8.20

tively be the tensions in them. Resolving horizontally for the
different weights in succession, we have,

$$T_1 \cos \theta_1 = T_2 \cos \theta_2$$
$$T_2 \cos \theta_2 = T_3 \cos \theta_3, \text{ etc.,}$$
$$\therefore T_1 \cos \theta_1 = T_2 \cos \theta_2 = \ldots = T_{n+1} \cos \theta_{n+1} = H \text{ (say)} \quad \text{(i)}$$

so that the horizontal component of the tension is the same
throughout.

Resolving vertically for the different weights in succession,
we have

$$T_2 \sin \theta_2 - T_1 \sin \theta_1 = W_1$$
$$T_3 \sin \theta_3 - T_2 \sin \theta_2 = W_2$$
$$\cdot \quad \cdot \quad \cdot \quad \cdot \quad \cdot \quad \cdot \quad \cdot \quad \cdot$$
$$T_{n+1} \sin \theta_{n+1} - T_n \sin \theta_n = W_n \quad \text{(ii)}$$

Substituting from equations (i) for T_1, T_2, etc., these equations give

$$\tan \theta_2 - \tan \theta_1 = \frac{W_1}{H}$$
$$\tan \theta_3 - \tan \theta_2 = \frac{W_2}{H}$$
$$\cdot \quad \cdot \quad \cdot \quad \cdot \quad \cdot \quad \cdot \quad \cdot$$
$$\tan \theta_{n+1} - \tan \theta_n = \frac{W_n}{H}.$$

Hence if the weights be all equal the tangents of the inclinations of the various portions of the string are in arithmetical progression. Now if h and k be the horizontal and vertical distances respectively between A and B

$$a_1 \cos \theta_1 + a_2 \cos \theta_2 + \ldots a_{n+1} \cos \theta_{n+1} = h \qquad \text{(iii)}$$

and $\quad a_1 \sin \theta_1 + a_2 \sin \theta_2 + \ldots a_{n+1} \sin \theta_{n+1} = k \qquad \text{(iv)}$

Equations (i), (ii), (iii), (iv), which amount to $2n+2$ equations, enable us to find the $(n+1)$ unknown tensions, and the $(n+1)$ unknown inclinations.

In many cases the horizontal distances h_1, h_2, . . . between A and A_1, A_1 and A_2, etc., are given, and then we have,

$$h_1 \tan \theta_1 + h_2 \tan \theta_2 \ldots + h_{n+1} \tan \theta_{n+1} = k.$$

The string may be replaced by a chain consisting of light links AA_1, A_1A_2, etc., freely jointed at A_1, A_2, etc., the weights being attached to the joints as in Example (iii), paragraph 8.9.

8.8. Graphical construction

If the inclinations of the different portions of the string are given we can easily obtain the ratios of W_1, W_2, . . . W_n graphically. Using the figure of the last paragraph, take any point O and draw OC horizontally, and a vertical CD at C (Fig. 8.21).

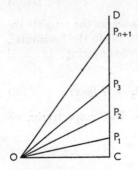

FIG. 8.21

Draw OP_1, OP_2, . . . OP_{n+1} parallel to the portions AA_1, A_1A_2, . . . A_nB to cut the vertical in P_1, P_2, . . ., so that the angles P_1OC, P_2OC, etc., are respectively θ_1, θ_2, etc.

Now $\qquad \dfrac{W_1}{H} = \tan \theta_2 - \tan \theta_1 = \dfrac{CP_2}{OC} - \dfrac{CP_1}{OC} = \dfrac{P_1P_2}{OC},$

$$\dfrac{W_2}{H} = \tan \theta_3 - \tan \theta_2 = \dfrac{CP_3}{OC} - \dfrac{CP_2}{OC} = \dfrac{P_2P_3}{OC},$$

and so on.

Hence H, W_1, W_2, ... W_n are respectively proportional to OC, P_1P_2, P_2P_3, ... P_nP_{n+1}, and their ratios are therefore determined.

If the weights are given, and the directions of any two portions of the string be also known, we can determine the directions of the others graphically. We proceed as follows.

Draw the vertical line CD, and on it mark off P_1P_2, P_2P_3, ... to represent W_1, W_2, ... to scale. If the directions of A_1A_2 and A_3A_4 are given, we draw P_2O and P_4O parallel to A_1A_2 and A_3A_4 respectively, to meet in O. The directions of the remaining portions of the string are obtained by joining O to P_1, P_3, etc.

8.9 EXAMPLE (i)

Five equal weights are attached to a light string which hangs from two points P and Q in the same horizontal. In equilibrium the horizontal projection of the six intervals of the string are all equal to a, and the depth below PQ of the lowest weight is $3a$. Show that the inclinations to the horizontal of the parts of the string are $\tan^{-1}\left(\frac{1}{3}\right)$, $\pi/4$, and $\tan^{-1}\left(\frac{5}{3}\right)$. Prove that the external angles of the polygon are $\tan^{-1}\left(\frac{1}{4}\right)$, $\tan^{-1}\left(\frac{1}{2}\right)$, and $\tan^{-1}\left(\frac{3}{4}\right)$.   (H.C.)

Let A, B, C, D, E (Fig. 8.22) be the points where the weights are attached, θ_1, θ_2, θ_3 the inclinations of PA, AB, BC to the horizontal, and T_1, T_2, T_3 the tensions in these portions of the string.

Resolving horizontally we have

$$T_1 \cos \theta_1 = T_2 \cos \theta_2 = T_3 \cos \theta_3 = H \text{ (say)} \qquad \text{(i)}$$

Resolving vertically for the weights at A, B, C respectively, we have

$$T_1 \sin \theta_1 - T_2 \sin \theta_2 = w$$
$$T_2 \sin \theta_2 - T_3 \sin \theta_3 = w$$
$$2T_3 \sin \theta_3 = w \qquad \text{(ii)}$$

Since the depth of C below PQ is $a(\tan \theta_1 + \tan \theta_2 + \tan \theta_3)$, we have

$$\tan \theta_1 + \tan \theta_2 + \tan \theta_3 = 3 \qquad \text{(iii)}$$

Substituting in equations (ii) for T_1, T_2, T_3, from (i) we have

$$\tan \theta_1 - \tan \theta_2 = w/H$$
$$\tan \theta_2 - \tan \theta_3 = w/H$$
$$2 \tan \theta_3 = w/H$$

∴ $\qquad\qquad \tan \theta_2 = 3 \tan \theta_3$ and $\tan \theta_1 = 5 \tan \theta_3$

Hence, from (iii), $\quad 9 \tan \theta_3 = 3$ or $\tan \theta_3 = \frac{1}{3}$

∴ $\qquad\qquad \tan \theta_2 = 1$ and $\tan \theta_1 = \frac{5}{3}$.

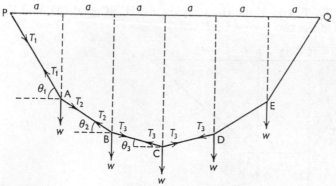

FIG. 8.22

The external angle of the polygon at C is $2\theta_3$ and

$$\tan 2\theta_3 = \frac{2 \tan \theta_3}{1 - \tan^2 \theta_3} = \frac{\frac{2}{3}}{1 - \frac{1}{9}} = \frac{3}{4}.$$

The external angle at B is $\theta_2 - \theta_3$ and

$$\tan (\theta_2 - \theta_3) = \frac{\tan \theta_2 - \tan \theta_3}{1 + \tan \theta_2 \tan \theta_3} = \frac{1 - \frac{1}{3}}{1 + \frac{1}{3}} = \frac{1}{2}.$$

Similarly the external angle at A is $\theta_1 - \theta_2$ and its tangent is $\frac{1}{4}$.

EXAMPLE (ii)

A *and* D *are two points in a horizontal line* 48 *cm apart; the ends of a light cord* 66 *cm long are attached to* A *and* D; B *is a point in the cord* 25 *cm from* A *and* C *a point in the cord* 29 *cm from* D. *A mass is suspended from* B. *Find the mass which must be suspended from* C, *so that the portion* BC *of the cord shall be horizontal.* (I.A.)

Draw the diagram as in Fig. 8.23, and let the verticals through B and C cut AD in E and F.

Let $\angle ABE = \theta_1$, $\angle FCD = \theta_2$, $AE = x$ cm, and $ED = y$ cm. Since $BE = CF$,

$$25^2 - x^2 = 29^2 - y^2,$$

$$\therefore \qquad y^2 - x^2 = 29^2 - 25^2 = 54 \times 4.$$

Since $AE + FD = 48 - 12 = 36$, we have

$$y + x = 36$$

$$\therefore \qquad y - x = 6$$

$$\therefore \qquad y = 21, \ x = 15.$$

Let T_1, T, and T_2 be the tensions in AB, BC, and CD, and the weights of the two masses be P and Q.

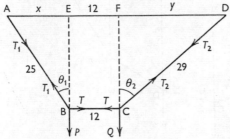

FIG. 8.23

Resolving horizontally for B and C, we get

$$T_1 \sin \theta_1 = T = T_2 \sin \theta_2$$

$$\therefore \qquad \frac{T_1}{T_2} = \frac{\sin \theta_2}{\sin \theta_1} = \frac{21}{29} \times \frac{25}{15}.$$

Resolving vertically for B, we get

$$T_1 \cos \theta_1 = P$$

and for C $\qquad T_2 \cos \theta_2 = Q$

$$\therefore \qquad \frac{Q}{P} = \frac{T_2 \cos \theta_2}{T_1 \cos \theta_1} = \frac{29 \times 15}{21 \times 25} \times \frac{25}{29} = \frac{5}{7}.$$

$$\therefore \qquad Q = \frac{5}{7}P.$$

Hence the mass at C is $\frac{5}{7}$ that at B.

EXAMPLE (iii)

Equal weights are suspended from the joints of a chain composed of five straight, light, smoothly jointed links. The extreme links are fastened to two points P, Q *in the same horizontal line by smooth joints. The projection of each link on the horizontal is equal to a, the distance* PQ *being 5a. The depth of the lowest link, which is horizontal, below* PQ *being 6a, find the inclination of each link to the horizontal.* (C.S.)

Let PBCDEQ (Fig. 8.24) represent the chain.

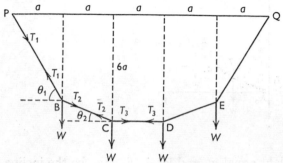

FIG. 8.24

Resolving horizontally for the joint B, we get

$$T_1 \cos \theta_1 = T_2 \cos \theta_2 \qquad \text{(i)}$$

Resolving vertically for the joint B, we get

$$T_1 \sin \theta_1 = W + T_2 \sin \theta_2$$

and for C $\qquad T_2 \sin \theta_2 = W$

$\therefore \qquad T_1 \sin \theta_1 = 2T_2 \sin \theta_2 \qquad \text{(ii)}$

Dividing (ii) by (i) $\tan \theta_1 = 2 \tan \theta_2$.

Also the depth of B below PQ is $a \tan \theta_1$, and the depth of C is $a \tan \theta_1 + a \tan \theta_2$.

$\therefore \qquad a \tan \theta_1 + a \tan \theta_2 = 6a$

$\therefore \qquad \tan \theta_1 + \tan \theta_2 = 6$

$\therefore \qquad 3 \tan \theta_2 = 6$

$\therefore \qquad \tan \theta_2 = 2$

and $\qquad \tan \theta_1 = 4.$

8.10. *A number of equal weights is attached at points of a light string, suspended from two points at the same level, in such a manner that the horizontal distances between the weights are equal.*

Show that the points of attachment of the weights and the points of suspension lie on a parabola.

We shall assume first that the number of weights is odd, and the middle weight will then be at the lowest point A_0 of the string, as in Fig. 8.25.

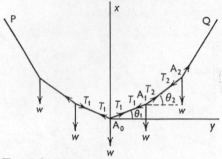

FIG. 8.25

Let each weight equal w, and let θ_1, θ_2, etc., be the inclinations of A_0A_1, A_1A_2, etc., to the horizontal and T_1, T_2, etc., the tensions in them. We have seen that $\tan \theta_1$, $\tan \theta_2$, etc., are in arithmetical progression, the common difference being w/H, where H is the horizontal tension.

Also $T_1 \cos \theta_1 = H$

and $2T_1 \sin \theta_1 = w,$

∴ $\tan \theta_1 = \tfrac{1}{2}(w/H).$

Let the horizontal distance between A_0 and A_1, A_1 and A_2 ... be a.

Take A_0 as origin, and the vertical and horizontal at A_0 as axes of x and y.

The y coordinates of A_1, A_2, etc., are a, $2a$, ... na.

The x coordinate of A_1 is $a \tan \theta_1$.

The x coordinates of A_2 is $a \tan \theta_1 + a \tan \theta_2$

$$= 2a \tan \theta_1 + a(w/H).$$

The x coordinate of A_3 is obtained by adding $a \tan \theta_3$ and

$$a \tan \theta_3 = a \tan \theta_1 + 2a(w/H).$$

The x coordinate of A_4 is obtained by adding $a \tan \theta_4$ and

$$a \tan \theta_4 = a \tan \theta_1 + 3a(w/H).$$

Hence the x coordinate of A_n is

$$na \tan \theta_1 + a\{1 + 2 + \ldots + (n-1)\}w/H$$

$$= \frac{1}{2}na\frac{w}{H} + \frac{n-1}{2}na\frac{w}{H} = \frac{n^2a}{2} \times \frac{w}{H}.$$

The coordinates of A_n are therefore

$$x = wan^2/2H, \; y = na.$$

Eliminating n these give,

$$x = \frac{w}{2aH}y^2$$

or

$$y^2 = \frac{2aH}{w}x$$

a parabola whose vertex is at A_0 and whose axis is vertical.

With an even number of weights the lowest portion of the string is horizontal, as in Fig. 8.26.

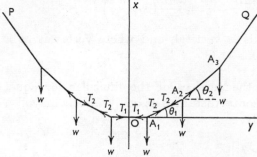

Fig. 8.26

Take the mid-point O of the lowest portion as origin, and the vertical and horizontal at O as axes of x and y.

The y coordinates of A_1, A_2, . . . A_n are

$$\frac{a}{2}, \frac{3a}{2}, \ldots \frac{2n-1}{2}a.$$

The x coordinate of A_2 is $a \tan \theta_1$.
The x coordinate of A_3 is

$$a \tan \theta_1 + a \tan \theta_2 = 2a \tan \theta_1 + a(w/H).$$

The x coordinate of A_4 is obtained by adding $a \tan \theta_3$ and

$$a \tan \theta_3 = a \tan \theta_1 + 2a(w/H).$$

The x coordinate of A_5 is obtained by adding $a \tan \theta_4$ and

$$a \tan \theta_4 = a \tan \theta_1 + 3a(w/H).$$

Hence the x coordinate of A_n is

$$(n-1)\, a \tan \theta_1 + \{1 + 2 + \ldots (n-2)\} a(w/H)$$
$$= (n-1)\, a \tan \theta_1 + \tfrac{1}{2}(n-2)(n-1)a(w/H).$$

In this case, $T_2 \cos \theta_1 = H$

and $T_2 \sin \theta_1 = w$

$\therefore$ $\tan \theta_1 = (w/H)$

$\therefore$ the coordinates of A_n are

$$x = \frac{n(n-1)}{2}a\frac{w}{H} \text{ and } y = \frac{2n-1}{2}a.$$

Eliminating n, $\left(\dfrac{y}{a} + \dfrac{1}{2}\right)\left(\dfrac{y}{a} - \dfrac{1}{2}\right) = \dfrac{2H}{aw}x$

$\therefore$ $y^2 = \dfrac{2aH}{w}x + \dfrac{a^2}{4} = \dfrac{2aH}{w}\left(x + \dfrac{wa}{8H}\right)$

a parabola whose axis is vertical, and whose vertex is at a depth $wa/8H$ below O.

8.11. The fact that the vertices of the funicular for equal weights at equal horizontal distances lie on a parabola may also be deduced graphically.

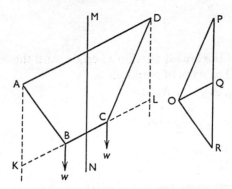

FIG. 8.27

Let A, B, C, D, E, ... (Fig. 8.27) be successive vertices, and let PQR be the portion of the force polygon for the weights at

B and C, the distance of the pole O from the vertical line PR being taken equal to the horizontal distance between successive weights.

The lines OP, OQ, OR representing the tensions in CD, BC, AB will then be equal to, as well as parallel to, the sides of the funicular.

If AK, DL be drawn vertically to meet BC produced in K and L, then, AKB is the triangle of forces for B (similar to OQR) and AK represents w. Similarly, DL represents w. Hence AK is equal and parallel to DL, and therefore AD is parallel to BC.

Also since the horizontal distances between A and B, B and C, C and D are equal, the vertical line MN bisecting BC will also bisect AD. Hence A, B, C, D lie on a parabola whose axis is vertical and of which MN is a diameter.

Similarly B, C, D, E lie on a parabola.

Now since the equation of a parabola is of the form

$$y = a+bx+cx^2,$$

three points on the curve are sufficient to determine a, b, c, i.e. the shape and size of the curve, and if the direction of the axis is known also the curve is completely determined.

Hence the successive parabolas ABCD, BCDE, . . . must coincide and all the vertices lie on one parabola.

8.12. Suspension bridges

In these bridges the horizontal roadway is suspended from a cable by a number of vertical ties, the ends of the cable being attached to the tops of rigid piers, or passing over the tops of these piers and attached to the ground on the other side.

The spaces between the vertical ties are called bays, and these are generally equal, i.e. the ties are equally spaced. Also the loading of the roadway is distributed uniformly.

We may then consider the roadway to be replaced by a number of equal weights acting on the ties, and the effect on the cable will be the same as that already considered when a number of weights are attached to a string.

With an even number of ties it is clear from symmetry that the lowest portion of the cable will be horizontal. With an odd number the lowest point of the cable will be at the middle tie.

A chain composed of light links jointed together may be used instead of a cable, the tie rods being attached to the joints.

EXAMPLE

The span of a suspension bridge is 18 m and the total load of the roadway is the weight of 60 Mg, uniformly distributed. There are five vertical tie rods, equally spaced, attached to a chain whose weight can be neglected, and the weight of the 1·5 m of roadway at each end is taken on the piers and not on the ties.

The chain hangs from points on the towers which are 9 m above each end of the roadway, and the middle tie is 3 m long. Show how to obtain the shape of the chain and the tension in each link.

Since the number of tie rods is odd the middle one will be at the lowest point of the chain, as in Fig. 8.28. The load taken by each tie rod equals the weight of 10 Mg.

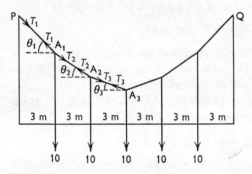

FIG. 8.28

Let the inclinations and tensions be as shown.
Resolving horizontally for the points A_1, A_2, A_3, in succession,

$$T_1 \cos \theta_1 = T_2 \cos \theta_2 = T_3 \cos \theta_3 \qquad \text{(i)}$$

Resolving vertically, we get

$$T_1 \sin \theta_1 - T_2 \sin \theta_2 = 10$$
$$T_2 \sin \theta_2 - T_3 \sin \theta_3 = 10$$

and $$2T_3 \sin \theta_3 = 10$$

$\therefore \qquad T_2 \sin \theta_2 = 15 \text{ and } T_1 \sin \theta_1 = 25.$

$\therefore \qquad \dfrac{T_2 \sin \theta_2}{T_3 \sin \theta_3} = 3.$

Using (i),

$$\cot \theta_3 \tan \theta_2 = 3 \text{ or } \tan \theta_2 = 3 \tan \theta_3.$$

Also
$$\frac{T_2 \sin \theta_2}{T_1 \sin \theta_1} = \frac{3}{5}$$

∴ using (i)

$$\cot \theta_1 \tan \theta_2 = \tfrac{3}{5} \text{ or } \tan \theta_1 = \tfrac{5}{3} \tan \theta_2 = 5 \tan \theta_3.$$

Now the depth of A_3 below PQ is 6m.

∴ $\qquad\qquad 3 (\tan \theta_1 + \tan \theta_2 + \tan \theta_3) = 6$

∴ $\qquad\qquad 9 \tan \theta_3 = 2 \text{ or } \tan \theta_3 = \tfrac{2}{9}.$

Hence $\qquad \tan \theta_1 = \tfrac{10}{9}, \tan \theta_2 = \tfrac{2}{3}, \tan \theta_3 = \tfrac{2}{9}.$

From these values we can calculate $\sin \theta_1$, etc., and obtain the

tensions from equations (ii), e.g. $\sin \theta_3 = \dfrac{2}{\sqrt{85}}$

∴ $\qquad\qquad T_3 = 2 \cdot 5 \sqrt{85}.$

EXAMPLES 8.3

1. ABCD is a light string attached to two fixed points A and D, and carrying weights at B and C. Determine the ratio of these weights if AB is inclined at 30° to the vertical, BC at 60° to the vertical, and CD is horizontal.

2. ABCD is a light string attached to two fixed points A and D, has two equal weights attached at B and C, and rests with the portions AB and CD inclined at 30° and 60° respectively to the vertical. Prove that the tension in the portion BC is equal to either weight, and that BC is inclined at 60° to the vertical.

3. The footway of a suspension bridge is horizontal, and is suspended by vertical rods attached at equal intervals along it. The upper ends of the rods are attached to a light cable. The tensions in all the vertical rods are equal. Show that the points of attachment to the cable lie on a parabola. (C.S.)

4. $PA_1A_2, \ldots A_{2n}Q$ is the chain of a suspension bridge. Each of the vertical bars $A_1B_1, A_2B_2, \ldots A_{2n}B_{2n}$ bears an equal portion of the weight of the roadway. P and Q are attached to the piers, and the ends of the roadway B_0 and B_{2n+1} are vertically below P and Q respectively. The distances $B_0B_1, B_1B_2, \ldots B_{2n}B_{2n+1}$ are all equal. The weights of the chains and bars may be neglected in comparison with the weight of the roadway. By means of a force diagram, or otherwise, show that the points $P, A_1, A_2, \ldots$

A_{2n}, Q lie on a parabola whose axis is vertical. If W is the total weight of the roadway supported by the bars, d the depth of $A_n A_{n+1}$ (the middle link) below PQ, and l the total span of the bridge, show that the tension in the chain at P or Q is

$$\tfrac{1}{2}W\left\{1+\frac{(n+1)^2 l^2}{4(2n+1)^2 d^2}\right\}^{\frac{1}{2}} \qquad\qquad \text{(C.S.)}$$

5. A suspension bridge has a span of 21 m, and there are six vertical tie rods. The roadway is horizontal, and has a load equal to the weight of 3 Mg per metre run. The towers at the ends are each 6 m high, and the chain comes down to the roadway at the centre. Find the tension in the middle link of the chain, and show how to determine the shape of the chain. Assume that at each end of the roadway the weight of 4·5 Mg are carried by the piers and not by the tie rods. (I.C.)

6. The span of a suspension bridge is 24 m, and the towers at the abutments are 15 m and 3 m high respectively. The chain carries seven vertical tie rods equally spaced, and the link in the bay next but one to the lower tower is horizontal. The load in the horizontal roadway is the weight of 80 Mg, uniformly distributed. Show how to determine the shape of the chain, neglecting the weight of it and that of the tie rods. Calculate the height of the horizontal link above the roadway. (I.E.)

7. Weights of mass 2, 4, and 3 kg are attached to points B, C, and D of a string of which the ends A and E are attached to fixed supports. If each of the portions of the string BC and CD makes an angle of 25° with the horizontal, prove that AB and DE make angles of approximately 43° and 49° 23′ with the horizontal.
(Q.E.)

8. An even number $2n$ of horizontally equidistant equal weights W are carried by a string. Show that the points from which the weights are suspended lie on a parabola, and calculate the tensions in the middle and extreme parts of the string respectively, given that a is the horizontal distance between successive weights, and k is the height of the highest above the lowest point of suspension. (Ex)

9. Four lamps of mass 1 kg are to be suspended across a road 15 m wide by strings attached at points B, C, D, E, of a string whose ends A, F are fixed at the same level to posts on opposite sides of the road. The strings supporting the lamps are to be 3 m apart, and the end lamps 3 m from the posts; CD is to be horizontal and 3·6 m below the line AF. Find (to 1 per cent) the lengths and tensions of the strings AB, BC. (C.W.B.)

10. Five particles of equal weight are attached to the ends and points of quadrisection of a light string. One of the end particles A is fixed and the other B is free to slide on a fixed smooth vertical rod. Prove that, in equilibrium, the tangents of the angles which the parts of the string make with the horizontal are in the ratios $1 : 2 : 3 : 4$.

Prove also that, if the lowest part of the string is inclined to the horizontal at an angle of $30°$, the ratio of the distance of A from the rod to the length of the string is

$$\sqrt{3}\left\{\frac{1}{8}+\frac{1}{4\sqrt{7}}+\frac{1}{8\sqrt{3}}+\frac{1}{4\sqrt{19}}\right\} : 1.$$

8.13. Uniform chain hanging under gravity

We shall now investigate the form of the curve in which a uniform heavy flexible cord or chain hangs when suspended from two points. In the cases so far considered the attached weights have been uniformly distributed horizontally. It is clear that when the weights are the actual particles of the cord itself this is not the case. The weight per unit length of the cord is constant, but except at the lowest point the cord is not horizontal.

The curve in which a uniform heavy cord hangs is called a *Catenary*, and its equation is obtained in the following paragraph. As we are dealing now with an infinite number of particles, we shall require the methods of the calculus.

8.14. The catenary

Let ACB (Fig. 8.29) be a uniform heavy flexible cord or chain attached to two points A and B at the same level, C being the lowest point of the cord.

Draw CO vertical, OX horizontal, and take OX as x-axis, OC as y-axis. Let ψ be the angle made by the tangent at any point P with OX. Consider the equilibrium of the portion of the cord CP. If the length of the arc CP equals s, and w is the weight of the cord per unit length, the weight of CP is ws.

The other forces acting on CP are the tensions at C and P along the tangents at these points, and the tangent at C is horizontal.

Let T_0 and T be the tensions at C and P respectively.
Resolving horizontally,

$$T \cos \psi = T_0 \qquad\qquad\text{(i)}$$

Resolving vertically,

$$T \sin \psi = ws \tag{ii}$$

If we put $T_0 = wc$, where c is some constant, then dividing (ii) by (i), we get

$$\tan \psi = s/c \text{ or } s = c \tan \psi \tag{iii}$$

This is called the *intrinsic* equation of the catenary. It gives

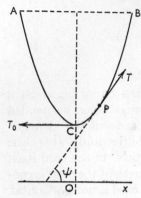

FIG. 8.29

the relation between the length of arc of the curve from the lowest point to any other point on the curve and the inclination of the tangent at the latter point.

We can find the Cartesian equation of the curve as follows.

Now $\qquad \tan \psi = \dfrac{dy}{dx}$ and $\sin \psi = \dfrac{dy}{ds}$

$\therefore \quad \dfrac{dy}{d\psi} = \dfrac{dy}{ds} \times \dfrac{ds}{d\psi} = \sin \psi \times c \sec^2 \psi = c \sec \psi \tan \psi.$

Integrating $\qquad y = c \sec \psi + \text{constant}.$

If $y = c$ when $\psi = 0$ the constant is zero and

$$y = c \sec \psi \tag{iv}$$

From (iii) and (iv)

$$y^2 = c^2(1 + \tan^2 \psi) = c^2 + s^2 \tag{v}$$

To obtain the relation between x and ψ, we have

$$\frac{dx}{d\psi} = \frac{dx}{ds} \times \frac{ds}{d\psi} = \cos \psi \times c \sec^2 \psi = c \sec \psi$$

Integrating we have

$$x = c \log (\sec \psi + \tan \psi) \qquad \text{(vi)}$$

the constant of integration being zero since $x = 0$ when $\psi = 0$.

To obtain the Cartesian equation we eliminate ψ from relations (iv) and (vi).

Alternatively we write

$$\frac{dy}{dx} = \tan \psi = \frac{s}{c} = \frac{\sqrt{(y^2 - c^2)}}{c}$$

Taking the positive square root, we have

$$\frac{dy}{\sqrt{(y^2 - c^2)}} = \frac{dx}{c}$$

$$\therefore \qquad \cosh^{-1} \frac{y}{c} = \frac{x}{c} + A.$$

Since $y = c$ when $x = 0$, $A = 0$ and therefore

$$y = c \cosh \frac{x}{c} \qquad \text{(vii)}$$

We can also obtain a connection between s and x.

We write

$$\frac{ds}{dx} = \sec \psi = (1 + \tan^2 \psi)^{\frac{1}{2}}$$

$$= \frac{1}{c} (c^2 + s^2)^{\frac{1}{2}}$$

$$\therefore \qquad \frac{ds}{(s^2 + c^2)^{\frac{1}{2}}} = \frac{dx}{c}$$

$$\therefore \qquad \sinh^{-1} \frac{s}{c} = \frac{x}{c} + B.$$

Since $x = 0$ when $s = 0$, $B = 0$,

$$\therefore \qquad s = c \sinh x/c \qquad \text{(viii)}$$

From (i) and (ii) we have

$$T \cos \psi = wc$$

and $$T \sin \psi = ws.$$

Squaring and adding these equations,

$$T^2 = w^2(c^2 + s^2) = w^2 y^2$$

$$\therefore \qquad T = wy \qquad \text{(ix)}$$

The constant c is called the parameter of the curve.

The axis of x when taken at a depth c below the lowest point C (as we have done) is called the *directrix*.

From equation (ix) we see that the tension at any point is equal to the product of the weight per unit length of the chain and the height of the point above the directrix.

The depth of the lowest point C below AB is the *sag*.

Since $wc = T_0$, the horizontal tension, the value of c will be large when the tension is large, i.e. when the curve is shallow and the sag in the middle is small.

Similarly, c is small when the tension is small and the sag comparatively large.

8.15. Approximate equations of catenary when c is large

We have
$$y = c \cosh \frac{x}{c} = \tfrac{1}{2}c\left(e^{\frac{x}{c}} + e^{-\frac{x}{c}}\right),$$
and expanding, we get

$$y = \tfrac{1}{2}c\left(1 + \frac{x}{c} + \frac{x^2}{2c^2} + \ldots + 1 - \frac{x}{c} + \frac{x^2}{2c^2} - \ldots\right)$$

$$= \tfrac{1}{2}c\left(2 + \frac{x^2}{c^2} + \ldots\right) = c + \frac{x^2}{2c}$$

since we can neglect x^4/c^3 and higher terms when c is large enough.

In this case the curve is approximately a parabola of latus rectum $2c$. This might have been expected, since, when the curve is flat the load is distributed horizontally in a manner very nearly uniform (cf. 8.10).

If k is half the span, then the length of half the chain is given approximately by

$$s = \frac{c}{2}\left(1 + \frac{k}{c} + \frac{k^2}{2c^2} + \frac{k^3}{6c^3} + \ldots - 1 + \frac{k}{c} - \frac{k^2}{2c^2} + \frac{k^3}{6c^3} - \ldots\right)$$

$$= \frac{c}{2}\left(\frac{2k}{c} + \frac{k^3}{3^3 c}\right)$$

$$= k + \frac{k^3}{6c^2},$$

$$\therefore \qquad\qquad\qquad s - k = \frac{k^3}{6c^2}.$$

Now if h is the sag in the middle, h is the difference in the value of y for $x = 0$ and $x = k$.

For $x = 0$, $y = c$, and for $x = k$, y is approximately $c + (k^2/2c)$,

$$\therefore \qquad h = \frac{k^2}{2c},$$

$$\therefore \qquad \frac{1}{c^2} = \frac{4h^2}{k^4},$$

$$\therefore \qquad s - k = \frac{k^3}{6} \times \frac{4h^2}{k^4} = \frac{4h^2}{6k}.$$

The span is $2k$ and

$$2(s - k) = \frac{8}{3} \times \frac{h^2}{2k}$$

$\therefore$ the difference between length and span

$$= \frac{8}{3} \times \frac{(\text{sag})^2}{\text{span}} \text{ approximately.}$$

8.16 EXAMPLE (i)

Show that the maximum tension in a wire which has a mass of 0·1 kg m⁻¹ and hangs with a sag of 0·5 m in a horizontal span of 50 m is about 613 N.

Since the catenary is shallow, c is large, and we have approximately $y - c = x^2/2c$.

The maximum tension is at the top and there $y - c = 0.5$ and $x = 25$.

$$\therefore \qquad 0.5 = 625/2c \text{ or } c = 625.$$

$$\therefore \qquad y = 625.5 \text{ at the top, } w = 0.1 \times 9.8 \text{ N m}^{-1}, \text{ and}$$

$$\therefore \qquad T = wy = 0.98 \times 625.5$$

$$= 612.99.$$

The maximum tension is therefore about 613 N.

EXAMPLE (ii)

A uniform flexible chain of length $2a(1+k)$ has its extremities attached to two points at the same level, distant $2a$ apart. Prove that, if k is so small that powers of k above the second may be neglected, the sag of the chain is $\frac{1}{2}a\left(1 + \frac{7k}{20}\right)(6k)^{\frac{1}{2}}$. (H.C.)

We have approximately

$$s = \frac{c}{2}\left(1 + \frac{x}{c} + \frac{x^2}{2c^2} + \frac{x^3}{6c^3} + \frac{x^4}{24c^4} + \frac{x^5}{120c^5} + \cdots\right.$$

$$\left. -1 + \frac{x}{c} - \frac{x^2}{2c^2} + \frac{x^3}{6c^3} - \frac{x^4}{24c^4} + \frac{x^5}{120c^5}\right)$$

$$= x + \frac{x^3}{6c^2} + \frac{x^5}{120c^4}$$

$$\therefore \qquad a + ak = a + \frac{a^3}{6c^2} + \frac{a^5}{120c^4} \text{ or } \frac{a^4}{120c^4} + \frac{a^2}{6c^2} - k = 0.$$

Writing this $\quad \dfrac{a^4}{c^4} + 20\dfrac{a^2}{c^2} - 120k = 0$, we get

$$\frac{a^2}{c^2} = -10 + 10\left(1 + \frac{6k}{5}\right)^{\frac{1}{2}} = 6k - \frac{9k^2}{5} \text{ approximately}$$

$$\therefore \qquad \frac{a}{c} = (6k)^{\frac{1}{2}}\left(1 - \frac{3k}{10}\right)^{\frac{1}{2}} = (6k)^{\frac{1}{2}}\left(1 - \frac{3k}{20}\right) \text{ approximately}$$

Also, $\quad \dfrac{a^3}{c^3} = 6k(6k)^{\frac{1}{2}}$ approximately

Now

$$y = \frac{c}{2}\left(1 + \frac{x}{c} + \frac{x^2}{2c^2} + \frac{x^3}{6c^3} + \frac{x^4}{24c^4} + \cdots\right.$$

$$\left. +1 - \frac{x}{c} + \frac{x^2}{2c^2} - \frac{x^3}{6c^3} + \frac{x^4}{24c^4} - \cdots\right)$$

$$= c + \frac{x^2}{2c} + \frac{x^4}{24c^3}.$$

The sag is the difference between the values of y when $x = 0$ and $x = a$.

$$\therefore \qquad \text{sag} = \frac{a^2}{2c} + \frac{a^4}{24c^3}$$

$$= \frac{a}{2}(6k)^{\frac{1}{2}}\left(1 - \frac{3k}{20}\right) + \frac{a(6k)^{\frac{1}{2}} \times 6k}{24}$$

$$= \frac{a}{2}(6k)^{\frac{1}{2}}\left(1 - \frac{3k}{20} + \frac{k}{2}\right)$$

$$= \frac{a}{2}(6k)^{\frac{1}{2}}\left(1 + \frac{7k}{20}\right).$$

EXAMPLE (iii)

O *is the lowest point of a heavy chain suspended from two points* A, B *whose coordinates referred to horizontal and vertical axes through* O *are* (6, 18) *and* (−4, 8). *If the weight of every element of chain varies as its horizontal projection, prove that the curve in which it hangs is a parabola; if the total mass of the chain be 50 kg, find the terminal tensions and the minimum tension.*   (Ex.)

Consider the equilibrium of the portion of the chain OP (Fig. 8.30) up to any point P whose coordinates are (x, y).

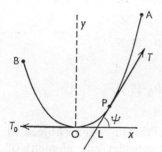

FIG. 8.30

If w is the weight per unit length of horizontal projection the weight of OP is wx.

The other forces acting on it are the horizontal tension T_0 at O, and the tension T at P acting along the tangent LP.

Let angle PLX = ψ.

Resolving vertically, $T \sin \psi = wx$ (i)

Resolving horizontally, $T \cos \psi = T_0$ (ii)

$\therefore$ $$\tan \psi = \frac{w}{T_0} x$$

$\therefore$ $$\frac{dy}{dx} = \frac{w}{T_0} x$$

$\therefore$ $$y = \frac{1}{2} \frac{w}{T_0} x^2 + c.$$

This is a parabola, and, as it passes through the origin, $c = 0$, and its equation is

$$y = \frac{1}{2} \frac{w}{T_0} x^2.$$

Since it passes through the points $(6, 18)$ and $(-4, 8)$

$$\frac{y}{x^2} = \frac{18}{36} \text{ or } \frac{8}{16} = \frac{1}{2}$$

$\therefore \qquad\qquad w = T_0.$

If the total weight of chain is $50 \times 9 \cdot 8$ N, we have

$$10w = 490 \text{ or } w = 49.$$

Since $T_0 = w$, the minimum tension is 49 N.
Also squaring and adding (i) and (ii),

$$T^2 = w^2 x^2 + T_0{}^2 = w^2(x^2 + 1).$$

At A, this gives

$$T = 49\sqrt{(36+1)} = 49\sqrt{37} \text{ N}.$$

At B, this gives

$$T = 49\sqrt{(16+1)} = 49\sqrt{17} \text{ N}.$$

8.17. Chain subject to forces

If a flexible chain or string is in equilibrium when subject to a system of external forces we may consider any element of it as maintained in equilibrium by the tensions at the two ends and the external forces acting on it. Relations between the form

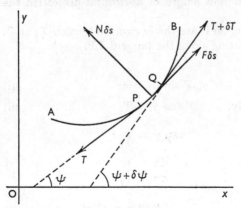

FIG. 8.31

of the chain, the external forces, and the tension at any point may be deduced from the conditions of equilibrium of such an element.

Suppose (Fig. 8.31) AB represents a flexible chain or string.

Any element PQ of length δs is acted on by the tension T at P at an angle ψ to the axis Ox, the tension $T+\delta T$ at Q at an angle $\psi+\delta\psi$ to Ox, and the external forces which are represented by $N\,\delta s$ parallel to the normal at P and $F\,\delta s$ parallel to the tangent at P.

Resolving along the tangent and normal at P we get

$$(T+\delta T)\cos\delta\psi - T + F\,\delta s = 0$$

and
$$(T+\delta T)\sin\delta\psi + N\,\delta s = 0$$

To the first order of small quantities, we have

$$\delta T + F\,\delta s = 0$$

and
$$T\,\delta\psi + N\,\delta s = 0.$$

These equations lead to

$$\frac{\mathrm{d}T}{\mathrm{d}s} = -F \text{ and } \frac{\mathrm{d}\psi}{\mathrm{d}s} = -\frac{N}{T}.$$

If F and N are known, T can be found. Also ψ can be expressed as a function of s, that is, the intrinsic equation of the curve can be derived.

8.18. Two special cases might be noted. First, suppose the chain is subject only to gravity as considered in 8.14 above.

The external forces are therefore $w\,\delta s$ vertically downwards, where w is the weight per unit length of the chain, so that

$$F = -w\sin\psi$$

and
$$N = -w\cos\psi.$$

It follows that

$$\frac{\mathrm{d}T}{\mathrm{d}s} = w\sin\psi$$

$\therefore$
$$\frac{\mathrm{d}T}{\mathrm{d}s} = w\frac{\mathrm{d}y}{\mathrm{d}s}.$$

$\therefore$
$$T = w(y - y_0)$$

if w is constant and $T = 0$ when $y = y_0$. This corresponds to relation (ix) of 8.14.

Further,
$$\frac{\mathrm{d}\psi}{\mathrm{d}s} = \frac{w\cos\psi}{T}$$

or
$$T = w\rho\cos\psi$$

where ρ is the radius of curvature of the chain at the point at which T is measured.

Also, if we write $T \cos \psi$, the constant horizontal component of the tension, as wc where c is a constant, we get

$$\frac{d\psi}{ds} = \frac{1}{c} \cos^2 \psi$$

or

$$\frac{ds}{d\psi} = c \sec^2 \psi$$

$\therefore \qquad\qquad\qquad\qquad s = c \tan \psi + A$

and $A = 0$ if $s = 0$ when $\psi = 0$.

Hence, $s = c \tan \psi$

as was derived in 8.14 (relation (iii)).

8.19. Second, suppose a chain is wrapped round a rough cylinder of any shape.

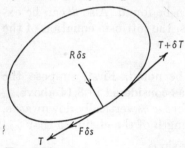

FIG. 8.32

Then indicating the forces on any element PQ of the chain (Fig. 8.32) of length δs by $R \, \delta s$ parallel to the normal at P and a frictional force $F \, \delta s$ parallel to the tangent at P we get

$$\delta T = F \, \delta s$$

and

$$T \, \delta \psi = R \, \delta s.$$

If the chain is on the point of slipping $F = \mu R$, and hence

$$\delta T = \mu T \, \delta \psi$$

$\therefore \qquad\qquad\qquad\qquad \dfrac{\delta T}{T} = \mu \, \delta \psi$

$\therefore \qquad\qquad\qquad\qquad \log T = \mu \psi + A$

$\therefore \qquad\qquad\qquad\qquad T = T_0 e^{\mu \psi}$

where $T = T_0$ when $\psi = 0$.

We note that T is only equal to T_0 if $\mu = 0$, that is, if the cylinder is smooth. Further, if T and T_0 are the tensions at the ends of a chain or string wrapped once completely round a rough post we have

$$T/T_0 = e^{2\pi\mu} = (535)^\mu.$$

Similarly, if the chain is wrapped n times round the post the ratio of the tensions at the two ends is

$$T/T_0 = e^{2\pi n\mu} = (535)^{n\mu}$$

Consequently, if n is large a very small force applied at one end of the chain can balance a very large force at the other. For example, the application of quite a small force to one end of a rope wrapped round a bollard on a quayside may be used to hold a boat in position.

8.20. EXAMPLE (i)

Weights W_1 and W_2 are suspended from the ends of a rope which passes over a rough circular cylinder fixed with its axis horizontal. The rope is in a plane perpendicular to the axis of the cylinder. If W_1 is on the point of moving downwards, find the weight that must be added to W_2 to cause W_1 to be on the point of moving upwards.

If W_1 is on the point of moving downwards, then

$$W_1/W_2 = e^{\mu\pi}.$$

If W is the weight added to W_2 to cause it to be on the point of moving downwards, then

$$(W_2+W)/W_1 = e^{\mu\pi}.$$
$$\therefore \qquad W_1{}^2 = W_2(W_2+W)$$
$$\therefore \qquad W = (W_1{}^2 - W_2{}^2)/W_2.$$

EXAMPLE (ii)

Three rough cylinders of equal radii a are fixed with their axes parallel in a horizontal plane at equal distances 4a apart. A rope passes over the outer two cylinders and under the other one, so that it is in a vertical plane perpendicular to the axes. Weights W_1 and W_2 are attached to the ends of the rope and the coefficient of function is $0\cdot3$; show that W_1 is on the point of descending if $W_1 = W_2 e^{\frac{1}{4}\pi}$.

The inclined portions of the rope must make angles of 30° with the horizontal (Fig. 8.33). Hence if T_1 and T_2 are the tensions in these portions of the rope we have

$$W_1 = T_1 e^{\mu(2\pi/3)}$$

and
$$T_1 = T_2 e^{\mu(\pi/3)}$$

and
$$T_2 = W_2 e^{\mu(2\pi/3)}$$

where μ is the coefficient of friction.

Hence
$$W_1 = W_2 e^{(5\pi/3)\mu} = W_2 e^{\frac{1}{2}\pi} = 4 \cdot 8 W_2.$$

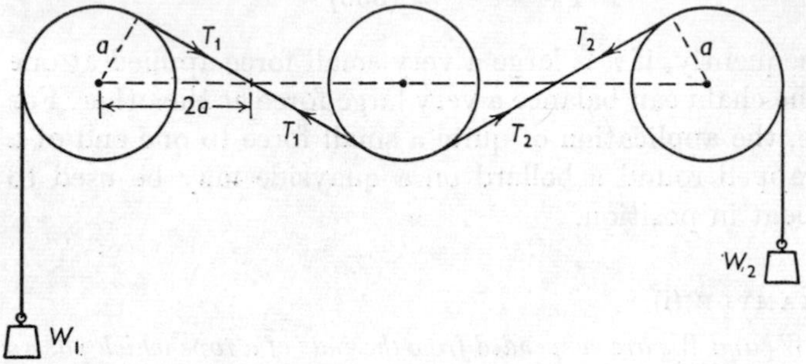

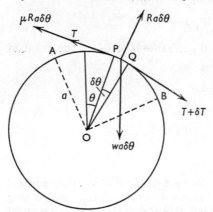

FIG. 8.33

EXAMPLE (iii)

A heavy uniform chain AB of length $\frac{1}{2}\pi a$ and of weight w per unit length rests in contact with a rough circular cylinder of radius a fixed

FIG. 8.34

with its axis horizontal. The chain is in a plane perpendicular to the axis of the cylinder, and is on the point of slipping when the radius OA *makes an angle of 30° with the upward vertical. Find the relation that the coefficient of friction μ must satisfy.*

In Fig. 8.34 PQ is an element of the chain, such that OP makes an angle θ with the vertical through O and the angle POQ $= \delta\theta$.

When the chain is on the point of slipping the forces acting on PQ are its weight $wa\,\delta\theta$, the tension T at P, the tension $T+\delta T$ at Q, the normal reaction $Ra\,\delta\theta$ and the limiting frictional force $\mu Ra\,\delta\theta$.

Resolving parallel to and perpendicular to the tangent at P we get:

$$\delta T = \mu Ra\,\delta\theta - wa\,\delta\theta \sin\theta$$

and
$$T\,\delta\theta = Ra\,\delta\theta - wa\,\delta\theta \cos\theta$$

Hence,
$$\frac{\mathrm{d}T}{\mathrm{d}\theta} = \mu Ra - wa \sin\theta$$

and
$$T = Ra - wa \cos\theta$$

Hence,
$$\frac{\mathrm{d}T}{\mathrm{d}\theta} - \mu T = wa(\mu \cos\theta - \sin\theta)$$

$\therefore$
$$Te^{-\mu\theta} = wa\int e^{-\mu\theta}(\mu \cos\theta - \sin\theta)\,\mathrm{d}\theta$$

$$= \frac{wa\,e^{-\mu\theta}}{1+\mu^2}(\cos\theta + 2\mu \sin\theta - \mu^2 \cos\theta) + A$$

Hence,
$$T = \frac{wa}{1+\mu^2}(\cos\theta + 2\mu \sin\theta - \mu^2 \cos\theta) + Ae^{\mu\theta}$$

Now at the end A, $\theta = -\pi/6$, and $T = 0$, since the end is free. Similarly at B, $\theta = \pi/3$ and $T = 0$.

$\therefore$
$$0 = \frac{wa}{1+\mu^2}(\tfrac{1}{2}\sqrt{3} - \mu - \tfrac{3}{4}\mu^2) + Ae^{-\mu\pi/6}$$

and
$$0 = \frac{wa}{1+\mu^2}(\tfrac{1}{2} + \mu\sqrt{3} - \tfrac{1}{4}^2) + Ae^{\mu\pi/3}$$

$\therefore$
$$(2\sqrt{3} - 4\mu - 3\mu^2) = e^{-\frac{1}{2}\mu\pi}(2 + 4\mu\sqrt{3} - \mu^2)$$

which is the relation μ must satisfy.

EXAMPLES 8.4

1. Obtain the equation $y = c \cosh x/c$ for the curve of a uniform chain hanging under gravity. If the chain is suspended from two points A, B on the same level, and the depth of the middle below AB is l/n, where $2l$ is the length of the chain, show that the horizontal span AB is equal to $l\left(n - \dfrac{1}{n}\right) \log \left(\dfrac{n+1}{n-1}\right)$. Find an

approximation to the difference between the arc AB and the span AB when n is large. (C.S.)

2. A uniform chain is suspended from one end and the other end hangs over a rough pulley. Prove that the friction brought into play at the pulley is the weight of a length of chain which would reach from the loose end to the directrix of the catenary formed by the chain. (C.S.)

3. Find the intrinsic and Cartesian equations of the curve in which a uniform heavy chain hangs when suspended from two points. One end of a rough uniform chain of length l is fastened to a point on a vertical wall at a height h above the ground. Show that the greatest distance from the wall at which the free end of the chain will rest on level ground is given by the expression

$$u\left\{1+\mu\,\log\left(\frac{h+l}{\mu u}+1-\frac{1}{\mu}\right)\right\}$$

where $\qquad u = l+\mu h-\{(\mu^2+1)h^2+2\mu lh\}^{\frac{1}{2}}$

and μ is the coefficient of friction. (C.S.)

4. Show that in a uniform chain at rest under gravity the tension at any point is proportional to the height above a certain level. A uniform chain of length l has its end link free to slide on a smooth vertical wire and passes over a smooth peg at a distance a from the wire. The other end is attached to a weight equal to n times the weight of a length a of the chain. Show that for equilibrium to be possible $n+l/a$ must not be less than e (the base of logarithms).

5. Prove from mechanical considerations the formulae $y^2 = s^2+c^2$ and $T = wy$ for the common catenary.

If the tangents of the points P and Q of a catenary are at right angles, prove that the tension at the mid-point of the arc PQ is equal to the weight of a length of the string equal to half the arc PQ. (O.C.)

6. A chain 9 m long is hung from two points in the same horizontal, and the sag in the middle is 3 m. Given $\log_e 5 = 1\cdot6094$, show that the span is 6·035 m and that the tension at the points of support is the weight of 4·875 m of chain. (H.C.)

7. A uniform string of length $64a$ rests symmetrically over two smooth pegs at the same level, the lowest point of the curved portion of the string between the pegs being at a depth a below the level of the pegs. Find the lengths of the portions of the string which hang vertically and in the form of a catenary respectively, and find the distance between the pegs. Prove also that the

reaction on either peg is equal to the weight of a length $40a$ of the string.   (H.C.)

8. A heavy uniform chain AB hangs freely under gravity, with the end A fixed and the other end B attached by a light string BC to a fixed point C at the same level as A. The lengths of the string and chain are such that the ends of the chain at A and B make angles of 60° and 30°, respectively, with the horizontal. Prove that the ratio of these lengths is $(\sqrt{3}-1):1$.

9. A uniform chain of length $2l$ is suspended from two points in the same horizontal line and has a sag h at the centre. Show that the span is

$$\frac{l^2-h^2}{h}\log_e\frac{1+h}{1-h}.$$

The length of the chain is 30 m, the sag is 3 m. Prove that the tension at each end is the weight of 39 m of chain and by expanding the logarithm show that the span is approximately 29·2 m.

10. A uniform chain of length $16l$ is hung symmetrically over two smooth pegs in the same horizontal so that the length $6l$ hangs in the form of the catenary and lengths $5l$ hang vertically. Show that the distance between the pegs is $5·545l$; and find the length of chain whose weight is equal to the reaction on either peg.

11. A uniform wire has a horizontal span of 40 m and the sag in the middle is 10 m. Find graphically or otherwise the length of the wire.

12. Prove that, for a catenary in which a heavy string hangs, $s = c \tan \psi$, where c is the tension at the lowest point divided by the weight per unit length. Deduce from this that, with suitable choice of axes.

$$y = c \cosh \frac{x}{c}, \quad s = c \sinh \frac{x}{c}.$$

A string of length $2l$ hangs in equilibrium over two smooth pegs which are at the same level and a distance $2a$ apart. Prove that the parameter c of the catenary in which the part between the pegs hangs is given by

$$ce^{a/c} = l.$$

Prove that this equation for c has no solution unless $l>ae$.   (O.C.)

13. A uniform flexible string hangs in equilibrium under gravity in the form of a catenary. Prove, with the usual notation, that

$$s = c \tan \psi, \, y = c \sec \psi.$$

A uniform heavy flexible string AB, of length $2a$, has its ends A, B attached to two light rings which can slide on a fixed horizontal wire. The coefficient of friction between each ring and the wire is $1/\sqrt{3}$. Show that, when equilibrium is limiting at A and B, the tangent to the string at A makes an angle $\frac{1}{3}\pi$ with the horizontal. Find the depth of the mid-point of the string below AB and show that the distance AB is

$$(2a/\sqrt{3}) \log_e (2+\sqrt{3}).$$ (N.U.)

14. A uniform flexible chain hangs in equilibrium under gravity in the form of a catenary. Prove, with the usual notation, that

$$T = wy, \ y^2 = s^2 + c^2.$$

The ends of a uniform flexible chain of length 16 m are attached to two points A and B, where B is 4 m higher than A. The tension at B is $39/25$ times the tension at A. If the lowest point C of the chain lies between A and B, find the lengths of chain CA and CB and show that the parameter c of the catenary in which the chain hangs is 4·3 m.   (N.U.)

15. Two pieces of dissimilar string, each uniform, are knotted together and a position of equilibrium is formed in which each string passes over a smooth peg with its free end hanging vertically while the tangent at the join is horizontal. Show that the depths of the free ends below the join are inversely proportional to the weights per unit length of the strings.

16. A uniform flexible string of length $4a$ and weight $4wa$ has one end attached to a fixed point A and passes over a smooth peg B fixed at the same level as A and distant $2a$ from it. A particle of weight wb is to be attached to the free end of the string so that there is an equilibrium position with a length a of the string hanging vertically from B and a length $3a$ in the form of a catenary between A and B. Using tables to obtain a graphical solution of the equation $2 \sinh u = 3u$, derive an approximation for the weight of the particle, determining the value of b/a correct to two significant figures.   (N.U.)

17. A uniform flexible string hangs in equilibrium in the form of a catenary. Prove, with the usual notation, that

$$s = c \tan \psi, \ y = c \sec \psi, \ T = wy.$$

A uniform heavy flexible string AB, of length a and weight wa, will break if subjected to a tension greater than $3wa$. The string is attached to a fixed point at A and hangs in equilibrium under the action of a horizontal force nwa applied at B. If the string is about to break, show that: (i) $n = 2\sqrt{2}$; (ii) the tangent to the string at A makes an angle $\sin^{-1}(\tfrac{1}{3})$ with the horizontal; and (iii) the height of A above the level of B is $(3-2\sqrt{2})a$.

(N.U.)

18. Prove that, if T denotes the tension at a point P of a chain which is at rest in a plane under the action of given forces, the forces exerted by the rest of the chain upon a short element PQ of itself are equivalent to forces δT along the tangent and $T\,\delta\psi$ along the inward normal at P, where $T+\delta T$ is the tension at Q and $\delta\psi$ is the angle between the tangents at P and Q.

    A uniform heavy chain rests symmetrically upon a smooth curve in the form of a catenary $s = c\tan\psi$ in a vertical plane, with its vertex *upwards* and directrix horizontal. Prove that the pressure per unit length at any point varies inversely as the square of the depth of the point below the directrix.   (O.C.)

19. Obtain the equations $s = c\tan\psi$, $y = c\cosh(x/c)$ for the common catenary.

    To the ends of a uniform string of length l are fastened smooth light rings which slide one on each of two smooth wires in the same vertical plane. If the wires make angles α and β with the downward vertical through their point of intersection on opposite sides of it, prove that the parameter of the catenary in which the string hangs is $l/(\tan\alpha+\tan\beta)$, and that the difference in the heights of the rings is $l\sin\tfrac{1}{2}(\alpha\sim\beta)\sec\tfrac{1}{2}(\alpha+\beta)$.

(O.C.)

20. A heavy flexible cord of variable density is hung up from two points not in the same vertical line. Prove that, if T_1, T_2, T_3, are the tensions at points A, B, C where the inclinations of the tangents to the horizontal are $\alpha-\beta$, α, $\alpha+\beta$, and if w_1, w_2 are the weights of the parts of the cord AB, BC respectively,

$$\frac{1}{T_1}+\frac{1}{T_3} = \frac{2\cos\beta}{T_2}; \quad T_1:T_3 = w_1:w_2. \qquad \text{(H.S.D.)}$$

21. A perfectly flexible non-uniform string hangs freely in a vertical plane (with its extremities fixed) in the form of an arc of a circle. If O is the centre of a circle, A its lowest point, and T the point in which the tangent at any point P meets the vertical OA, show that: (i) the weight of the portion AP of the string varies as TP; (ii) the tension at P varies as TO; and (iii) the weight of the string per unit length at P varies as TO^2.   (Ex.)

22. Obtain equations for determining the tension and the shape of a non-uniform chain hanging under gravity.

    Show that, if a light chain is loaded in such a way that, when it hangs freely, the load per unit horizontal distance is constant, then its form is a parabola.

It is required to put a parabolic suspension bridge over a river 6 m wide, the sag in the suspension cables being 1·8 m. If the cables can withstand a tension equal to the weight of 10 Mg, find the maximum weight of the bridge. (The weights of the cables and the vertical ties are to be neglected.)   (O.C.)

23. Prove that if a chain hangs freely in the form of a parabola the tension at any point of the chain varies as the square root of the height of the point above the directrix of the parabola.

24. The weight of a heavy chain is so distributed that, when it is suspended from two fixed points at the same level distant $2a$ apart, the load per unit length of the horizontal projection is constant and equal to w, and the tension at either end is then T. Prove that the chain hangs in the form of a parabola; that the tension at a point at a distance x from the axis of the parabola is

$$\{T^2 - w^2(a^2 - x^2)\}^{\frac{1}{2}},$$

and that the lowest point of the chain is at a depth

$$\tfrac{1}{2}wa^2(T^2 - w^2a^2)^{-\frac{1}{2}}.$$

below the line of the supports.

25. A weight W is suspended from a fixed point by a uniform rope of length l and weight w per unit length. It is drawn aside by a horizontal force P. Show that in the position of equilibrium the distance of W from the vertical through the fixed is

$$\frac{P}{W}\left\{\sinh^{-1}\left(\frac{W+lw}{P}\right) - \sinh^{-1}\left(\frac{W}{P}\right)\right\}.$$

26. A uniform chain of length l and weight W_1 hangs between two fixed points at the same level and a weight W_2 is attached at the mid-point, and the sag at the middle is d. Show that the pull on each support is

$$\left(\frac{d}{2l} + \frac{l}{8d}\right)W_1 + \frac{l}{4d}W_2.$$

27. A rope is given three complete turns round a rough cylindrical post, and a force of 200 N is applied at one end. What force must be applied at the other end of the rope in order to cause it to slip round the post, given that the coefficient of friction between the rope and the post is 0·5?

28. Find the number of times a hauling rope must be wound round a rotating capstan in order to haul twelve trucks, each of mass 20 Mg, up a gradient of 1 in 20. The trucks are subject to a

resistance equal to $\frac{1}{200}$ of their weight, the coefficient of friction is 0·3 and the pull given to the free end of the rope is 200 N.

29. Two rough cylinders of radii a_1, a_2 are fixed with their axes at the same horizontal level and at a distance d apart. A string passes over the cylinders in a vertical plane perpendicular to their axes and has weights W_1 and W_2 attached to its ends. The coefficients of friction are μ_1 and μ_2. Find the ratios of W_1/W_2 corresponding to the two cases of limiting equilibrium.

30. A heavy uniform rope rests in limiting equilibrium on the inside of a rough rod in the form of the cycloid $s = 4a \sin \psi$ fixed in a vertical plane with the vertex downwards. One end of the rope is at the vertex and the other is at the cusp. Show that the coefficient of friction μ satisfies the relation $(1+\mu^2)e^{\frac{1}{2}\pi\mu} = 3$.

31. Prove the formulae for the common catenary, with the usual notation,

$$s = c \tan \psi, \; y^2 = s^2 + c^2.$$

One end of a length l of a uniform flexible chain is attached to one end of a length l of a second uniform flexible chain, whose density is twice that of the first, so as to form a perfectly flexible chain of length $2l$, which is suspended by attaching its two free ends to two points at the same level. If the heavier portion hangs in a catenary of parameter c, and s is the distance along the chain from the joint to the lowest point of the chain, prove that

$$\sqrt{\{(l+2s)^2+4c^2\}} = \sqrt{\{(l-s)^2+c^2\}}+\sqrt{(s^2+c^2)}.$$

From this equation obtain l in terms of s and c, and show that the inclination of the chain to the horizontal at the joint must be less than $\sin^{-1}(1/3)$. (O.C.)

32. A chain is such that the weight per unit length w varies as the tension T, so that $T = \lambda w$. Prove that, if the origin be chosen at the lowest point of the chain, the equation of the curve in which the chain hangs is

$$y = \lambda \log \sec(x/\lambda).$$

Prove that for the above chain: (i) the tension varies as the radius of curvature of the curve in which the chain hangs, and (ii) $T = T_0 \cosh(s/\lambda)$, where T_0 is the tension at the lowest point. (C.W.B.)

CHAPTER NINE

VIRTUAL WORK—STABILITY—MISCELLANEOUS EXAMPLES

9.1. Virtual work

If a particle is in equilibrium under the action of any number of forces, and the particle is displaced in any direction while the forces remain constant in magnitude and direction, the total work done by the forces is zero, since their resultant is zero (see 6.4).

It is often possible to solve problems in Statics very easily by imagining a body at rest to be displaced through a *small* distance, finding the resulting small distances moved by the forces acting on it and equating the total work done by the forces to zero. Since the displacement is not necessarily a *real* one, the principle used in this method is called the *Principle of Virtual Work*.

If a force P acts on a particle at a point A and we suppose that the particle is displaced to A′, then the virtual work done by P is $P \, \mathrm{d}p$ where $\mathrm{d}p$ is the projection of AA′ on the direction of P.

The principle of virtual work states that if any set of forces $P_1, P_2, \ldots P_n$ act on a particle and maintain it in equilibrium, then in any virtual displacement $\Sigma P \, \mathrm{d}p = 0$. We shall now consider the extension of this principle to a system of rigid bodies. The general principle of virtual work may be stated as follows.

If a system of forces $P_1, P_2, \ldots P_n$, acting at the points $A_1, A_2, \ldots A_n$ of a body or system of bodies, these bodies being connected together in any manner so as either to allow or exclude relative motion, and if the forces $P_1, P_2, \ldots P_n$ are in equilibrium, then if the system undergoes any small displacement consistent with its geometrical connections the total virtual work done by all the forces is zero.

This means that if the points $A_1, A_2, \ldots A_n$ move to $A'_1, A'_2, \ldots A'_n$, and $\mathrm{d}p_1, \mathrm{d}p_2, \ldots \mathrm{d}p_n$ are the projections of the displacements $A_1A'_1, A_2A'_2, \ldots A_nA'_n$, on the directions of $P_1, P_2, \ldots P_n$, respectively, then

$$P_1 \, \mathrm{d}p_1 + P_2 \, \mathrm{d}p_2 \ldots + P_n \, \mathrm{d}p_n = 0.$$

Conversely, if $\sum P \, dp$ is zero for any small displacement the system is in equilibrium.

9.2. *The total virtual work done by a system of coplanar forces which is in equilibrium is zero.*

Take any two rectangular axes Ox, Oy (Fig. 9.1) in the plane of the forces, and let the rigid body on which the forces act undergo a slight displacement.

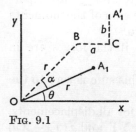

FIG. 9.1

It is clear that this can be done by turning the body through a suitable small angle α (radians) above O, and then moving it through suitable distances a and b parallel to Ox and Oy (see Dynamics, 10.26).

Let A_1 be the original position of the point of application of one of the forces P, and let (x, y) be the Cartesian and (r, θ) be the polar coordinates of A_1. Then $OA_1 = r$, $\angle xOA_1 = \theta$.

Let A'_1 be the position of A after displacement, performed as described above by rotating the body about O through an angle α which brings A_1 to B, then a displacement through a distance a parallel to Ox bringing it to C, and finally a displacement b parallel to Oy bringing it to A'_1.

It is clear that the Cartesian coordinates of A'_1 are

$$r \cos (\theta + \alpha) + a \text{ and } r \sin (\theta + \alpha) + b$$

or $\qquad r \cos \theta - \alpha r \sin \theta + a \text{ and } r \sin \theta + \alpha r \cos \theta + b.$

retaining only the first power of the small angle α.

The displacements of A_1 parallel to the axes are therefore

$$a - \alpha r \sin \theta \text{ and } b + \alpha r \cos \theta$$

or $\qquad a - \alpha y \text{ and } b + \alpha x.$

Hence, if X and Y are the components of P parallel to the axes, the virtual work done by P, which is equal to the virtual work done by its components, is

$$X(a-\alpha y)+Y(b+\alpha x) = aX+bY+\alpha(Yx-Xy).$$

Since a, b, and α are the same for each force of the system, the total virtual work done by all the forces is

$$a\sum X+b\sum Y+\alpha\sum(Yx-Xy) \qquad\qquad\text{(i)}$$

But since the forces are in equilibrium, $\sum X$ and $\sum Y$ are separately zero. Also $\sum(Yx-Xy)$ is equal to the sum of the moments of the forces about O, and is therefore zero.

Hence the total virtual work is zero if the forces are in equilibrium.

Strictly speaking, we should say that this sum is zero to the first order of small quantities.

Conversely, if the expression (i) is zero for all displacements, that is, for all values of a, b, and α, then $\sum X$, $\sum Y$, and $\sum(Yx-Xy)$ must all be zero. Hence the forces must be in equilibrium.

In fact, if the total work done by a system of coplanar forces acting on a rigid body is zero for *three* independent small displacements in the plane of the forces it follows that the forces must be in equilibrium.

For suppose the total work done is zero for a small displacement a parallel to the axis of x, a small displacement b parallel to the axis of y, and a small rotation α about O. Then from (i) it follows that

$$a\sum X = 0, \ b\sum Y = 0 \ \text{and} \ \alpha\sum(Yx-Xy) = 0,$$

that is, $\sum X = 0, \ \sum Y = 0 \ \text{and} \ \sum(Yx-Xy) = 0.$

Consequently the forces are in equilibrium.

It should also be noted that expression (i) equals the work done by the resultant of the forces in the displacement of the body, and hence establishes *that the total work done by the forces in any small displacement of the body equals the work done by their resultant* (cf. 6.4).

9.3. *Forces which may be omitted in using the principle of virtual work*

When a body is not free but can move either under the guidance of certain constraints or is subject to the actions of other rigid bodies it is important to know which actions and reactions can be omitted in writing down the equation of virtual work.

The following commonly occur.

(i) *The tension of an inextensible string, when the displacement which we suppose the system to undergo does not involve a change in the length of the string.*

For let AB (Fig. 9.2) be such a string whose tension is T, and

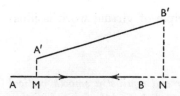

FIG. 9.2

let A'B' be the displaced position of the string. Draw A'M, B'N perpendicular to AB.

Since the angle made by A'B' with AB is very small, A'B' = MN to the first order.

Hence since AB = A'B', we have

$$AM = BN.$$

The virtual work done by the tensions is

$$T \times AM + T(-BN) = 0.$$

(ii) *The reaction R of any smooth surface with which the body is in contact.*

For as the surface is smooth the reaction R is normal to the surface at the point of contact, and is therefore at right angles to the displacement of this point. Hence the virtual work done by R is zero. This case includes the reaction of a smooth hinge.

(iii) *The reaction at any point of contact with a fixed surface on which the body rolls without sliding.*

For the point of contact is instantaneously at rest. Hence the normal reaction and the friction at this point have zero displacement.

(iv) *The reactions between any two bodies of the system considered.*

For the action and reaction between two bodies are equal

and opposite. Hence provided we consider *both* bodies in writing
down the equation of virtual work, the expressions for the
virtual work done by the equal and opposite reactions will
cancel.

This does not, of course, apply to the case where there is
motion of the point of contact of the two bodies and the bodies
are rough, e.g. a rough hinge.

9.4. Examples

The method of applying the principle of virtual work is illustrated in the following examples.

EXAMPLE (i)

*A uniform ladder rests with its upper end against a smooth vertical
wall and its foot on rough horizontal ground; to find the force of friction
at the ground.*

Let AB (Fig. 9.3) represent the ladder. Let l be its length, θ its

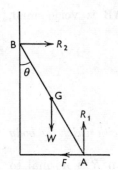

FIG. 9.3

inclination to the vertical, G its centre of gravity, W its weight, and
F the horizontal component of the reaction at the ground.

The height of G above the ground is $\frac{1}{2}l \cos \theta$, and the distance of
A from the foot of the wall is $l \sin \theta$.

Let the ladder be slightly displaced so that it remains in contact
with the wall and ground and θ changes by $d\theta$. The normal reactions
at A and B do not work, and the equation of virtual work reduces to

$$-W \, d(\tfrac{1}{2}l \cos \theta) - F \, d(l \sin \theta) = 0$$

$$\therefore \qquad +\tfrac{1}{2}Wl \sin \theta \, d\theta - Fl \cos \theta \, d\theta = 0$$

$$\therefore \qquad F = \tfrac{1}{2}W \tan \theta.$$

EXAMPLE (ii)

AB *and* BC *are two rods of equal length freely jointed at* B, *the weight of* AB *is* W *and that of* BC *is* 2W. *They are placed in a vertical plane inclined to one another at* 90° *with the ends* A *and* C *on a smooth horizontal plane. The ends* A *and* C *are joined by an inextensible string. Find the tension in this string.*

Let D, E (Fig. 9.4) be the mid-points of the rods, l the length of each, T the tension in the string. Let angle BAC $= \theta$.

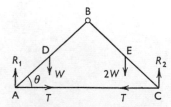

FIG. 9.4

Suppose the system to be slightly displaced so that θ is increased by dθ, A and C remaining in contact with the plane.

The normal reactions at A and C, and the reactions at the hinge do not work.

The height of D and E above AC is $\frac{1}{2}l \sin \theta$, the length of AC is $2l \cos \theta$.

Hence the equation of virtual work is

$$-3W \, d(\tfrac{1}{2}l \sin \theta) - T \, d(2l \cos \theta) = 0$$

$$\therefore \qquad -\tfrac{3}{2}Wl \cos \theta \, d\theta + 2Tl \sin \theta \, d\theta = 0$$

$$\therefore \qquad T = \tfrac{3}{4}W \cot \theta$$

But we are given that $\theta = 45°$ in the equilibrium position.

$$\therefore \qquad T = \tfrac{3}{4}W.$$

Note. It is very important to remember that all lengths of strings and heights of weights *must be expressed in terms of one variable only.*

Also heights must be measured from a *fixed level*, i.e. one which does not alter with the displacement we suppose the system to undergo.

EXAMPLE (iii)

Four equal uniform rods are smoothly jointed to form a rhombus ABCD, *which is placed in a vertical plane with* AC *vertical and* A *resting on a horizontal plane. The rhombus is kept in shape, with the angle BAC equal to θ, by a light string joining B and D. Find the tension in the string.*

Let G_1, G_2, G_3, G_4 (Fig. 9.5) be the mid-points of the rods, W the weight and $2l$ the length of each rod, and let T be the tension in the string.

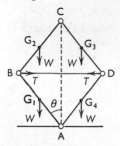

FIG. 9.5

The height of G_1 or G_4 above A is $l \cos \theta$, and that of G_2 or G_3 is $3l \cos \theta$.

The length of BD is $4l \sin \theta$.

Suppose the rhombus displaced so that θ increases by $d\theta$, C descending in the vertical direction CA.

The work done by the weights at G_1 and G_3 is $-2W \, d(l \cos \theta)$.

,, ,, ,, ,, ,, ,, ,, G_2 and G_3 is $-2W \, d(3l \cos \theta)$.

,, ,, ,, ,, ,, tension is $-T \, d(4l \sin \theta)$.

,, ,, ,, ,, ,, reactions at the hinges and at A is zero.

$$\therefore \quad -2W \, d(l \cos \theta) - 2W \, d(3l \cos \theta) - T \, d(4l \sin \theta) = 0$$

$$\therefore \quad +2Wl \sin \theta \, d\theta + 6Wl \sin \theta \, d\theta - 4Tl \cos \theta \, d\theta = 0$$

$$\therefore \quad 4T \cos \theta = 8W \sin \theta$$

$$\therefore \quad T = 2W \tan \theta.$$

It might be noted that we could consider the total weight $4W$ of the four rods concentrated at their centre of gravity which is at a distance $2l \cos \theta$ above A. The work done by the total weight in the small displacement is $-4W \, d(2l \cos \theta)$, that is, $8Wl \sin \theta \, d\theta$, as obtained by treating each rod separately.

Many of the problems in Examples 3.3 can be solved quite easily by the method of virtual work.

This method is especially useful in cases where we require the tension or thrust in a light string or rod which is keeping a jointed framework in shape, as in Examples (ii) and (iii) above.

EXAMPLE (iv)

The ends of a uniform rod AB *of length* 2l *and weight* w *rest on two smooth planes inclined at* 45° *to the horizontal and with their line of intersection horizontal. A light rod* BC *of length* h *is fixed at* B *so that the angle* ABC *is* β, BC *being above* AB. *A weight* W *is applied at* C, *and the frame* ABC *remains always in a vertical plane normal to both of the smooth planes. If the inclination of* AB *to the horizon is* θ, *show that in the position of equilibrium*

$$\tan \theta = \frac{W(h \cos \beta - l)}{wl + Wl + Wh \sin \beta}.$$

Let O (Fig. 9.6) be the intersection of the planes, G the centre of AB, and AD, BE perpendiculars on the horizontal plane.

$$AD = DO \text{ and } BE = OE$$

Also $$AD + BE = DE = AB \cos \theta.$$

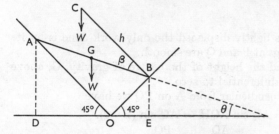

Fig. 9.6

Hence the height of G above OE is $\frac{1}{2}AB \cos \theta = l \cos \theta$.
The height of B above OE is

$$\frac{OB}{\sqrt{2}} = \frac{2l \sin (45 - \theta)}{\sqrt{2}} = l(\cos \theta - \sin \theta)$$

and the height of C above B is $h \sin (\beta + \theta)$,
∴ the height of C above O is $l \cos \theta - l \sin \theta + h \sin (\beta + \theta)$.

For a small displacement of the system in which θ increases by dθ, we have

work done by w is $-wd (l \cos \theta) = +wl \sin \theta \, d\theta$

work done by W is $+Wl \sin \theta \, d\theta + Wl \cos \theta \, d\theta - Wh \cos (\beta + \theta) \, d\theta$

∴ $wl \sin \theta + Wl \sin \theta + Wl \cos \theta - Wh \cos \beta \cos \theta + Wh \sin \beta \sin \theta = 0$

∴ $wl \tan \theta + Wl \tan \theta + Wh \sin \beta \tan \theta = W(h \cos \beta - l)$

∴ $$\tan \theta = \frac{W(h \cos \beta - l)}{wl + Wl + Wh \sin \beta}.$$

EXAMPLE (v)

A uniform isosceles triangle ABC rests with its plane vertical and its two equal sides AB, AC in contact with two smooth fixed pegs P and Q. PQ is horizontal. Prove that the angle between BC and PQ is either zero or $\cos^{-1}\left[\dfrac{BC}{6PQ}(1+\cos A)\right]$.

Let CB produced (Fig. 9.7) meet the horizontal PQ in E, and let angle BEQ $= \theta$.

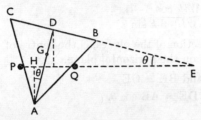

FIG. 9.7

If the triangle is lightly displaced the only work done is by its weight, since the pegs at P and Q are smooth.

We therefore find the height of the centre of gravity, G, above PQ, and equate its differential to zero.

If AH is the perpendicular from A on PQ we have

$$AH = AQ \sin AQH = AQ \sin (B-\theta).$$

Also
$$\frac{AQ}{\sin APQ} = \frac{PQ}{\sin A}$$

or
$$AQ = \frac{PQ}{\sin A}\sin (B+\theta).$$

∴
$$AH = \frac{PQ}{\sin A}\sin (B+\theta) \sin (B-\theta)$$

$$= \frac{PQ}{2 \sin A}(\cos 2\theta - \cos 2B).$$

The height of G above A is $\frac{2}{3}$AD $\cos \theta$, so that the height of G above PQ is

$$\tfrac{2}{3}AD \cos \theta - \frac{PQ}{2 \sin A} (\cos 2\theta - \cos 2B).$$

Hence by the principle of virtual work we get

$$-\tfrac{2}{3}AD \sin \theta \, d\theta + \frac{PQ}{\sin A} \sin 2\theta \, d\theta = 0$$

∴
$$\frac{PQ}{\sin A} \sin \theta \cos \theta = \frac{1}{3} \times \frac{BC}{2}\cot \tfrac{1}{2}A \sin \theta.$$

Hence $\sin \theta = 0$, and then $\theta = 0$, or

$$\cos \theta = \frac{BC}{6PQ}\sin A \cot \tfrac{1}{2}A$$

$$= \frac{BC}{6PQ} \times 2 \cos^2 \tfrac{1}{2}A$$

$$= \frac{BC}{6PQ}(1 + \cos A).$$

EXAMPLE (vi)

Four equal uniform rods, each of weight w, are connected at one end of each by means of a smooth joint, and the other ends rest on a smooth table and are connected by equal strings. A weight W is suspended from the joint. Show that the tension in the strings is

$$\tfrac{1}{4}(W + 2w)a(4l^2 - 2a^2)^{-\frac{1}{2}}$$

where l is the length of each rod and a is the length of each string. (C.S.)

Let OA, OB, OC, OD (Fig. 9.8) represent the rods.
ABCD is a square, and if E is its centre OE is vertical.

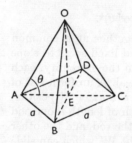

FIG. 9.8

Let angle $OAE = \theta$, then $AE = l \cos \theta$, and the length of a side of the square is $\sqrt{2}l \cos \theta$.

The height of O is $l \sin \theta$, and that of the mid-points of the rods is $\tfrac{1}{2}l \sin \theta$.

Suppose the system displaced so that θ is increased by $d\theta$. Then O ascends and so do the mid-points of the rods, while the strings are shortened by equal small amounts.

The equation of virtual work, if T is the tension in each string, is

$$-4T \, d(\sqrt{2}l \cos \theta) - W \, d(l \sin \theta) - 4w \, d(\tfrac{1}{2}l \sin \theta) = 0.$$

$$\therefore \quad +4T\sqrt{2}l \sin \theta d\theta - Wl \cos \theta d\theta - 4w\tfrac{1}{2}l \cos \theta d\theta = 0$$

$$\therefore \qquad T = \frac{W+2w}{4} \times \frac{\cos\theta}{\sqrt{2}\sin\theta}.$$

But
$$l\sqrt{2}\cos\theta = a$$

$$\therefore \qquad \cos\theta = \frac{a}{l\sqrt{2}} = \frac{a\sqrt{2}}{2l}$$

and
$$\sin\theta = \frac{(4l^2-2a^2)^{\frac{1}{2}}}{2l}$$

$$\therefore \qquad T = \frac{W+2w}{4} \times \frac{a}{2l} \times \frac{2l}{(4l^2-2a^2)^{\frac{1}{2}}}$$

$$= \left(\frac{W+2w}{4}\right)\frac{a}{(4l^2-2a^2)^{\frac{1}{2}}}$$

EXAMPLES 9.1

1. A uniform rod AB of length a and weight W, free to turn about one end A, is supported, in a position in which B is below the level of A, and AB makes an angle θ with the horizontal, by a light string attached to B, passing over a smooth peg, at the same level at A and distant b ($>a$) from it, and carrying a weight W'. Prove that

$$2bW'\tan\theta = W(a^2+b^2-2ab\cos\theta)^{\frac{1}{2}}.$$

2. Two smooth straight wires are joined together at a common extremity so as to contain an angle 2α, and fixed so as to slope downwards in the same vertical plane from the extremity, each of them making an angle α with the vertical. A uniform rod of length $2a$ and weight W is supported in a horizontal position between the wires by two light strings, each of length l, one end of each string being attached to an end of the rod, and the other end being fastened to a small ring of weight W' which can slide on one of the wires. Prove that the distance of either ring from the junction of the wires is

$$\left\{a+l\frac{(W+2W')\cos\alpha}{(W^2+4W'^2\cos^2\alpha+4WW'\cos^2\alpha)^{\frac{1}{2}}}\right\}\operatorname{cosec}\alpha.$$

3. Two equal rods, each of weight wl and length l, are hinged together and placed astride a smooth horizontal cylindrical peg of radius r. Then the lower ends are tied together by a string, and the rods are left at the same inclination θ to the horizontal. Find the tension in the string. If the string is slack show that θ satisfies the equation

$$\tan^3\theta + \tan\theta = l/2r.$$

4. An endless chain of weight W rests in the form of a circular band round a smooth vertical cone which has its vertex upwards. Show that the tension in the chain due to its own weight is $(W/2\pi) \cot \alpha$, where 2α is the vertical angle of the cone.

5. A uniform ladder of length l and weight W is held with its upper end resting against a smooth vertical wall, and with its lower end on a smooth horizontal surface; a man of weight W' stands on the ladder at a distance l' from the lower end. Show that if the ladder is kept from slipping by means of a couple the moment of the couple is equal to

$$(\tfrac{1}{2}Wl + W'l') \sin \theta,$$

where θ is the inclination of the ladder to the vertical. Draw a diagram to indicate the sense of the couple. (H.C.)

6. AB, AC are two equal rods, each of length $2a$ and weight W, smoothly jointed at A, while B and C are connected by a light string, of length $4a$, to whose mid-point is fastened a weight w. The system is placed with AB, AC resting symmetrically over two smooth pegs which are in a horizontal line at a distance $2c$ apart. Show, by the method of virtual work or otherwise, that in the position of equilibrium the rods are inclined to the vertical at an angle θ, where

$$\sin^3 \theta = (2W+w)c/(4w+2W)a.$$

7. Four equal heavy uniform rods are freely jointed to form a rhombus ABCD which is freely suspended from A, and kept in the shape of a square by an inextensible string connecting A and C. Find the tension in the string.

8. A regular hexagon ABCDEF consists of six equal uniform rods, each of weight W and freely jointed together. The hexagon rests in a vertical plane and AB is in contact with a horizontal table; if C and F are connected by a light string, prove that its tension is $W\sqrt{3}$.

9. Six equal uniform rods AB, BC, CD, DE, EF, FA, each of weight W, are freely jointed to form a regular hexagon. The rod AB is fixed in a horizontal position, and the shape of the hexagon is maintained by a light rod joining C and F. Find the thrust in the rod.

10. A regular pentagon ABCDE is formed of five uniform rods, each of weight W, freely jointed at their extremities. It is freely suspended from A, and is maintained in its regular pentagonal form by a light rod joining B and E. Prove that the stress in the rod is $W \cot 18°$. (C.S.)

11. A uniform lamina in the shape of an equilateral triangle ABC of side a has its vertices attached to the vertices of a fixed horizontal equilateral triangle of side b by equal strings AA′, BB′, CC′. A couple of moment M in a horizontal plane acts on the lamina and holds it turned through an angle θ from its undisturbed position. Prove that $\sin \theta = 3hM/abW$, where W is the weight of the lamina and h the distance between the planes ABC and A′B′C′ (in the *disturbed* position). (C.S.)

12. A weightless tripod, consisting of three legs of equal length l, smoothly jointed at the apex, stands on a smooth horizontal plane. A weight W hangs from the apex. The tripod is prevented from collapsing by three inextensible strings, each of length $l/2$, joining the mid-points of the legs. Show that the tension in each string is $W\sqrt{6}/9$. (C.S.)

13. ABCD is a rhombus formed by four light rods smoothly jointed at their ends, and PQ is a light rod smoothly jointed at one end to a point P in BC and at the other end to a point Q in AD. Two forces each equal to F are applied at A and C in opposite directions along AC. Prove that the stress in PQ is

$$\frac{F \times AB \times PQ}{AC(AQ \sim BP)}.$$ (C.S.)

14. A hexagon ABCDEF, consisting of six equal heavy rods of weight W freely jointed together, hangs in a vertical plane with AB horizontal, and the frame is kept in the form of a regular hexagon by a light rod connecting the mid-points of the rods CD and EF. Determine the thrust in the light rod. (C.S.)

15. A uniform lamina in the form of a parallelogram rests with two adjacent sides on two smooth pegs in the same horizontal line at a distance c apart. $2h$ is the length of the diagonal through the intersection of the two sides; α, β, θ are the angles which this diagonal makes with the sides and with the vertical $(\alpha > \beta)$. Prove that $h \sin \theta \sin (\alpha + \beta) = c \sin (2\theta - \alpha + \beta)$. (C.S.)

16. Two heavy uniform bars AB, BC of weights W, W' are connected at B by a rough pivot at which there is a small frictional couple G. The bar AB can turn in a vertical plane about a smooth pivot at A, and the end C of the bar BC can slide in a smooth horizontal groove whose direction passes through A. Prove that the smallest horizontal force at C which will maintain equilibrium is

$$\frac{(W+W') \cos A \cos C}{2 \sin B} - \frac{G \times AC}{AB \times BC \sin B}$$ (N.U.)

17. Six equal heavy beams are freely jointed at their ends to form a

hexagon, and are placed in a vertical plane with one beam resting on a horizontal plane; the mid-points of the two upper slant beams, which are inclined at angle θ to the horizontal, are connected by a light cord. Show that its tension is $6W \cot \theta$, where W is the weight of each beam. (C.S.)

18. A triangle ABC of any shape is formed of light rods smoothly jointed to each other at their ends. It is placed in a vertical plane with A downwards and the rods AB, AC resting on two smooth pegs on a horizontal line. A weight W is suspended from A; prove that the stress in the rod BC is

$$\frac{1}{2} \frac{Wl}{p} \operatorname{cosec}^2 \tfrac{1}{2}\text{A},$$

where $2l$ is the distance between the pegs and p is the perpendicular from A on BC. (C.S.)

19. A tripod of three equal light rods of length l, loosely jointed together at the top, rest on a smooth table, their lower ends being held together by three equal horizontal strings of length a, which join them in pairs. A weight W is hung from the top. Find the tensions in the strings. (C.S.)

20. Six equal uniform rods freely jointed at their extremities form a tetrahedron. If this tetrahedron is placed with one face on a smooth horizontal table, prove that the thrust along a horizontal rod is $w/2\sqrt{6}$, where w is the weight of a rod. (C.S.)

21. A rhombus ABCD of smoothly jointed rods rests on a smooth table with the rod BC fixed in position. The mid-points of AD, DC are connected by a string which is kept taut by a couple L applied to the rod AB. Prove that the tension in the string is $2L/\text{AB} \cos \tfrac{1}{2}\text{ABC}$. (C.S.)

22. Three equal rods, each of weight w, are freely jointed together at one extremity of each to form a tripod and rest with their other extremities on a smooth horizontal plane, each rod inclined at an angle θ to the vertical, equilibrium being maintained by three equal light strips each joining two of these extremities. Prove by means of the principle of virtual work, or in any other manner, that the tension in each string is $W \tan \theta/2\sqrt{3}$. (H.C.)

23. A uniform isosceles triangular lamina is supported vertically with its vertex downwards upon two smooth pegs in a horizontal line. Prove that, if p is the distance of the centre of gravity from the vertex, α the vertical angle, and q the distance between the pegs, there will be one or three positions of equilibrium according as p is $>$ or $<2q \operatorname{cosec} \alpha$. (C.S.)

24. Show that, if a step ladder consists of two equal halves, each of weight W, and a load W' is placed on top of the ladder, the tension in the cord is equal to $(W+W')b^2/2ah$, where $2a$ is the length of the cord, h is the height of the ladder, and $2b$ the distance between its ends. The cord is fastened to the two halves at equal distances from the top.

25. A rhombus of four equal light rods of length a smoothly jointed has its opposite corners joined by elastic strings of the same material and of natural lengths $a\sqrt{2}$ and $\frac{1}{2}a\sqrt{2}$, and rests on a horizontal plane. Show that in equilibrium an angle of the rhombus is 30°.

26. Four equal uniform rods AB, BC, CD, DE, each of weight W, are suspended from fixed points A and E in the same horizontal line. A string connects the mid-points of BC and CD. Show that, if α and β are the angles made by AB and BC with the horizontal, the tension of the string is $W(3 \cot \alpha - \cot \beta)$.

27. Four freely jointed bars form a framework having the form of a parallelogram ABCD. The points B and C are attached by strings to two points X and Y respectively of the bar AD. Apply the principle of virtual work to prove that, the system being in equilibrium, the tensions in the strings will be equal if the angle ABX is equal to the angle DCY.

28. Four equal uniform rods of weight W are freely jointed so as to form a square ABCD which is suspended from A, and is prevented from collapsing by an inextensible string joining the middle points of AB and BC. Prove that the tension of the string is $4W$, and find the magnitude and direction of the reaction at B.
(C.S.)

29. A parallelogram ABCD is formed of uniform heavy rods freely jointed at their extremities. AB is held fixed in a horizontal position, and the parallelogram is maintained in its form so that ADC is an acute angle α by means of a string joining A to a point P in DC. Prove that the tension of the string is $W \times AP \cot \alpha/DP$ where W is half the weight of the parallelogram. (C.S.)

30. Four uniform rods of equal lengths are smoothly jointed together to form a rhombus ABCD. The rods AB and DC can turn in the plane of the rhombus, which is vertical, about smooth pivots at their mid-points, which are in the same vertical line. The system is kept in equilibrium, in the position in which the angle ABC is 2θ by a light rod smoothly jointed to A and C. Prove that the stress in this rod is $(W \sim W') \sin \theta$, where W and W' are the weights of AD and BC respectively.

9.5. Potential energy of a mechanical system

The concept of potential energy can be extended to any mechanical system. We define the *potential energy* V of the system as such that the work done against the internal and external forces acting on the system in any displacement equals the difference of the potential energy of the system in the two positions. More precisely, if the potential energy in one position is V_1 and in the other position is V_2, then the work done *against* the internal and external forces in the displacement is $V_2 - V_1$.

This applies only to *conservative* systems, that is, systems for which the work done in any displacement is independent of the way in which the displacement takes place, and therefore depends only on the initial and final positions of the system. Systems involving frictional or dissipative forces are, in general, not conservative.

If a body of weight mg moves from a height h_1 to a height h_2 above a fixed plane the work done against its weight is $mg(h_2 - h_1)$. The potential energy of the weight at a height h above the fixed plane is therefore $V = mgh + A$, where A is some constant. If we choose $V = 0$ when $h = 0$, then $A = 0$ and $V = mgh$.

This value of V also holds when forces other than its weight act on the body, provided they are such that they do no work in any displacement of the body, for example, reactions between the body and any smooth surfaces with which it is in contact or tensions in inextensible strings attached to it.

The potential energy of any system can be used to find the positions of equilibrium of the system and to examine whether such positions are stable or unstable.

9.6. Stability of equilibrium

Suppose the potential energy of the system is a function of a single variable θ, that is, $V = f(\theta)$. Suppose there is a position of equilibrium corresponding to $\theta = \alpha$.

Then the work done against the forces acting on the system in some small displacement equals δV, and if the displacement is from an equilibrium position this is zero to the first order of small quantities by the principle of virtual work.

Hence at any position of equilibrium the potential energy V

must have a stationary value, that is, $dV/d\theta$ must be zero for the value of θ corresponding to an equilibrium position.

If the position is stable, then the forces acting on the system in any slightly disturbed position must tend to bring the system back to the equilibrium position. This means that a positive amount of work must be done against the forces in any small displacement from the equilibrium position; in other words, V must have a minimum value at a position of stable equilibrium.

Hence, the position $\theta = \alpha$ is stable if $dV/d\theta = 0$ and $d^2V/d\theta^2$ is positive when $\theta = \alpha$.

Similarly, the position $\theta = \alpha$ is unstable if $dV/d\theta = 0$ and $d^2V/d\theta^2$ is negative when $\theta = \alpha$.

If when $\theta = \alpha$, $dV/d\theta = 0$ and $d^2V/d\theta^2 = 0$ the potential energy may have a maximum or a minimum value, but it may have neither. Such cases require further investigation. The equilibrium may be neutral, but frequently it is unstable.

In many cases the only force acting on a body (other than the reactions of smooth constraints) is gravity, in which case $V = Wz$, where W is the weight of the body and z is the height of the centre of gravity above some fixed level.

Hence if $dz/d\theta = 0$ and $d^2z/d\theta^2$ is positive when $z = h$, the position corresponding to $z = h$ is stable.

9.7. The problem of paragraph 7.16 can be dealt with by this method. It exemplifies the power of the technique.

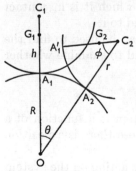

FIG. 9.9

Let O (Fig. 9.9) be the centre of the surface of the fixed body, C_1 that of the upper body, and G_1 its centre of gravity, and A_1, the point of contact.

Let $OA_1 = R$, $A_1C_1 = r$ and $A_1G_1 = h$.

Let C_2, G_2, A'_1 be the displaced positions of C_1, G_1, A_1, and A_2 the new point of contact. Let $\angle A_1OA_2 = \theta$ and $\angle A_2C_2A'_1 = \phi$. Then as before

$$r\phi = R\theta.$$

The height of G_2 above O is

$$z = (R+r) \cos \theta - (r-h) \cos (\theta+\phi)$$

$$= (R+r) \cos \theta - (r-h) \cos\left(\frac{R+r}{r}\right)\theta$$

$$\therefore \qquad \frac{dz}{d\theta} = -(R+r) \sin \theta + \frac{(r-h)(R+r)}{r} \sin \left(\frac{R+r}{r}\right)\theta$$

$$\text{and} \frac{d^2z}{d\theta^2} = -(R+r) \cos \theta + \frac{(r-h)(R+r)^2}{r^2} \cos \left(\frac{R+r}{r}\right)\theta$$

It is clear that $\theta = 0$ makes $dz/d\theta = 0$, and therefore gives a position of equilibrium.

This is stable if $d^2z/d\theta^2$ is positive when $\theta = 0$, i.e. if

$$\frac{(r-h)(R+r)^2}{r^2} > (R+r),$$

or $$Rr+r^2-hR-hr > r^2$$

or $$Rr > h(R+r)$$

or $$\frac{1}{h} > \frac{1}{r} + \frac{1}{R}.$$

If $1/h = 1/r + 1/R$, then $d^2z/d\theta^2 = 0$, and we must examine the value of the higher differential coefficients.

We have

$$\frac{d^3z}{d\theta^3} = (R+r) \sin \theta - \frac{(r-h)(R+r)^3}{r^3} \sin \left(\frac{R+r}{r}\right)\theta$$

and this vanishes when $\theta = 0$.

$$\frac{d^4z}{d\theta^4} = (R+r) \cos \theta - \frac{(r-h)(R+r)^4}{r^4} \cos \left(\frac{R+r}{r}\right)\theta.$$

When $\theta = 0$, this becomes

$$(R+r)\left[1-\frac{(r-h)(R+r)^3}{r^4}\right]$$

$$=\left(\frac{R+r}{r^4}\right)\left[r^4-\left(r-\frac{rR}{R+r}\right)(R+r)^3\right]$$

$$=\frac{(R+r)}{r^4}\left[r^4-r^2(R+r)^2\right]$$

and this is negative, so that z is a maximum and the equilibrium is unstable.

9.8. EXAMPLE (i)

Three equal uniform bars AB, BC, CD, each of length 2a, are smoothly jointed at B, C and rest with BC horizontal and AB, CD each on small smooth pegs at the same level at a distance $2(a+b)$ apart.

Show that, if $2a>3b$, there are two positions of equilibrium, and determine which of them is stable.

If $2a = 3b$, show that there is only one position of equilibrium and that it is unstable. (N.U.)

Let P, Q (Fig. 9.10) be the positions of the pegs, and let $\angle BPQ = \theta$.

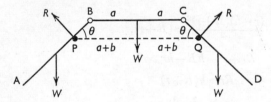

FIG. 9.10

The height of the centre of gravity of AB and CD above PQ is

$$b \tan \theta - a \sin \theta$$

and that of BC is $b \tan \theta$.

For a small displacement $d\theta$, we have

$$-2W\, d(b \tan \theta - a \sin \theta) - W\, d(b \tan \theta) = 0$$

since the reactions on the pegs at P and Q and at the hinges B and C do no work.

$$\therefore \qquad 2b \sec^2 \theta - 2a \cos \theta + b \sec^2 \theta = 0$$

$$\therefore \qquad 2a \cos \theta = 3b \sec^2 \theta$$

$$\therefore \qquad \cos^3 \theta = \frac{3b}{2a} \text{ or } \cos \theta = \sqrt[3]{\frac{3b}{2a}}.$$

Hence for equilibrium to be possible $2a > 3b$.

There are two possible values of θ, and if one is α the other is $-\alpha$.

To determine which is stable we differentiate the height of the centre of gravity of the system twice.

This height is

$$z = \tfrac{1}{3}(2b \tan \theta - 2a \sin \theta + b \tan \theta)$$

$$\therefore \qquad 3\frac{dz}{d\theta} = 2b \sec^2 \theta - 2a \cos \theta + b \sec^2 \theta.$$

$$\therefore \qquad 3\frac{d^2z}{d\theta^2} = +\frac{6b}{\cos^3 \theta} \sin \theta + 2a \sin \theta.$$

We note that $dz/d\theta = 0$ when $\cos^3\theta = 3b/2a$, in which case $d^2z/d\theta^2 = 2a \sin \theta$.

If θ is negative $d^2z/d\theta^2$ is negative and the equilibrium is unstable.

If θ is positive $d^2z/d\theta^2$ is positive and the equilibrium is stable.

The positive shown in the figure is the stable one.

If $2a = 3b$, $\cos \theta = 1$, $\theta = 0$, and there is only one position of equilibrium; all the rods being horizontal.

In this case $d^2z/d\theta^2 = 0$, and the usual test for a maximum or minimum fails.

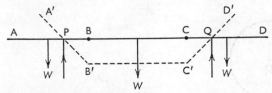

FIG. 9.11

Let the system be displaced into the position A'B'C'D' (Fig. 9.11) through a small angle θ.

$$PB = \tfrac{2}{3}a, \text{ and } BB' = \tfrac{2}{3}a \tan \theta.$$

$PB' = \tfrac{2}{3}a \sec \theta$, and the distance of the centre of gravity of A'B' from P is therefore

$$a - \tfrac{2}{3}a \sec \theta.$$

The centres of gravity of AB and CD have therefore risen a distance

$$a \sin \theta - \tfrac{2}{3}a \tan \theta.$$

The centre of gravity of BC has descended a distance $\tfrac{2}{3}a \tan \theta$.

Hence the loss of potential energy is

$$\tfrac{2}{3}Wa \tan \theta - 2Wa \sin \theta + \tfrac{4}{3}Wa \tan \theta = 2Wa(\tan \theta - \sin \theta).$$

Now this expression is always positive for small values of θ.

Hence for this displacement the potential energy decreases and the equilibrium is unstable.

EXAMPLE (ii)

A smooth cylinder of radius a is fixed with its axis horizontal and at a distance c from a smooth vertical wall. A uniform rod of length l rests at right angles to the axis of the cylinder with one end in contact with the cylinder and the other end in contact with the wall. Show that in the position of equilibrium the inclination ϕ of the rod to the vertical satisfies the equation

$$a \sin \phi = (c - l \sin \phi)\sqrt{(1 + 3 \cos^2 \phi)}.$$

Also show that the equilibrium is unstable.   (O.C.)

Suppose AB (Fig. 9.12) indicates the position of the rod, and O

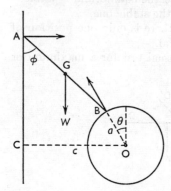

Fɪɢ. 9.12

the axis of the cylinder. Then $AB = l$, $OB = a$, and $OC = c$. Suppose OB is inclined at an angle θ to the vertical.

The height of the centre of gravity G of the rod above OC is

$$\tfrac{1}{2}l \cos \phi + a \cos \theta.$$

Therefore the potential energy of the rod is

$$V = W(\tfrac{1}{2}l \cos \phi + a \cos \theta)$$

for the smooth reactions acting on the rod at A and B do no work in any displacement of the rod.

Differentiating with respect to ϕ we get

$$\frac{\mathrm{d}V}{\mathrm{d}\phi} = W\left(-\tfrac{1}{2}l \sin \phi - a \sin \theta \frac{\mathrm{d}\theta}{\mathrm{d}\phi}\right)$$

But $\qquad\qquad c = l\sin\phi + a\sin\theta$

$\therefore\qquad\qquad 0 = l\cos\phi + a\cos\theta\,\dfrac{d\theta}{d\phi}$

Hence $\qquad\dfrac{dV}{d\phi} = W(-\tfrac12 l\sin\phi + l\tan\theta\cos\phi)$

which is zero when $\tan\phi = 2\tan\theta$.

This relation might also have been derived from Fig. 9.12 so drawn that the three forces acting on the rod AB met at a point. We note too that θ has the same sign as ϕ and is less than ϕ.

It follows that there is a position of equilibrium given by the equations

$$\tan\phi = 2\tan\theta$$

and $\qquad\qquad c - l\sin\phi = a\sin\theta$

Eliminating θ we get

$$c - l\sin\phi = a\sin\phi/\sqrt{(1+3\cos^2\phi)}.$$

To examine the stability of the position of equilibrium we examine the sign of $d^2V/d\phi^2$.

We have

$$\dfrac{dV}{d\phi} = -Wl(\tfrac12\sin\phi - \tan\theta\cos\phi)$$

$$\therefore\quad \dfrac{d^2V}{d\phi^2} = -Wl\left(\tfrac12\cos\phi + \tan\theta\sin\phi - \sec^2\theta\cos\phi\dfrac{d\theta}{d\phi}\right)$$

$$= -Wl[\tfrac12\cos\phi + \tan\theta\sin\phi + (l/a)\cos^2\phi\sec^3\theta]$$

which is negative since ϕ and θ both lie between $0°$ and $90°$.

Hence the position of equilibrium is unstable.

EXAMPLES 9.2

1. A uniform solid hemisphere hangs with its plane face towards a smooth vertical wall suspended by a string of length equal to the diameter, one end being fastened to the wall and the other to the rim of the hemisphere. Find the inclination of the string to the wall in equilibrium, and determine whether the equilibrium is stable or not.   (C.S.)

2. A uniform heavy rod of length $2l$ rests with its ends on a fixed smooth parabola with axis vertical and vertex downwards (latus rectum $= 4a$). Show that if $l>2a$ there are three positions of equilibrium and that the horizontal position is then unstable, but that if $l\le 2a$ the only position of equilibrium is horizontal.
                                                              (C.S.)

3. The ends A, B of a uniform rigid rod of length $2l$ are constrained to move on two fixed smooth wires OA, OB at right angles, each of which is inclined at $45°$ to the vertical. O is the highest point. Prove that when the rod is horizontal the equilibrium is stable.
(C.S.)

4. A solid cylinder, of height h and radius r, is suspended from a point in a smooth wall by means of a string of length l which is attached to a point on the circumference of the middle section of the cylinder. Show that, when the string is inclined at an angle θ to the wall, and the axis of the cylinder is in the same vertical plane as the string, the centre of gravity is at a distance $l \cos \theta + (2lr/h) \sin \theta$ below the level of the upper end of the string. Thence find the value of θ in the equilibrium position.

5. A uniform sphere of weight W has a particle of weight w attached to its surface. Show there are two distinct positions of equilibrium on a rough inclined plane of inclination α provided $(W+w) \sin \alpha < w$. Prove that the position in which the particle is below the centre is the stable position.

6. A heavy sphere, with its centre of gravity at a distance from the centre equal to one-third of the radius, rests on an inclined plane which is sufficiently rough to prevent slipping; find the greatest possible inclination of this plane to the horizon.

   If the inclination is less than this, show that there are two possible positions of equilibrium, and that one of these is stable and the other unstable.

7. A rigid triangular frame consisting of equal heavy rods hangs in equilibrium with the two upper rods in contact with smooth pegs in the same horizontal line. Prove that, if the distance between the pegs is greater than a quarter the length of each rod, there is an oblique position of equilibrium which is stable.
(C.S.)

8. Two small rings P, Q can slide on the upper part of a smooth circular wire in a vertical plane, and are attached by strings of equal length to a third ring R which is free to slide along the vertical diameter of the circle. The weights of the three rings are equal. Prove that, if the lengths of the strings are less than the radius of the circle, there is a stable position of equilibrium in which R is at the centroid of the triangle POQ, where O is the centre of the circle.
(C.S.)

9. AB is the horizontal diameter of a circular wire whose plane is vertical. A load of mass M at the lowest point C can slide on the wire and is attached to two strings which pass through small

fixed rings at A, B. To the other ends of the strings are attached equal particles m which hang freely. Find the potential energy of the system when it is displaced so that the radius to M makes an angle θ with the vertical. Deduce that the equilibrium with M at C is stable if $m < M\sqrt{2}$. (C.S.)

10. A light lever AOB of length $2a$ can turn freely about its mid-point O. A weight $2w$ hangs from the end A. A light rod BC of length b is smoothly jointed to AB at B, and the end C supports a weight w and is restricted by a frictionless constraint to move in the downward vertical through O. Investigate the positions of limiting equilibrium and their stability according as a is greater than, equal to, or less than b. Also show that if, in the case $b > a$, turning at the joint B is resisted by a constant friction couple F there is a position of limiting equilibrium in which the lever makes an angle θ with the vertical given by

$$4wa^3\, F \sin \theta \cos^2 \theta = (b^2 - a^2)(wa \sin \theta - F)^2. \qquad \text{(C.S.)}$$

11. A thin wire has the form of a circle in a vertical plane with centre C. A, B are pegs attached to the wire so that CA, CB make angles α on opposite sides of the downward vertical through C. A small ring of mass M can slide on the wire, and is attached to two strings passed over the pegs with masses m hanging from their ends. Write down the potential energy of the system when the radius to M makes an angle θ with the vertical. Hence discuss the stability of equilibrium positions in the cases $M > m \sin \frac{1}{2}\alpha$ and $M < m \sin \frac{1}{2}\alpha$. (C.S.)

12. A rectangular block of height $2h$ rests with two faces vertical and its base in contact with a fixed rough cylinder of radius a whose axis is horizontal, the base of the block making an angle α with the horizontal plane. Find the change in the potential energy when the block is rolled on the cylinder through a small angle θ; and show that if $h = a \cos^2 \alpha$, the block is in neutral equilibrium to a first approximation, but is actually in unstable equilibrium. (C.S.)

13. A circular cylinder of radius a and weight W having its centre of gravity at a distance c from its axis rests in stable equilibrium on a horizontal plane. A uniform plank of thickness $2b$ and weight w is placed on the cylinder, so as to rest in a horizontal position with its length perpendicular to the axis of the cylinder. Prove that the system is in stable equilibrium for small rolling displacements if $(a - b) > wa^2/Wc$. (C.S.)

14. A uniform rod of length $2l\sqrt{3}$ has attached to its ends a string of length $4l$, which passes through a small smooth ring fixed in a

smooth vertical wall. Prove that there is a position of equilibrium in which *one* end of the rod presses against the wall, and that the length of the string in contact with the wall is then l. Show also that the position of equilibrium is unstable.   (H.C.)

15. The cross-section of a uniform homogeneous prism is a rectangle ABCD. It is placed between the two planes which have a common horizontal straight line and are each inclined to the horizontal at an angle α, so that the edge through A is in contact with one plane and the edge through B with the other. If θ is the inclination of AB to the horizontal, prove that the height of the centre of gravity of the prism above the common line of the planes is $\frac{1}{2}(AB \tan \alpha + BC) \cos \theta$. Hence show that if the planes are smooth the position in which AB is horizontal is a position of *unstable* equilibrium. What other positions of equilibrium are there with the prism between the planes with the face ABCD vertical, and which, if any, of them are stable?   (H.S.C.)

16. A uniform rough plank of weight W and thickness $2b$ rests horizontally in equilibrium across a fixed rough cylinder of radius a, and a particle of weight w is fixed to the plank vertically above the axis of the cylinder. Prove that, if $(W+w)a > b(W+2w)$ equilibrium is stable, and that if $(W+w)a < b(W+2w)$ it is unstable. Show that in the former case there are two oblique positions of equilibrium which are unstable, provided that the friction is great enough to prevent slipping.   (C.S.)

17. Explain what is meant by the stability of a position of equilibrium. How may it be determined? Illustrate your answer by considering the case of a uniform solid right circular cone resting on a fixed spherical surface, the surfaces considered being rough.   (C.S.)

18. A uniform regular hexagonal lamina ABCDEF rests in a vertical plane with the sides AB and CD in contact with two fixed parallel smooth horizontal rods in the same horizontal plane. Show that the only position of equilibrium is the one in which BC is horizontal and that it is stable.   (C.S.)

19. A uniform hemisphere of weight W and radius a is placed symmetrically on top of a fixed sphere of radius b, the curved surfaces being in contact and sufficiently rough to prevent sliding. Show that if the hemisphere is rolled through a small angle θ the gain of potential energy is approximately

$$\frac{1}{16} \frac{W(3b-5a)a\theta^2}{(a+b)},$$

and deduce the condition for stability.   (C.S.)

20. The two ends of a uniform rod of length l and weight w are attached to the two ends of a light elastic string of natural length l and modulus E. The rod is suspended in a horizontal position by hanging the string over a smooth nail. Write down the potential energy of the system when the two portions of the string make an angle θ with the rod, and prove that in the position of equilibrium

$$\tan \theta - \sin \theta = w/2E. \qquad \text{(N.U.)}$$

21. A cubical box whose sides are of length $2a$ is in equilibrium with a pair of adjacent sides resting one on each of two parallel horizontal smooth supports at the same level, whose distance apart is $c(>2a)$. Show that, if $a^2 > 2c^2$, there is only one position of equilibrium, the symmetrical position. Show further that, if $a^2 < 2c^2$, other positions of equilibrium exist, determined by $2c \cos \theta = a\sqrt{2}$, where θ is the angle that the plane joining the centre of the box to the lowest edge makes with the vertical.

(H.S.C.)

22. Two equal uniform rods are fastened at right angles to one another at a common end, and, with that end uppermost, are free to slide each in contact with one of two smooth rails in the form of right circular cylinders, of equal radius a, which have their axes parallel, in the same horizontal plane, and at a distance c apart. Prove that, if the length of either rod lies between $4(a+c)$ and $4(a+c\sqrt{2})$, there are three configurations of equilibrium, that in which the rods are equally inclined to the vertical being stable and the other two unstable. (C.S.)

23. A rigid square framework of rods is hung over two smooth pegs in the same horizontal line. Prove that if the distance between the pegs is greater than $\frac{1}{4}$ (diagonal) and less than $\frac{1}{2}$ (side) of the square an unsymmetrical position of equilibrium exists, and show that it is stable. (H.C.)

24. A uniform rod AB, of weight W and length a, is free to turn about its end A, which is fixed. A fine string is attached to B and, passing over a smooth pulley vertically above A and at a distance $h(>a)$ from A, carries a weight w. Show that if

$$\frac{h+a}{h} > \frac{2w}{W} > \frac{h-a}{h},$$

the two positions in which AB is vertical are stable positions of equilibrium. (H.C.)

25. A uniform square trap-door ABCD, of weight W, can turn freely about the edge AB, which is horizontal. One end of an elastic

cord, of length $\frac{1}{2}$BC and modulus of elasticity $\frac{1}{2}W$, is fastened to a hook which is fixed at a point vertically above the mid-point of AB and at a height equal to BC; the other end is attached to the mid-point of CD. Prove that the door can rest in equilibrium at an angle of 60° to the vertical; and prove that the equilibrium is stable. (H.C.)

26. Explain how the potential energy of a system may be used to find positions of equilibrium and to determine their stability.

A uniform hemisphere of mass m and radius r is placed with its plane face horizontal and its curved surface in contact with the highest point of a fixed rough sphere of radius R. Prove that, when the hemisphere is displaced by rolling so that the radius from the centre of the fixed sphere to the point of contact is inclined at an angle θ to the vertical, the potential energy is

$$V = mg\left\{(R+r)\cos\theta - \tfrac{3}{8}r\cos\frac{(R+r)}{r}\theta\right\} + \text{constant.}$$

Prove that the equilibrium in the position $\theta = 0$ is unstable if $5r \geqslant 3R$.

If $R = 2r$, find another position of equilibrium and discuss its stability. (O.C.)

27. A uniform rod of length l and weight W has one end A freely jointed to one end of a *light* rod of length b and passes through a smooth ring which is fixed at C. The other end of the light rod is freely jointed to a fixed point O which is vertically above C at a distance a from it. Prove that the potential energy of the system is

$$W\left\{\frac{l(x-a)}{2(a^2+b^2-2ax)^{\frac{1}{2}}}-x\right\} + \text{constant,}$$

where x is the depth of A below the level of O.

Show that, if an oblique position of equilibrium exists, then x satisfies the equation

$$l^2(b^2-ax)^2 = 4(a^2+b^2-2ax)^3. \tag{O.C.}$$

28. Explain how the potential energy of a system may be used to find positions of equilibrium and to determine their stability.

A uniform rod AOB of length $2a$ is free to turn in a vertical plane about its mid-point O. A weight $3w$ is suspended from A and a light rod BC of length $b(>a)$ is smoothly jointed to AB at B. The end C of the rod BC carries a weight $2w$ and is restricted by a frictionless constraint to move along the downward vertical through O. Show that, if A is above the level of O and C is below the level of B, the potential energy V of the system is given by

$$V = wa \cos \theta - 2w\sqrt{(b^2 - a^2 \sin^2 \theta)} + \text{constant},$$

where θ is the angle COB.

Show that, if $b < 2a$, there is a position of equilibrium with the rod inclined to the vertical, and discuss its stability. (O.C.)

29. A smooth rigid wire is in the form of a parabola and its axis, and is fixed with the axis vertical and the vertex downwards. A uniform rod PQ, of length l, is constrained by small strings at its ends to slide on the wire, so that one end P is on the parabolic portion of the wire and the other end Q on the axis. Show that, if $l > a$, where $4a$ is the latus-rectum of the parabola, there is a position of equilibrium in which Q is higher than P; and examine its stability.

30. A cylinder rests in equilibrium on a table. Show that if the radius of curvature of any cross-section at its point of contact is greater than the height of the centre of gravity the equilibrium is stable. Show that the stable equilibrium of an elliptic cylinder lying with its generators horizontal on a table cannot be rendered unstable by loading it at the top if $e > 1/\sqrt{2}$. (C.S.)

9.9. Miscellaneous examples

The following examples illustrate the methods of this and the preceding chapters. Some of them are of a more difficult type than those already considered.

EXAMPLE (i)

A solid hemisphere rests with its base in an inclined position at an angle θ to the horizontal, its curved surface resting on a horizontal plane (coefficient of friction μ) and against a vertical plane (coefficient of friction μ'). If the hemisphere is on the point of slipping, show that $c \sin \theta/a = \mu(1+\mu')/(1+\mu\mu')$, where a is the radius of the hemisphere and the centre of gravity is on the axis of symmetry at a distance c from the centre.

If $\mu = \mu'$ and $c/a = 3/8$, show that there is no position of limiting equilibrium for the hemisphere if $5\mu > \sqrt{31} - 4$. (C.S.)

Let O (Fig. 9.13) be the centre of the hemisphere, G its centre of gravity, C and D the points of contact with wall and ground. Let λ, λ' be the angles of friction at the ground and wall.

The normals at C and D pass through O, and the resultant reactions at C and D (making angles λ' and λ with CO and DO) must meet at E on the vertical through G. The required relation must follow from the geometry of the figure.

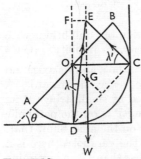

FIG. 9.13

Draw EF perpendicular to DO produced.

Then, $\qquad \tan \lambda = \text{EF}/\text{FD} = c \sin \theta/\text{FD}$

But, $\qquad \text{FD} = \text{FO}+a$

$\qquad\qquad\qquad = (a-c \sin \theta) \tan \lambda' + a$

$\therefore \qquad c \sin \theta = (a-c \sin \theta) \tan \lambda \tan \lambda' + a \tan \lambda$

$\qquad\qquad\qquad = (a-c \sin \theta)\mu\mu' + a\mu$

$\therefore \qquad \dfrac{c \sin \theta}{a} = \dfrac{\mu(1+\mu')}{1+\mu\mu'}.$

When $\mu = \mu'$ and $\dfrac{c}{a} = \dfrac{3}{8}$, we have

$$\sin \theta = \frac{8\mu(1+\mu)}{3(1+\mu^2)}$$

Hence for limiting equilibrium to be possible we must have

$$8\mu+8\mu^2<3+3\mu^2$$

or $\qquad\qquad\qquad 5\mu^2+8\mu-3<0$

or $\qquad\qquad\qquad (\mu+\tfrac{4}{5})^2<\tfrac{31}{25}$

or $\qquad\qquad\qquad 5\mu<\sqrt{31}-4.$

EXAMPLE (ii)

An aeroplane rests on the ground and is supported in front by a pair of wheels of radius a and behind by a tail skid which touches the ground at a distance l from the line joining the points of contact of the wheels. It is found that, if the aeroplane is tilted up through an angle θ, the vertical force required to support the skid is half the original pressure on the ground. Prove that the aeroplane will tip on its nose if tilted through a total angle greater than

$$\cot^{-1}\tfrac{1}{2}\left(\cot\theta-\frac{a}{l}\right). \qquad \text{(C.S.)}$$

Fig. 9.14 represents a vertical section through the centre of the plane, C the axle, A the line of contact of the wheels, B the point of contact of the tail skid.

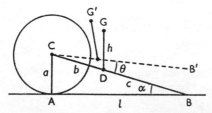

Fɪɢ. 9.14

Let G be taken as the centre of gravity, and let the vertical through G meet CB in D, where $CD = b$, $DB = c$; also $GD = h$, $\angle ABC = \alpha$.

If R is the pressure on the ground at B, taking moments about A,

$$Rl = Wb \cos \alpha.$$

When tilted through an angle θ the distance of the vertical through G′ from A becomes $b \cos (\alpha-\theta)-h \sin \theta$, and the distance of the vertical through B′ from A becomes $(b+c) \cos (\alpha-\theta)$.

Taking moments about A, we get

$$\tfrac{1}{2}R(b+c) \cos (\alpha-\theta) = W[b \cos (\alpha-\theta)-h \sin \theta]$$

$$\therefore \quad h \sin \theta = b \cos (\alpha-\theta)-\frac{b \cos \alpha}{2l}(b+c) \cos (\alpha-\theta)$$

$$= b \cos (\alpha-\theta)\left[1-\frac{(b+c) \cos \alpha}{2l}\right] = \tfrac{1}{2}b \cos (\alpha-\theta).$$

Now the plane will tip over when tilted through an angle ϕ so that G comes vertically over C, i.e. when $b \cos (\alpha-\phi) = h \sin \phi$.

Hence ϕ is given by

$$b \cos (\alpha-\phi) = \frac{b \cos (\alpha-\theta) \sin \phi}{2 \sin \theta}$$

$$\therefore \quad \cos \alpha \cot \phi+\sin \alpha = \tfrac{1}{2} \cos \alpha \cot \theta+\tfrac{1}{2} \sin \alpha$$

$$\therefore \quad \cot \phi = \tfrac{1}{2} \cot \theta-\tfrac{1}{2} \tan \alpha = \tfrac{1}{2}\left(\cot \theta-\frac{a}{l}\right).$$

EXAMPLE (iii)

A light string of length l has a heavy ring of weight W threaded on it, and has one end fixed to the upper end O of a rough rod fixed at an angle α to the horizontal, the other end of the string being attached to another ring P of weight w which slides along the rod. If OP = x, and the two portions of the string are inclined at an angle 2θ, show that the positions of limiting equilibrium are given by x = l sin θ/cos α, where θ has either of the values given by

$$W \tan \theta = (2w+W) \tan (\alpha \pm \lambda),$$

λ being the angle of friction.   (Ex.)

Let A (Fig. 9.15) be the position of the weight W, OAP the string, and P the position of the ring w on the rod.

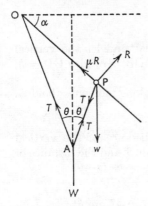

Fig. 9.15

Since W is smooth and free to move on the string, AO and AP make equal angles θ with the vertical.

The projections of OA and AP on the horizontal are together equal to $l \sin \theta$, and the projection of OP is OP cos α.

∴ $x \cos \alpha = l \sin \theta.$

Let R be the normal reaction at P, T the tension in the string, and μ the coefficient of friction. If we assume the ring is on the point of slipping down the rod the frictional force μR is in the direction shown in the figure.

Resolving vertically for W,

$$2T \cos \theta = W \qquad\qquad (i)$$

Resolving along the rod for w, we get

$$\mu R = w \sin \alpha + T \sin (\alpha - \theta) \tag{ii}$$

Resolving perpendicular to the rod for w, we have

$$R = w \cos \alpha + T \cos (\alpha - \theta) \tag{iii}$$

$$\therefore \qquad \mu w \cos \alpha + \mu T \cos (\alpha - \theta) = w \sin \alpha + T \sin (\alpha - \theta)$$

$\therefore$ using (i)

$$\mu w \cos \alpha + \frac{\mu W \cos (\alpha - \theta)}{2 \cos \theta} = w \sin \alpha + \frac{W \sin (\alpha - \theta)}{2 \cos \theta}$$

$$\therefore \ 2w\left[\sin \alpha - \frac{\sin \lambda \cos \alpha}{\cos \lambda} \right]$$

$$= W(\mu \cos \alpha + \mu \sin \alpha \tan \theta - \sin \alpha + \cos \alpha \tan \theta)$$

$$\therefore \ 2w\frac{\sin (\alpha - \lambda)}{\cos \lambda} = W\left[\frac{\sin \lambda \cos \alpha}{\cos \lambda} - \sin \alpha + \tan \theta\left(\frac{\sin \lambda \sin \alpha}{\cos \lambda} + \cos \alpha \right) \right]$$

$$= W\left[\tan \theta\frac{\cos (\alpha - \lambda)}{\cos \lambda} - \frac{\sin (\alpha - \lambda)}{\cos \lambda} \right]$$

$$\therefore \qquad W \tan \theta \cos (\alpha - \lambda) = (W + 2w) \sin (\alpha - \lambda)$$

$$\therefore \qquad W \tan \theta = (W + 2w) \tan (\alpha - \lambda).$$

The sign of λ in this result is changed if we assume that the ring is on the point of slipping in the other direction, that is, towards O.

EXAMPLE (iv)

A cylinder rests inside a fixed hollow cylinder whose axis is horizontal and subtends an angle 2α at this axis. A cylinder equal to the former is placed so as to rest in contact with both without disturbing the former. Show that if the surfaces are equally rough the angle of friction must be greater than each of $\frac{1}{4}\pi - \frac{1}{2}\alpha$ and α. (C.S.)

Let A, B, C (Fig. 9.16) be the centres of the cylinders, and D, E, F the points of contact.

Consider the equilibrium of the lower cylinder.

The weight and the resultant reaction at D both act at this point so that for equilibrium the resultant reaction at F must also pass through D.

Hence the angle of friction at F must be not less than BFD, which is equal to $\frac{1}{4}\pi - \frac{1}{2}\alpha$.

(The resultant reaction at D acts between DB and DF.)

Consider the equilibrium of the upper cylinder.

The resultant reaction at F meets the vertical through C at H on the circumference. Hence the resultant reaction at E must also pass through H.

Now $\angle\mathrm{HCA} = \angle\mathrm{BAC} = 2\alpha,$

and $\angle\mathrm{HEC} = \tfrac{1}{2}\angle\mathrm{HCA} = \alpha.$

Hence the angle of friction at E must be not less than α. Since the surfaces are equally rough, the angle of friction must not be less than either $\tfrac{1}{4}\pi - \tfrac{1}{2}\alpha$ or α.

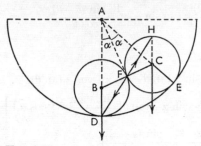

Fig. 9.16

EXAMPLE (v)

Show that two cylindrical logs, of equal radius but unequal weights W and W', where W' > W, can rest in contact on an inclined plane with their axes horizontal and the heavier log uppermost, if the coefficient of friction μ (supposed the same at each line of constant) exceeds $\dfrac{W'+W}{W'-W}$ and the inclination of the plane is less than

$$\tan^{-1}\frac{2\mu W'}{(\mu+1)(W'+W)}.$$ (C.S.)

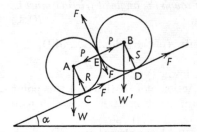

Fig. 9.17

Let A, B (Fig. 9.17) be the centres of the logs, C, D their points of contact with the plane, and E the point of contact with each other. Let R, S, P be the normal reactions at C, D, E respectively.

By taking moments about the centres it is clear that the forces of friction at each contact must be equal; let each be F.

If slipping can take place at E the lower log will roll down whatever the inclination of the plane.

To prevent the lower log rolling about C, we must have

$$Fa = Pa + Wa \sin \alpha \qquad \text{(i)}$$

and to prevent the upper log rolling about D,

$$Fa + Pa = W'a \sin \alpha \qquad \text{(ii)}$$

These two equations give

$$F = \tfrac{1}{2}(W + W') \sin \alpha$$

and
$$P = \tfrac{1}{2}(W' - W) \sin \alpha.$$

But μP must not be less than F, and hence

$$\mu \nless \frac{W' + W}{W' - W}.$$

If μ is greater than this value no slipping can take place at E. In this case, however, it is possible that the lower log may roll about C while the upper one rotates in such a way that the two roll on each other at E while the point of contact at D slips downwards.

This will happen if the friction at D is insufficient.

Now taking moments about E for the upper log, we have

$$Sa = W'a \cos \alpha - Fa$$

$$\therefore \qquad S = W' \cos \alpha - \tfrac{1}{2}(W + W') \sin \alpha.$$

And if there is to be no slipping downwards at D,

$$\mu S + P \nless W' \sin \alpha.$$

$$\therefore \qquad \mu W' \cos \alpha - \tfrac{1}{2}\mu(W + W') \sin \alpha + \tfrac{1}{2}(W' - W) \sin \alpha \nless W' \sin \alpha$$

$$\therefore \qquad \tan \alpha [W' - \tfrac{1}{2}(W' - W) + \tfrac{1}{2}\mu(W + W')] \ngtr \mu W'$$

$$\therefore \qquad \tan \alpha \left[(W + W')\frac{1 + \mu}{2}\right] \ngtr \mu W'$$

$$\therefore \qquad \tan \alpha \ngtr \frac{2\mu W'}{(W + W')(1 + \mu)}.$$

It should be noticed that unless W' is greater than W, equations (i) and (ii) cannot both hold, and equilibrium is impossible.

EXAMPLE (vi)

Two light rods, AC, BC, *each 1 m long, are freely jointed to fixed points A and B in a horizontal plane 1 m apart; from C hangs a weight of mass* 10 *kg the system being kept in equilibrium, with the plane* ABC *inclined to the vertical, by a string* CD 1·5 *m long, attached to a point* D *in the same horizontal plane as* A *and* B, *the triangle* ABD *being equilateral. Find the thrusts in the rods and the tension in the string.* (Ex.)

Let E (Fig. 9.18) be the mid-point of AB.

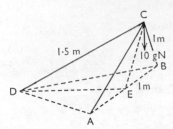

FIG. 9.18

$$DE = CE = \sqrt{3}/2$$

∴ $$\cos DEC = \frac{\frac{3}{4}+\frac{3}{4}-\frac{9}{4}}{2\times\frac{3}{4}} = -\frac{1}{2}$$

∴ $$\angle DEC = 120°.$$

By symmetry, the thrusts in BC and AC are equivalent to a thrust along EC, say R N; let T N be the tension in the string.

Resolving vertically,

$$R\cos 30° - T\cos 60° = 10\times 9·8 \tag{i}$$

Resolving horizontally,

$$R\cos 60° = T\cos 30° \tag{ii}$$

∴ $$R = \sqrt{3}T$$

and from (i) $$3T - T = 196$$

∴ $$T = 98$$

∴ $$R = 98\sqrt{3}$$

If T' N is the thrust in AC and BC,

$$2T'\cos 30° = R = 98\sqrt{3}$$

∴ $$T' = 98.$$

EXAMPLE (vii)

A rhombus is formed of rods each of weight W *and length* l *with smooth joints. It rests symmetrically with its two upper sides in contact with two smooth pegs at the same level and at a distance apart* 2a. *A weight*

*W' is hung at the lowest point. If the sides of the rhombus make an angle
θ with the vertical, show that*

$$\sin^3 \theta = \frac{a(4W + W')}{l(4W + 2W')}.$$ (C.S.)

Let ABCD (Fig. 9.19) represent the rhombus, P and Q the
positions of the pegs.

The depth of the centre of gravity of AB or AD below PQ is

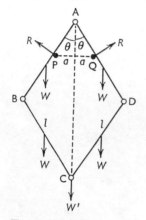

FIG. 9.19

$\frac{1}{2}l \cos \theta - a \cot \theta$, and the depth of the centre of gravity of BC or AD
below PQ is $\frac{1}{2}l \cos \theta - a \cot \theta + l \cos \theta$, or $\frac{3}{2}l \cos \theta - a \cot \theta$.

The depth of C is $\frac{1}{2}l \cos \theta - a \cot \theta + \frac{3}{2}l \cos \theta$ or $2l \cos \theta - a \cot \theta$.

The equation of virtual work for a small displacement $d\theta$ is

$$2W \, \mathrm{d}(\tfrac{1}{2}l \cos \theta - a \cot \theta) + 2W \, \mathrm{d}(\tfrac{3}{2}l \cos \theta - a \cot \theta)$$
$$+ W' \, \mathrm{d}(2l \cos \theta - a \cot \theta) = 0$$

$$\therefore -Wl \sin \theta + 2Wa \, \mathrm{cosec}^2 \, \theta - 3Wl \sin \theta + 2Wa \, \mathrm{cosec}^2 \, \theta$$
$$-2W'l \sin \theta + W' a \, \mathrm{cosec}^2 \, \theta = 0$$

$$\therefore \qquad (4Wl + 2W'l) \sin^3 \theta = 4Wa + W'a$$

$$\therefore \qquad \sin^3 \theta = \frac{a(4W + W')}{l(4W + 2W')}.$$

An alternative method is as follows.

The forces acting on the rods AB and BC are, from symmetry,
as shown in Fig. 9.20.

Considering the equilibrium of AB we get

$$Y + R \sin \theta = W$$ (i)

and $$R \cos \theta = X_1 - X_2$$ (ii)

Taking moments about A

$$R\,a/\sin\theta + Yl\sin\theta + X_2 l\cos\theta = W\,\tfrac{1}{2}l\sin\theta \qquad\qquad\text{(iii)}$$

Similarly for BC we get

$$Y + W + \tfrac{1}{2}W' = 0 \qquad\qquad\text{(iv)}$$

and taking moments above C

$$Y\,.\,l\sin\theta + \tfrac{1}{2}Wl\sin\theta = X_2 l\cos\theta \qquad\qquad\text{(v)}$$

From (iv) $Y = -W - \tfrac{1}{2}W'$

From (v) $X_2 = -\tfrac{1}{2}(W + W')\tan\theta$

From (i) $R = (2W + \tfrac{1}{2}W')/\sin\theta$

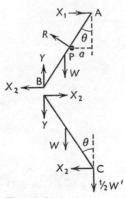

Fig. 9.20

Substituting in (iii) we get

$$(2W + \tfrac{1}{2}W')a/\sin^2\theta - (W + \tfrac{1}{2}W')l\sin\theta - \tfrac{1}{2}(W + W')l\sin\theta = W\,\tfrac{1}{2}l\sin\theta$$

which gives

$$\sin^3\theta = a(4W + W')/l(4W + 2W').$$

EXAMPLE (viii)

A uniform plank is to be lowered to the ground from a vertical position by one man, who places the lower end against a smooth vertical step and then walks backwards exerting a force on the plank perpendicular to the length of the latter at a point which is always 1·8 m above the ground. Show that the plank will slip if its length is greater than $5\cdot4\sqrt{3}$ m.

(C.S.)

Let AB (Fig. 9.21) represent the plank when inclined at an angle θ to the ground, let G be its centre of gravity, and C the point where the man applies the force.

If R is the vertical reaction of the ground at A the plank will slip when R vanishes. X is the reaction of the step at A.

Resolving vertically, we have

$$P \cos \theta + R = W.$$

Taking moments about A, we have

$$P \frac{1 \cdot 8}{\sin \theta} = Wl \cos \theta, \text{ where } 2l = \text{ the length of AB}$$

$$\therefore \qquad R = W - \frac{1}{1 \cdot 8} Wl \sin \theta \cos^2 \theta.$$

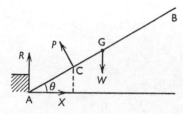

Fig. 9.21

Hence R will vanish when $l \sin \theta \cos^2 \theta = 1 \cdot 8$.

$$\therefore \qquad l \not> \frac{1 \cdot 8}{\sin \theta \cos^2 \theta}.$$

For the maximum value of $\sin \theta \cos^2 \theta$ we have, by differentiation,

$$\cos^3 \theta - 2 \sin^2 \theta \cos \theta = 0$$

$$\therefore \qquad \cos \theta = 0 \text{ or } \tan \theta = 1/\sqrt{2}.$$

The former value is impossible, and the latter value gives

$$\cos \theta = \sqrt{(2/3)} \text{ and } \sin \theta = 1/\sqrt{3}.$$

With these values

$$\frac{1 \cdot 8}{\sin \theta \cos^2 \theta} = \frac{1 \cdot 8 \sqrt{3}}{\frac{2}{3}} = 2 \cdot 7 \sqrt{3}.$$

Also l must not be greater than this minimum value. The maximum length is therefore $5 \cdot 4 \sqrt{3}$ m.

EXAMPLE (ix)

A particle C is attached to two light elastic strings CA and CB of the same modulus of elasticity, which are fastened to points A and B in the same horizontal plane as C, and the strings are just taut. The particle is then repelled from M, *the mid-point of* AB, *by a force which moves it to a new position* C'. *Prove that*

$$\frac{1}{AC} - \frac{1}{AC'} = \frac{1}{BC} - \frac{1}{BC'}.$$ (N.U.)

Let Fig. 9.22 represent the original and displaced positions, and let T_1, T_2 be the tensions in AC', BC'.

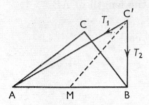

F<small>IG.</small> 9.22

If E is the modulus of elasticity we have

$$T_1 = E\frac{AC'-AC}{AC} \text{ and } T_2 = E\frac{BC'-BC}{BC}.$$

Resolving perpendicular to MC',

$$T_1 \sin AC'M = T_2 \sin BC'M$$

since the force acting on C' is along MC'.

But $\dfrac{\sin BC'M}{\sin AC'M} = \dfrac{AC'}{BC'}$

∴ $\dfrac{AC'-AC}{AC} = \dfrac{BC'-BC}{BC} \times \dfrac{AC'}{BC'}$

∴ $\dfrac{1}{AC} - \dfrac{1}{AC'} = \dfrac{1}{BC} - \dfrac{1}{BC'}.$

E<small>XAMPLE</small> (x)
Prove that if a transversal cuts the lines of action OA, OB, OC, *of three forces* P, Q, R *in equilibrium, at the points* A, B, C, *then with a convention regarding sign*

$$\frac{P}{OA \times BC} = \frac{Q}{OB \times CA} = \frac{R}{OC \times AB},$$

and deduce that $\dfrac{P}{OA} + \dfrac{Q}{OB} + \dfrac{R}{OC} = 0.$

Show also that if any number of forces P, Q, R, *etc., acting at a*

point in the lines OA, OB, OC, . . . *are in equilibrium, and if K be any*
point upon a transversal cutting the lines at A, B, C . . . *then*

$$\frac{P}{OA}+\frac{Q}{OB}+\frac{R}{OC}\cdots=0, \, and \, \frac{P\times AK}{OA}+\frac{Q\times BK}{OB}+\frac{R\times CK}{OC}+\ldots=0.$$

(Ex.)

In Fig. 9.23 it is clear that if P and R act in the direction OA,

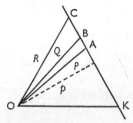

FIG. 9.23

OC, then Q must act from B to O, since the forces are in equilibrium.
By Lami's Theorem,

$$\frac{P}{\sin BOC}=\frac{Q}{\sin AOC}=\frac{R}{\sin AOB}.$$

Also in the triangle OBC,

$$\frac{\sin OCB}{OB}=\frac{\sin BOC}{BC}$$

and in the triangle OAC,

$$\frac{\sin OCA}{OA}=\frac{\sin AOC}{AC}$$

$$\therefore \quad \frac{OA}{OB}=\frac{AC}{BC}\times\frac{\sin BOC}{\sin AOC}$$

$$\therefore \quad \frac{P}{Q}=\frac{\sin BOC}{\sin AOC}=\frac{OA\times BC}{OB\times AC}$$

$$\therefore \quad \frac{P}{OA\times BC}=\frac{Q}{OB\times CA}$$

where CA is in the opposite direction to BC, or negative, since Q is in
the opposite direction to P. Similarly for the other equality.

Putting each fraction equal to k, we have

$$\frac{P}{OA}+\frac{Q}{OB}+\frac{R}{OC}=k(BC+CA+AB)=0$$

since CA $=$ $-$AC.

Let p be the perpendicular from O on the transversal.

The sum of the resolved parts of the forces perpendicular to the transversal must be zero.

Hence $\qquad P \sin A + Q \sin B + R \sin C \ldots = 0,$

$$\therefore \qquad \frac{Pp}{OA} + \frac{Qp}{OB} + \frac{Rp}{OC} + \ldots = 0,$$

$$\therefore \qquad \frac{P}{OA} + \frac{Q}{OB} + \frac{R}{OC} + \ldots = 0.$$

Let K be any point in the transversal, and join OK.

$$\frac{\sin AOK}{\sin K} = \frac{AK}{OA} \text{ and } \frac{\sin BOK}{\sin K} = \frac{BK}{OB}, \text{ etc.}$$

The sum of the resolved parts of the forces perpendicular to OK must be zero.

$$\therefore \qquad\qquad P \sin AOK + Q \sin BOK + \ldots = 0,$$

$$\therefore \qquad \frac{P \times AK}{OA} + \frac{Q \times BK}{OB} + \frac{R \times CK}{OC} + \ldots = 0.$$

EXAMPLE (xi)

A thin smooth elliptic tube, of axes $2a$, $2b$ ($a>b$), is attached by light spokes to a horizontal axis which passes through the centre of the ellipse and is perpendicular to its plane. The weight of the tube is W, and its centre of gravity is on the major axis at a distance d from the centre; and a particle of weight w is placed in the tube. Prove that there are 2 or 4 positions of equilibrium according as $d>$ or $< w(a^2-b^2)/aW$. (C.S.)

Let ACA' (Fig. 9.24) be the major axis of the ellipse, C its centre,

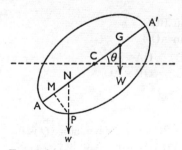

FIG. 9.24

G the position of its centre of gravity, and let the axis ACA' be inclined at an angle θ to the horizontal.

The particle w can only rest at a point P where the normal to the ellipse is vertical.

Let this normal meet AA' in N and let the coordinates of P (referred to the axes of the ellipse) be $a \cos \phi$ and $b \sin \phi$.

The equation of the normal at P is

$$\frac{x - a \cos \phi}{\dfrac{\cos \phi}{a}} = \frac{y - b \sin \phi}{\dfrac{\sin \phi}{b}}$$

$\therefore$ $\qquad$ $\text{CN} = a \cos \phi - \dfrac{b^2 \cos \phi}{a} = \dfrac{(a^2 - b^2) \cos \phi}{a}.$

If PM is perpendicular to AA' we have

$$\text{MN} = PM \tan \theta.$$

$\therefore$ $\qquad$ $a \cos \phi - \text{CN} = b \sin \phi \tan \theta$

$\therefore$ $\qquad$ $\dfrac{b^2 \cos \phi}{a} = b \sin \phi \tan \theta$

$\therefore$ $\qquad$ $\tan \phi = \dfrac{b}{a} \cot \theta$

and $\qquad$ $\cos \phi = a/(a^2 + b^2 \cot^2 \theta)^{\frac{1}{2}}$

Taking moments about C, we get

$$W \times d = w \times \text{CN}$$

$\therefore$ $\qquad$ $W \times d = w(a^2 - b^2)/(a^2 + b^2 \cot^2 \theta)^{\frac{1}{2}}$

$\therefore$ $\qquad$ $a^2 + b^2 \cot^2 \theta = \left[\dfrac{w(a^2 - b^2)}{Wd} \right]^2$

$\therefore$ $\quad$ $b \cot \theta = \{[w(a^2 - b^2) - aWd][w(a^2 - b^2) + aWd]\}^{\frac{1}{2}}/Wd$

Now if $d > w(a^2 - b^2)/aW$ the expression under the root sign is negative and there is no real value of θ.

The tube can, however, obviously rest with the major axis vertical and G above or below C.

There are two positions of equilibrium in this case.

If $d < w(a^2 - b^2)/aW$ there are two real values for $b \cot \theta$, one positive and one negative.

Hence in this case, besides the two positions with the major axis vertical, there are two others in which the major axis is inclined to the horizontal.

In one position G is to the right of C and in the other to the left of it (θ is an obtuse angle).

EXAMPLE (xii)

Four equal rods, each of length l, freely jointed at their ends, form a rhombus ABCD. The opposite corners A, C are connected by an elastic *string whose unstretched length is b($<2l$). The system is kept in equilibrium by forces at B and D acting inwards along the line BD. Show that these forces have a maximum value when*

$$\frac{AC}{BD} = \frac{\left(\dfrac{b}{2l}\right)^{\frac{1}{3}}}{\left[1-\left(\dfrac{b}{2l}\right)^{\frac{2}{3}}\right]^{\frac{1}{2}}}$$

(C.S.)

Let T be the tension in the string, P the value of either of the forces acting at B and D, and θ the angle DAC (Fig. 9.25).

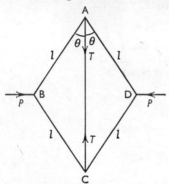

Fig. 9.25

$$AC = 2l \cos \theta$$
$$BD = 2l \sin \theta$$

Also,
$$T = \lambda(2l \cos \theta - b)/b$$

where λ is the modulus of elasticity of the string.

The equation of virtual work gives ·

$$-T\, d(2l \cos \theta) - 2P\, d(l \sin \theta) = 0$$

∴
$$2Tl \sin \theta - 2Pl \cos \theta = 0.$$

∴
$$P = T \tan \theta = \lambda \tan \theta(2l \cos \theta - b)/b.$$

This is a maximum when $2l \cos \theta - b \sec^2 \theta = 0$

∴
$$\cos \theta = \left(\frac{b}{2l}\right)^{\frac{1}{3}}$$

∴
$$\sin \theta = \left[1 - \left(\frac{b}{2l}\right)^{\frac{2}{3}}\right]^{\frac{1}{2}}$$

Also
$$\frac{AC}{BD} = \frac{\cos \theta}{\sin \theta} = \frac{\left(\dfrac{b}{2l}\right)^{\frac{1}{3}}}{\left[1-\left(\dfrac{b}{2l}\right)^{\frac{2}{3}}\right]^{\frac{1}{2}}}$$

EXAMPLE (xiii)

Of three equal discs in the same vertical plane, two rest on a horizontal table, not necessarily in contact with each other, and the third rests on the first two. Show that the least coefficient of friction between two of the discs for which this is possible is three times the least possible between a disc and the table. Can three pennies rest like the discs? The coefficient of friction between the edges of two pennies is about $\frac{1}{8}$ and between a penny and the table about $\frac{1}{4}$. (C.S.)

Let A, B, C (Fig. 9.26) be the centres of the discs; D, E, F, H the points of contact with the ground and of the discs respectively.

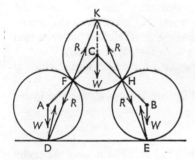

FIG. 9.26

If the disc A is not to roll about D the resultant reaction at F must pass through D, the point where the weight of A and the resultant reaction of the ground on disc A meet.

The resultant reaction at D will then act somewhere between DA and DF.

The angle of friction λ at F must not be less than AFD, and, if this condition holds, no slipping can take place at F. The condition may be written $\lambda \not< \theta$ where θ is the angle AFD, or $\mu \not< \tan \theta$. where μ is the coefficient of friction between two of the discs.

The resultant reaction at F meets the vertical through C at K on the circumference of the upper disc, and the reaction at H also passes through this point.

If W is the weight of each disc, and R the resultant reaction at F or H, we have

$$2R \cos \theta = W.$$

Now although there can be no slipping at F, it is possible that the lower discs may slip at D and E.

The normal reaction at D is $\frac{3}{2}W$, and the maximum friction is

$\mu'\left(\dfrac{3}{2}W\right)$, where μ' is the coefficient of friction between a disc and the table.

This must be sufficient to balance the horizontal component of R, in which case

$$\mu'\frac{3W}{2} = R\sin\theta = \frac{1}{2}W\tan\theta$$

that is, $$\mu' = \frac{1}{3}\tan\theta \text{ or } \mu = 3\mu'.$$

Hence the coefficient of friction between two of the discs must be three times that between a disc and the table.

It is clear that the angle θ is greater when the lower discs are farther apart. The least value of θ is when the lower discs are touching and its value is then 15°. The coefficient of friction between two discs must then be not less than tan 15° $(2-\sqrt{3}$ or 0·268).

The least coefficient between two discs for which equilibrium is possible is therefore 0·268.

If the coefficient is $\frac{1}{5}$ or 0·2 equilibrium is impossible; the lower pennies would roll apart.

EXAMPLE (xiv)

The distance between the axles of a railway truck is a, and the centre of gravity is half-way between them and at a perpendicular distance h from the rails. With the lower wheels locked the greatest incline on which the truck can rest is α. Show that the coefficient of friction between the wheels and rails is $2a/(a\cot\alpha + 2h)$.   (C.S.)

Let A, B (Fig. 9.27) be the axles, G the centre of gravity, and C, D the points of contact of wheels and ground.

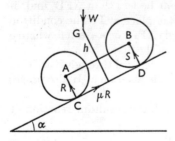

FIG. 9.27

Let R, S be the normal reactions at C and D, W the weight of the truck. Limiting friction at C is μR up the incline. Assuming that the upper wheels are free, any friction at D would rotate the wheel.

Resolving perpendicular to the incline, we get

$$R+S = W \cos \alpha.$$

Taking moments about G, we get

$$\mu Rh+S\frac{a}{2} = R\frac{a}{2}$$

$$\therefore \qquad 2\mu R\frac{h}{a}+S = R$$

$$\therefore \qquad R-\frac{2\mu h}{a}R = W \cos \alpha-R$$

$$\therefore \qquad 2R\left(1-\frac{\mu h}{a}\right) = W \cos \alpha.$$

Resolving parallel to the incline, we get

$$\mu R = W \sin \alpha$$

$$\therefore \qquad \frac{2W \sin \alpha}{\mu}\left(1-\frac{\mu h}{a}\right) = W \cos \alpha$$

$$\therefore \qquad \frac{2}{\mu}-\frac{2h}{a} = \cot \alpha$$

$$\therefore \qquad \frac{2}{\mu} = \frac{2h+a \cot \alpha}{a}$$

$$\therefore \qquad \mu = \frac{2a}{2h+a \cot \alpha}.$$

EXAMPLES 9.3

1. A uniform cylinder rests on two fixed planes as in Fig. 9.28; the Plane AB is smooth and the coefficient of friction between the

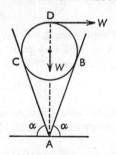

Fig. 9.28

cylinder and the plane BC is μ. A horizontal force equal to the weight of the cylinder acts at D, the mid-point of the highest generator of the cylinder. Show that equilibrium is impossible unless α is greater than $\pi/4$, and that if $\alpha = \tan^{-1} 2\cdot4$ there will be equilibrium if μ is not less than $4/7$. (C.S.)

2. Two uniform circular cylinders of equal radii rest in contact on an inclined plane with their axes horizontal. The weight of the upper cylinder is three times that of the lower, and all pairs of surfaces have the same frictional coefficient μ. Show that equilibrium is not possible whatever the inclination of the plane to the horizontal when $\mu<2$, and that when $\mu>2$ equilibrium is possible for all inclinations less than

$$\tan^{-1}\frac{3\mu}{2(\mu+1)}.$$ (C.S.)

3. A uniform rod of length c rests with one end on a smooth elliptic arc whose major axis is horizontal, and with the other end on a smooth vertical plane at a distance h from the centre of the ellipse, the ellipse and the rod being in a vertical plane. Prove that, if θ is the angle which the rod makes with the horizontal, and $2a$, $2b$ are the axes of the ellipse,

$$2b\tan\theta = a\tan\phi$$

where $\qquad a\cos\phi+h = c\cos\theta.$

In the special case when $a = 2b = c$, $h = 0$, prove that there are an infinite number of positions of equilibrium. (C.S.)

4. A uniform rod AB of weight W rests horizontally on two equally rough supports at A and C. Prove that the least horizontal force applied at B in a direction perpendicular to BA which is able to move the rod is $\frac{1}{2}\mu W$ or $\mu W\dfrac{b-a}{2a-b}$, according as $3b$ is greater or less than $4a$, where AB $= 2a$, AC $= b$, and μ is the coefficient of friction. (C.S.)

5. A uniform wire in the form of an ellipse of semi-axes a, b is hung over a peg. If the wire can rest in equilibrium with any point of it in contact with the peg, show that the coefficient of friction cannot be less than $\dfrac{1}{2}\left(\dfrac{a}{b}-\dfrac{b}{a}\right).$ (C.S.)

6. The end links of a uniform chain can slide on a fixed rough horizontal rod. Prove that the ratio of the extreme span on the rod to the length of the chain is

$$\mu\sinh^{-1}\left(\frac{1}{\mu}\right),$$

where μ is the coefficient of friction. (C.S.)

7. A thin uniform rod of length $2l$ rests in limiting equilibrium inside a rough vertical circular hoop of radius a. Prove that the inclination of the rod to the horizontal is

$$\cot^{-1}\left[\frac{a^2-l^2-l^2\mu^2}{a^2\mu}\right],$$

where μ is the coefficient of friction. (C.S.)

8. A car rests on four equal weightless wheels of diameter 0.9 m, which can rotate about central bearings of diameter 5 cm. If the angle of friction between each wheel and its bearing is $18°$, show that the car will not rest on a rough inclined plane if the inclination of the plane to the horizontal be greater than $1°$ approximately, assuming that a wheel and its bearings are in contact along a single generator. (C.S.)

9. Out of a circular disc of metal a circle is punched whose diameter is a radius OA of the disc. The disc is then placed vertically resting on two rough parallel rails in the same horizontal plane, the plane of the disc being perpendicular to the rails. The chord of contact subtends an angle 2α at the centre of the disc. Show that if the angle which OA makes with the vertical is greater than $\sin^{-1}(3\sin 2\epsilon/\cos\alpha)$, where ϵ is the angle of friction, the disc will slip. (C.S.)

10. A drawer of depth b (from back to front) is jammed by pulling at a handle at a distance c from the centre of the front. Prove that the coefficient of friction must be at least $b/2c$. (C.S.)

11. A uniform rectangular board is supported with its plane vertical and with two edges, of length a, horizontal, by pressures applied at two points, one in each of its vertical edges, at which the coefficient of friction is μ. Prove that the vertical distance between the points of support cannot exceed μa. (C.S.)

12. A uniform plank of length l and thickness $2h$ rests symmetrically across a fixed rough cylinder of radius a. Taking λ to be the angle of friction between the bodies, find the relation between λ, a, and h in order that if the plank be slowly tilted another position of equilibrium may be reached, and show that if a were less than h no amount of friction would make this possible.

(C.S.)

13. A strong wire AOB is bent at its mid-point O to form an angle of 2α, the parts OA and OB being straight, and is hung up at O. A weight W is suspended by two equal strings (length a), whose other ends are tied to rings, one of which slides on OA, the other on OB. Find the highest position of equilibrium of W if the

coefficient of friction between the rings and wire is tan λ, and if the weight of the rings is negligible. If the rings are heavy, each of weight P, and θ is the angle between a string and the adjacent lower end of the wire, prove that, for the higher position of equilibrium of the rings,

$$\tan{(\theta+\lambda)} = -\frac{W+2P\,\cos^2(\alpha+\lambda)}{P\,\sin{(2\alpha+2\lambda)}}. \qquad \text{(Ex.)}$$

14. A uniform rectangular signboard ABCD of weight W is supported at right angles to a wall, with AB the upper horizontal edge and AD vertical, by three equal rods AP, AQ, DR of negligible weight, each inclined to the horizontal at an angle θ, AQ, DR being parallel and sloping down to the wall, while AP slopes upwards, P, Q, R being the points where they are hinged to the wall in the same vertical line. If AB $= a$, AD $= b$, determine the stresses in the rods, and the conditions under which the stress in AQ is: (i) tensile; (ii) compressive; (iii) zero.   (Ex.)

15. A triangle ABC, formed of three rods freely jointed together, is lying on a horizontal plane; B is connected to any point Q in AC by a tight string, and C to a point R in AB, the strings crossing at E. If the stress at A is parallel to BC, prove that the tensions in BQ and CR are proportional to BE and CE respectively.   (Ex.)

16. A rhombus ABCD is formed of four uniform freely jointed rods, each of length $2a$ and weight W. It is placed symmetrically with AC vertical and A uppermost, and BC and CD resting one on each of two smooth pegs at the same level and distant $2c$ apart, the rhombus being prevented from collapsing by a rod of negligible weight joining B and D. Find the stress in the rod when the angle BCD is 2θ. If $2c = a$, and the framework is held on the pegs in the form of a square, without the rod BD, and is then released, how will it begin to move?   (H.S.C.)

17. Two smooth balls of different weights and sizes rest inside a smooth hemispherical bowl which is placed on a smooth table. Prove that the bowl cannot remain in equilibrium if the plane of its rim is inclined to the horizontal. If the balls are made of the same material of radii a, b, and the radius of the bowl is R, prove that the line joining the centres of the balls will be horizontal if

$$R = \frac{(a+b)(a^2+b^2)}{(a^2-ab+b^2)}. \qquad \text{(Ex.)}$$

18. Five equal uniform rods AB, BC, CD, DE, EA, each of weight W, are freely jointed together at their extremities and hang from the joint A with the rods AB and AE making the same angle θ with the vertical, and the rods BC and ED making the same angle ϕ with the vertical, this configuration being maintained by a rod of negligible weight connecting B and E. Prove that the vertical and horizontal components of the reaction at the joint C are $\frac{1}{2}W$ and $W \tan \phi$, and that the stress in the rod BE is $W(2\tan \theta + \tan \phi)$. (H.S.C.)

19. Forces lBC, mCA, nAB, where l, m, n are positive, act along the sides BC, CA, AB respectively of a triangle ABC in the senses indicated by the order of the letters. Show that the line of action of their resultant divides BC, CA, AB respectively externally in the ratios $m:n$, $n:l$, $l:m$. Discuss the case when $l = m = n$. (H.S.C.)

20. Show that forces acting along the sides of a closed quadrilateral ABCD, in the same sense and proportional to the sides, will form a couple. Show also that forces pAB, qBC, rCD, and sDA will form a couple if the following conditions are satisfied:

If s is the greatest of the multiples p, q, r, s, divide AB, BC, and CD at K_1, K_2, K_3, so that $AK_1 = (p/s)AB$, $K_2C = (q/s)BC$, $K_3D = (r/s)CD$, then the forces form a couple if K_1K_2 is equal and parallel to K_3C. (H.S.C.)

21. Two barges, one following the other, are moving steadily at the same rate along a canal, the second being pulled by a rope attached to its bow at A and to the stern of the first barge at B, the points A and B being on the same level. The rope has a mass of 2·5 kg and is 9 m long. The sag or dip in the middle of the rope below AB is 15 cm, and it is known that the tension at A and B exceeds the tension at the lowest point by the weight of a rope of length equal to the sag. By considering the horizontal and vertical components of the tension at A or B, calculate the resistance of the water to the second barge.

22. An acute-angled and isosceles triangular prism stands on a rough horizontal plane, and one of its side faces is subjected to increasing uniform normal pressure. Show that equilibrium will be broken by sliding or tumbling as the angle of friction is less or greater than the vertical angle of the prism, supposed less than 60°. (C.S.)

23. A heavy cubical block of edge $2a$ is placed on a rough table with one face parallel to the edge of the table at a distance $a \cot \alpha$ from it; to the centre of this face a light smooth rod of length l

is freely jointed; it passes over the smooth edge of the table and carries a weight W at its end. Show that as W is increased the equilibrium of the block is broken by its tilting about an edge if

$$\mu > \frac{l \cos \alpha \sin^2 \alpha}{a + l \sin \alpha \cos \alpha (\sin \alpha - \cos \alpha)}.$$ (C.S.)

24. If forces lPA, mPB, nPC acting along the chords PA, PB, CP of a circle are in equilibrium, show that

$$l\text{PA}: m\text{PB}: n\text{PC} = \text{BC}: \text{CA}: \text{AB}.$$

Show also that

$$l\text{PA}^2 + m\text{PB}^2 = n\text{PC}^2$$

and $\qquad n(l+m-n)\text{PC}^2 = lm\text{AB}^2.$ (Ex.)

25. ABC is a triangle of which O is the circumcentre and H the orthocentre, and D, E, and F are the feet of the perpendiculars from A, B, C respectively on the opposite sides. If forces acting along OA, OB, OC are in equilibrium, show that they are proportional to the sides of the triangle DEF. If equal forces represented by OA, OB, OC act at O, prove that their resultant is represented by OH. (Ex.)

26. To the mid-point C of an unstretched string of length $2a$, secured at its ends by two pegs A and B (also $2a$ apart), is attached another elastic string CD of length a. The other end D of this string is pulled at right angles to AB until C is displaced a distance $a/10$. Find the shift of D, assuming the strings to have the same modulus of elasticity. (N.U.)

27. A particle of weight w lying on a rough inclined plane is attached to one end of an elastic string which passes through a hole in the plane. The other end of the string is attached to a fixed point underneath the plane, which is at such a distance from the hole that the string is just taut when the particle on the plane is at the position of the hole. Find the locus of the possible positions of equilibrium of the particle on the plane, taking the inclination of the plane to the horizontal as α, and the coefficient of friction for the particle on it as μ. (N.U.)

28. A uniform heavy rod is supported by two elastic strings, with the same modulus of elasticity, attached to its ends, the other ends of the strings being attached to a peg. Prove that if l_1, l_2 are the natural lengths of the strings and x_1, x_2, their extensions in the equilibrium position, then

$$\frac{l_2 x_1}{l_1 + x_1} = \frac{l_1 x_2}{l_2 + x_2}.$$ (N.U.)

29. A, B, C (Fig. 9.29) are three toothed wheels which revolve in contact at their edges, A and C being free to turn about their common centre, and the teeth of C being turned inwards from a projecting rim so as to catch on the teeth of B. Couples K_a,

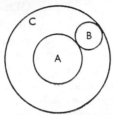

FIG. 29

K_b, K_c are applied to them respectively in the same sense round their axes, and when the wheel B is free to move round in contact with the wheels A and C. Prove that the wheels will be in equilibrium if

$$\frac{K_a}{a} = \frac{K_b}{2b} = -\frac{K_c}{c}.$$ (N.U.)

30. Two equal rods, each of weight w, are rigidly jointed together to make two arms of a right angle. This angle rests with one foot on a rough horizontal ground, and the other against a rough vertical wall, the plane of the angle being vertical and perpendicular to the plane of the wall, and the arm against the wall makes an angle θ with the wall. If the friction between the foot on the wall and the wall in this position is limiting (coefficient μ), prove that the force of friction exerted by the ground on the other foot is

$$\frac{w(\sin \theta + 3 \cos \theta)}{2[(\mu+1) \sin \theta + (\mu-1) \cos \theta]}.$$ (C.W.B.)

31. A circular disc of weight w and radius a stands in a corner between a vertical wall and a horizontal floor with its plane perpendicular to both the plane of the wall and the floor. The centre of gravity of the disc is at a distance c from its centre of figure, and the line joining these centres makes an angle θ with the vertical on the wall side of the centre. If in this position the friction between the disc and the wall is limiting (coefficient μ), prove that the force of friction between the disc and the floor is

$$\frac{wc \sin \theta}{(1+\mu)a},$$

and find the normal reaction also at this point.   (C.W.B.)

32. Six equal light rods form the edges of a regular tetrahedron, which is placed with one face on a horizontal table. A weight W is hung from the top vertex. Prove that the thrust in each horizontal rod is one-third of that in an inclined rod, and calculate these thrusts. (H.S.D.)

33. Two smooth spheres of radii a and b, of equal density, are connected by a light string of length l, the ends of the string being attached to points on the surfaces of the spheres. The string is slung over a fixed smooth peg, and the spheres hang freely in contact with one another. Show that in equilibrium the peg divides the length of the string in the ratio

$$b^3(b+l)-a^4 : a^3(a+l)-b^4.$$ (H.S.D.)

34. A regular hexagon ABCDEF is formed of six equal uniform heavy rods freely jointed to each other at their ends. It is suspended freely from the angular point A and the regular hexagonal form is maintained by a light horizontal rod PQ, freely jointed to a point P in BC and to a point Q in FE. Prove that

$$BP : PC = FQ : QE = 1 : 5.$$ (C.S.)

35. A trolley wire is carried on poles round a curve of 360 m radius. The poles are 36 m apart, and in the middle of each span the wire sags 18 cm below the points of support. If the wire has a mass of 0·8 $kg\,m^{-1}$, show that the resultant horizontal pull on each pole is very nearly 700 N.

36. Four equal uniform rods of length a are jointed so as to form a square. Two adjacent sides rest in contact with two smooth pegs which are on the same level at a distance $2c$ apart. The square is kept rigid by two strings stretched between opposite corners, and in the position of equilibrium these two strings are horizontal and vertical respectively. If T_1, T_2 are the tensions of the two strings and W is the weight of each rod, show that

$$T_1 - T_2 = 4w\left(\frac{c\sqrt{2}}{a} - \frac{1}{2}\right).$$ (C.S.)

37. ABCD is a rhombus of smoothly jointed rods resting on a smooth horizontal table, to which CD is fixed. The points A, C are joined by an inelastic string which is kept taut by a force F applied at right angles to AB at its mid-point. Show that the forces along the rods BC, AD are in the ratio $\sin \alpha : \sin 3\alpha$, where the angle CAB $= \alpha$. (C.S.)

38. Two uniform spheres of equal weight but unequal radii a, b are connected by a cord of length l attached to a point on each

surface. They rest in contact, the string hanging over a smooth peg. Show that the two portions of the string make equal angles

$$\sin^{-1} \frac{a+b}{a+b+l}$$

with the vertical. (C.S.)

39. Five equal uniform rods AB, BC, CD, DE, EA are hinged together, and the framework is supported with AB and BC in a horizontal line resting on two smooth pegs, and DE also horizontal. Show that the distance between the pegs is $1\frac{3}{5}$ times the length of a rod. (C.S.)

40. A rhombus of smoothly jointed rods rests with two sides in contact with a smooth circular disc all in the same vertical plane. Show that, if the diameter of the disc be one-fifth of the length of a rod, the reactions at the highest and lowest joints are in the ratio 15: 1. (C.S.)

41. A number of weights are to be hung on a light string so that the vertical lines drawn through them are at equal horizontal distances apart, and so that the particles lie on a curve of the form $a^2y = x^3$, where Ox is a horizontal axis and Oy an upward vertical axis (only positive values of x are to be considered). Show that the weights of the particles must be in arithmetical progression. (C.S.)

42. OA is a slightly compressible vertical rod of height h and negligible mass (modulus of compressibility μ) freely pivoted at its lowest point O. AB is a slightly extensible cord of natural length l (modulus λ). B is a point in the horizontal plane through O distant a from O, where $a^2 = l^2 - h^2$. A horizontal force P is applied at A in the direction BO. Show that the horizontal and vertical components of the displacement of A are approximately (neglecting x^2 and y^2)

$$x = \frac{P}{a^2}\left(\frac{h^3}{\mu} + \frac{l^2}{\lambda}\right) \text{ and } y = \frac{Ph^2}{a\mu}.$$ (C.S.)

43. Six equal rigid weightless bars, freely jointed at their ends, form a regular tetrahedron ABCD. It is suspended from A, and three weights, each equal to W, are hung from B, C, and D respectively. Find the stresses in all the bars. (C.S.)

44. A hemisphere, whose weight is W and radius a, is placed with its curved surface on a smooth horizontal table, and a string of length $l(l < a)$ is attached to a point on its rim and to a point on the table. Prove that the tension in the string is

$$\tfrac{3}{8}W(a-l)/(2al-l^2)^{\frac{1}{2}}$$ (C.S.)

45. A uniform rod of length $2a$ and weight W hangs in an oblique position supported by a light inextensible string of length $2l(l>a)$ whose ends are fastened to the ends of the rod and which passes over a small peg, and a weight w is attached to the rod at a distance d from its mid-point. Prove that the lengths of the string on the two sides of the peg are

$$l\left(1-\frac{d}{an}\right) \text{ and } l\left(1+\frac{d}{an}\right),$$

where $nw = W+w$.   (C.S.)

46. A picture hangs in a vertical plane from a smooth nail by a cord of length $2a$, fastened to two symmetrically placed rings at a distance $2c = 2\sqrt{(a^2-b^2)}$ apart. Show that if the depth of the centre of gravity below the line of the rings is greater than c^2/b the symmetrical position of equilibrium is the only one, and it is stable. Prove that if the depth is c^2/b the symmetrical position is unstable.   (C.S.)

47. A light framework of three rods BC, CA, AB, freely jointed together to form an equilateral triangle of side a, is suspended by three strings OA, OB, OC, each of length l, from a point O, and a weight W is suspended by three equal strings, each of length l', connecting it to A, B, C. Show that the thrust in each of the rods is

$$\frac{Wa}{3\sqrt{3}}\left[\left(3l^2-a^2\right)^{-\frac{1}{2}}+\left(3l'^2-a^2\right)^{-\frac{1}{2}}\right]$$ (C.S.)

48. A uniform solid cube of edge $2c$ rests on two parallel horizontal bars placed under one face parallel to the edges of that face at distances b from the centre of it. The plane containing the bars makes an angle θ with the horizontal. Show that, if equilibrium exists and $b>\mu c$, then

$$\tan\theta<\frac{(\mu'+\mu)b}{2b+(\mu'-\mu)c},$$

where μ, μ' are the coefficients of friction between the cube and the lower and upper rails respectively.   (C.S.)

49. Three smooth equal cylinders of radius a and weight w have their axes parallel and horizontal. Two of them rest on a smooth horizontal table; the third rests between them, and equilibrium is maintained by means of a string of length $(2\pi a+l)$ passing round all three and everywhere perpendicular to the direction of their axes. If $l<8a$, show that the tension in the string is

$$\frac{w(l-4a)}{2[l(8a-l)]^{\frac{1}{2}}}.$$ (C.S.)

50. Three equal uniform rods, of length l and weight w, are smoothly jointed together to form a triangle ABC. This triangle is hung up by the joint A, and by two strings each of length $l/\sqrt{2}$ a weight W is attached to B and C. The system hangs under gravity, show that the stress in the rod BC is

$$\frac{1}{\sqrt{3}}\left[w+\frac{1}{2}W(1+\sqrt{3})\right].$$ (C.S.)

51. A smooth sphere is suspended from a fixed point by a string of length equal to its radius. To the same point a second string is attached which, after passing over the sphere, supports a weight equal to that of the sphere. Show that the first string then makes an angle $\sin^{-1}\left(\frac{1}{4}\right)$ with the vertical. (C.S.)

52. A tetrahedron ABCD is formed of light rods smoothly jointed at their extremities, and X, Y, the mid-points of AB, CD, are jointed by a string in which there is a tension T. Prove that the tension in AB is $\frac{1}{4}T$ AB/XY, and write down the stresses in the other rods, stating in each case whether the stress is a tension or a thrust. (C.S.)

53. AB, BC, DE, and EF are four equal rods; a hinge at B connects AB and BC, and a hinge at E connects DE and EF; also AB and DE are hinged together at their mid-points, and BC and EF are hinged together at their mid-points. Equal forces, applied at A and D in directions perpendicular to AB and DE, are balanced by forces applied at C and F. Show that these forces must be equal and their directions must make angles θ with CF, such that

$$\cot\theta = 4\tan\alpha+\cot\alpha,$$

where 2α is the angle between the rods AB and DE.

54. $(n+1)$ bricks of the same size are piled one above another in a vertical plane so that they rest each one overlapping the one below by as much as possible. Prove that, if $2a$ is the length of a brick, the lowest but one overlaps the lowest by a length a/n. Show also that, if each brick overlaps the one next below by a length a/n, the greatest number of bricks that may be piled up is $(2n-1)$. (C.S.)

55. A tripod consists of three equal uniform rods AO, BO, CO, rigidly connected at O so that they are at right angles to one another. If the tripod be hung from the point A, show that the plane ABC makes an angle $\tan^{-1}2\sqrt{2}$ with the horizontal. (C.S.)

56. Two equal smooth circular cylinders of radius c are fixed with

their axes parallel and in the same horizontal plane at a distance b apart. A cube of side $2a$ rests with two adjacent faces touching the cylinders. Show that, if $a+c<b\sqrt{2}$ and $a^2+c^2>b^2$, there are two positions of equilibrium in which the plane through the highest and lowest edges of the cube makes an angle $\cos^{-1}\left[\dfrac{a+c}{b\sqrt{2}}\right]$ with the vertical. Also show that these positions are unstable.

(C.S.)

57. Fig. 9.30 shows a plate gripped by two cylinders which lean against it, the cylinders being hinged at A, B to fixed supports. The coefficient of friction between each cylinder and the plate

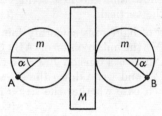

FIG. 9.30

is μ. The masses are m, M as shown. Show that the plate will not slip if

$$\frac{M}{m}<\frac{2\mu\cos\alpha}{\sin\alpha-\mu(1+\cos\alpha)}.$$ (C.S.)

58. ABC is a triangle, O the centre of its circumcircle. Forces P, Q, R act along BC, CA, AB and forces P', Q', R' along OA, OB, OC. Show that if the forces are in equilibrium,

$$P\cos A+Q\cos B+R\cos C = 0,$$

and $$\frac{PP'}{\sin A}+\frac{QQ'}{\sin B}+\frac{RR'}{\sin C} = 0.$$ (C.S.)

59. Two equal rectangular blocks of length a having square ends of side b are placed on a horizontal table with two square faces in contact, and a third block of the same size is placed symmetrically on top of them. Equal forces are then applied to the centres of the end faces of the lower blocks. Prove that provided that the horizontal components of these forces are greater than $3Wa/2b$ the table may be removed without disturbing equilibrium, W being the weight of each block.   (C.S.)

60. A frame, formed of four light rods of equal length, freely jointed at A, B, C, D, is suspended at A. A particle of mass m is suspended from B and D by two strings each of length l. The frame

is prevented from collapsing by a string AC. Show that the tension of the string is equal to $\frac{1}{2}mg$ AP/PN, where P is the particle and N is the centre of the rhombus ABCD.   (C.S.)

61. A smooth right circular cone, of semi-vertical angle α, has its axis vertical and vertex upwards; a heavy elastic string of weight W, modulus of elasticity λ, and natural length $2\pi l$ is placed round it and allowed to sink gradually to rest. Find the position of equilibrium. A second string of equal weight and equal natural length, but of modulus λ', $<\lambda$, is placed, without stretching, round the cone, and the two strings are allowed to sink gradually to rest. Show that in the position of equilibrium both strings will be at a depth h below the vertex if

$$\pi(\lambda+\lambda')(h \tan \alpha-l) \tan \alpha = Wl. \qquad \text{(C.S.)}$$

62. On a fixed circular wire (radius r) in a vertical plane slide two small smooth rings, each of weight W. The rings are joined by a light inextensible string of length $2a(<2r)$ on which slides a small smooth ring of weight P. Prove that for equilibrium either both parts of the string are vertical or else P is at a distance from the centre of the wire equal to

$$\left[\frac{W(r^2-a^2)}{W+P}\right]^{\frac{1}{2}}.$$

63. A uniform rod AB of length l is constrained without friction so that A moves on the circumference and B on the vertical diameter (not produced) of a circle in a vertical plane. The radius of the circle is a. Find the positions of equilibrium given that l is between $\frac{1}{2}a$ and a, and discuss separately the stability of the upper and lower positions. What happens if l is less than $\frac{1}{2}a$?   (C.S.)

64. A triangle ABC formed of uniform rods of the same material and thickness rests in a vertical plane with each rod in contact with a smooth circular cylinder whose axis is horizontal and whose section is the inscribed circle of the triangle. Show that in equilibrium the intersection of the medians of the triangle ABC must be in the vertical plane through the axis of the cylinder.   (C.S.)

65. A number of equal uniform rectangular blocks, each of length $2a$, are placed in a pile with each block projecting over the end of the block below it, the lowest block resting on a fixed horizontal plane. If the number of blocks is n, and the distance which each projects over the one below it is so adjusted that the total overhang (the horizontal distance between the vertical

ends of the highest and lowest blocks) is a maximum, prove that this overhang is

$$a\left(1+\tfrac{1}{2}+\tfrac{1}{3}+ \ldots +\frac{1}{n-1}\right).$$

If, however, the distance which each block projects over the one below it are all equal, prove that the greatest overhang is $2a(n-1)/n$ and verify that this never exceeds the maximum overhang. (H.C.)

66. Three equal uniform spheres of radius a are placed in contact with each other in a fixed spherical bowl of radius b. A fourth equal sphere is placed gently on top of them. Prove that the three lower spheres do not separate if b is less than $a(1+2\sqrt{11})$. (H.C.)

67. A horizontal bridge truss is supported at its ends, and consists of a number of congruent panels numbered in order $1, 2, 3, \ldots$ from one end. Each panel consists of a rectangle of members braced by a diagonal, and the diagonals are all parallel. If equal loads are placed at each of the bottom joints, prove that the stresses in the diagonal members form an arithmetical progression while those in the upper members vary according to a quadratic formula with the order, r, of the panel. (N.U.4)

68. A car with four equal wheels has a wheel-base a (i.e. a is the distance between the points of contact with the ground of the front and back wheels); its centre of gravity is equidistant from all four wheels and at a height h above the ground. Find the pressures on the wheels when the car rests with the hind wheels locked on a slope of inclination α facing: (i) uphill; (ii) downhill. Show that, if the hill is slippery, the car may be able to rest in position (i) but not in position (ii). If $a = 3$ m, $h = 0.6$ m, find the least coefficient of friction between the wheels and the road that the car may be able to rest in either position on a slope of $10°$. (H.C.)

69. A light frame ABC in the form of half an equilateral triangle having the angles A, B, C equal to $30°$, $60°$, and $90°$ respectively, is freely hinged at A to a wall and attached by a light string from B to a point D in the wall vertically above A; it is given that $AD = AB$, and $\angle DAB = 30°$, and a weight W is suspended from C. Find the tension in BD by a force diagram. If the length of the string is altered so that the angle $DAB = \alpha$, show that the tension in the string is

$$W\sqrt{3} \cos (60°-\alpha)/2 \cos \tfrac{1}{2}\alpha. \qquad \text{(H.S.C.)}$$

70. Two uniform equal heavy rods, each of weigat w, are smoothly jointed at A and the rest across a smooth horizontal cylinder. From A a weight W is suspended. If $2l$ is the length of each rod, r the radius of the cylinder, and θ the inclination of each rod to the horizontal, show that

$$Wr \sec \theta \tan \theta = 2w(l \cos \theta - r \sec \theta \tan \theta). \quad \text{(H.S.C.)}$$

71. A plank rests symmetrically across two cylinders on the ground whose axes are parallel, the plank being perpendicular to the axes of the cylinders, and making an angle α with the ground. The centre of mass of the plank divides the part of it between cylinders in the ratio $p : q$. The larger cylinder is fixed to the ground. If θ is the angle between the vertical and the total resistance at the point of contact of the smaller cylinder with the ground, show that

$$\tan \theta = \frac{\tan \frac{1}{2}\alpha}{1 + \left(\dfrac{p+q}{q} \sec \alpha\right) \dfrac{W}{w}}$$

where W is the weight of the small cylinder and w that of the plank.   (H.S.C.)

72. A sphere of 10 cm radius and of mass 6 kg is supported on three smooth pegs A, B, C in the same horizontal plane; BC = 8 cm, AB = AC, and the angle BAC = 30°. Find the thrust on each peg.   (H.S.C.)

73. A straight uniform rod lies on a rough horizontal table and a small horizontal force is applied perpendicular to the rod at one end and gradually increased in intensity until equilibrium is broken. Prove that equilibrium can be broken only by the rod turning about a point in its length, and determine the particular point about which the rod first turns.   (N.U.4.)

74. A uniform heavy string of length l and weight w per unit length hangs vertically from a fixed end, and a weight W is hung on the lower end. Find the expression for the tension T at the point of the string distant x from the upper end. If the weight per unit length instead of being uniform varies uniformly from w at the top end to $2w$ at the bottom, prove that the tension T is given by

$$T = w(l-x) + \frac{w}{2l}(l^2 - x^2) + W. \quad \text{(C.W.B.)}$$

75. A tetrahedron is formed by six light rods jointed together, and the mid-points of a pair of opposite rods are connected by a tight string. By using the method of virtual work, or otherwise,

show that the stresses in the remaining rods are proportional to their lengths and that, if the rods are equal, these stresses are $\frac{1}{4}\sqrt{2}$ times the tension of the string.   (O.C.)

76. A rough heavy uniform sphere of radius a and centre C rests in contact with a horizontal floor at D. A uniform rod AB of length $2b$ and weight W is smoothly hinged at A to a fixed point on the floor and rests on the sphere, touching it at E. The rod is inclined at an angle 2θ to the horizontal (with $2b > a \cot \theta$) and is in the vertical plane ACD. If the contacts at D and E are rough enough to prevent slipping, prove that the mutual action and reaction at E act in the line ED and are each of magnitude $Wb \sin \theta(1-\tan^2 \theta)/a$.

The angle of friction at both D and E is λ. Prove that if $\lambda > \theta$ the friction is not limiting at either contact, but that if $\lambda = \theta$, then the friction is limiting at E and not at D.   (N.U.)

77. Three equal rods BC, CA, AB, each of length $2a$ and weight W, are freely jointed at A, B, and C. The triangle is suspended by two strings BD, CE, each of length $2a$, where DE is horizontal and of length $4a$. Show that the tension in the rod BC is $W/\sqrt{3}$.

78. A heavy uniform circular disc of radius r rests in a vertical plane on two equally rough fixed horizontal pegs at the same level and at a distance $2a$ apart $(a < r)$. The plane of the disc is perpendicular to the axes of the pegs. The coefficient of friction at each point of contact is μ. A gradually increasing horizontal force is applied at the highest point of the disc and in its plane. Show that the disc will slip at both points of contact before it will turn about one of them if

$$\mu\{r+\sqrt{(r^2-a^2)}\}\{r-\mu a\} < a\sqrt{(r^2-a^2)}.   \text{(L.A.)}$$

79. Show that the centre of gravity of a uniform solid hemisphere of radius r is a distance $3r/8$ from the centre.

The hemisphere is placed with its curved surface on an inclined plane which makes an angle α with the horizontal. The plane is sufficiently rough to prevent sliding. Show that for a certain range of values of α the hemisphere can rest in either of two positions of equilibrium.

Determine: (a) the greatest value of α consistent with equilibrium; (b) the value of α below which only one position of equilibrium is possible.   (L.A.)

80. A uniform plane lamina OAB is in the shape of a sector of a circle of radius a. The angle between the two radii OA and OB is 2α. The centroid of the lamina is on the chord AB. Show that $2 \tan \alpha = 3\alpha$.

The lamina is bent along AB so that the two parts are perpendicular to each other, and it is then freely suspended from O. Show that the angle of inclination of the triangular part with the horizontal is $\tan^{-1}(\tan^2 \alpha + \sec^2 \alpha)$. (L.A.)

81. A uniform rod AB of length $2a$ and of weight W is freely hinged to a wall at A. The end B is attached by means of two light elastic cords, each of natural length $2a$ and of the same modulus of elasticity, to fixed points C and D of the wall. C is vertically below A, and D is vertically above A. $AC = AD = 2a$. The tension in one cord is three times the tension in the other. Find these tensions and the direction of the action at the hinge.

(L.A.)

82. Prove that the potential energy of a stretched light elastic string of natural length l and modulus λ is $\frac{1}{2}\lambda x^2/l$, when x is the extension.

Two particles, each of mass m, are attached one at an end and the other at the mid-point of such a string. If the string lies in a small coil on a horizontal table and the other end is slowly lifted until the lower mass is just clear of the table, show that the work done during the operation will be

$$mgl\left[\frac{1}{2} + \frac{7mg}{4\lambda}\right].$$ (L.A.)

83. Three smooth rods form the sides of a triangle ABC which lies on a smooth horizontal table. Three equal elastic strings whose natural length is one-half the radius of the inscribed circle are knotted together at one end of each and the other ends are fixed to smooth rings one on each of the rods. Prove that in equilibrium the actual lengths of the strings are $\frac{1}{2}r + ka$, $\frac{1}{2}r + kb$, $\frac{1}{2}r + kc$, where $k(a^2 + b^2 + c^2)$ is the area of the triangle.

(O.C.)

84. Two uniform rods AB and BC, each of length $2a$ and weight W, are maintained at right angles to each other by a stiff hinge at B. The friction at the hinge is such that the greatest couple which either rod can exert on the other has moment K. The end A is smoothly pivoted to a fixed point from which the rods hang. A vertical force F acting at C holds the rods in equilibrium with AB inclined at an angle ϕ to the downward vertical at A, and with C higher than B. Find F, and show that the moment of the reaction couple at B is $2Wa \sin \phi/(1 + \tan \phi)$.

By considering the maximum value of this function of ϕ, show that equilibrium is possible for any value of ϕ between 0 and $\pi/2$, provided that $K \geqslant Wa/\sqrt{2}$. (N.U.)

85. One end of an elastic string of natural length a and modulus $2W$ is attached to a particle A of weight W resting on a rough horizontal table. The other end is attached to a fixed point O distant a vertically above A. The coefficient of friction between A and the table is μ. A horizontal force P is now applied to A and is always just sufficient to make it move in a straight line on the table. Find the normal reaction between the particle and the table when the string is inclined at an angle θ to the vertical, and show that

$$P = W(2\mu \cos \theta - \mu + 2 \tan \theta - 2 \sin \theta).$$

Show that the particle is about to leave the table when $\theta = 60°$ and that the total work done by P is then

$$Wa\{2\mu \log_e(2 + \sqrt{3}) + 1 - \mu\sqrt{3}\}. \qquad \text{(N.U.)}$$

86. A uniform ladder AB, of weight W and length $2a$, is placed at an inclination θ to the horizontal with A on a rough horizontal floor and B against a rough vertical wall. The latter is in a vertical plane perpendicular to the wall, and the coefficients of friction at A and B are μ and μ' respectively. Show that the ladder will be in equilibrium if

$$\tan \theta \geqslant (1 - \mu\mu')/2\mu.$$

If this condition is not satisfied and a couple of moment K is applied in the vertical plane so that the ladder is just prevented from slipping down, show that

$$K = Wa \cos \theta(1 - \mu\mu' - 2\mu \tan \theta)/(1 + \mu\mu'). \qquad \text{(N.U.)}$$

87. Two uniform rods AB and BC, each of weight W and length $2l$, are smoothly hinged together at B. At A and C small light rings are attached to the rods and are threaded on a rough fixed horizontal bar below which the rods hang in equilibrium, the inclination of each rod to the horizontal being denoted by θ. The coefficient of friction between each ring and the bar is μ. Show that $\theta \geqslant \alpha$ where $\cot \alpha = 2\mu$.

Two equal and opposite couples in the plane ABC are now applied, one to each rod, of moment just sufficient to cause A and C to slip towards each other. Find the moment K of either couple as a function of θ. If the couples are first applied when the rods are in the position given by $\theta = \alpha$ and continue to act so as just to cause slipping, show that the work they do in moving the system to the position $\theta = \frac{1}{2}\pi$ is

$$2Wl\{\sqrt{(4\mu^2 + 1)} - 1\}. \qquad \text{(N.U.)}$$

88. A number of forces acting in the plane of rectangular axes Ox, Oy have a resultant whose x and y components are X and Y

respectively. The sum of the moments of the forces about O in the sense from Ox to Oy is G. Show that the equation of the line of action of the resultant is $Yx-Xy-G=0$.

A rigid lamina lying on a smooth horizontal table is freely pivoted to a fixed point P on the table and is in equilibrium under forces whose components are $(4, 2)$, $(6, 7)$, and $(-5, -3)$ acting at the points $(a, 0)$, $(0, a)$, and $(-a, 0)$ respectively, the coordinate axes Ox, Oy being rectangular and in the plane of the lamina. Show that P must lie on the line $6x-5y+a=0$. The direction of each force is now turned through a right angle in the sense from Ox to Oy, without changing its magnitude and point of application, and the lamina is still found to be in equilibrium. Show that P must be the point $(74a/61, 101a/61)$. Show also that, if the directions of the forces are similarly turned through any angle θ, the lamina will still be in equilibrium.

(N.U.)

89. The rectangle ABCD with its diagonal AC represents the vertical sections of two similar wedges ACD and ABC, whose weights are W and W_1: the wedge ABC rests with its face represented by AB on a horizontal table, and the wedge ACD rests on the wedge ABC. If a gradually increasing horizontal force acts on the upper wedge in a direction such that the wedges remain in contact, and slipping occurs first between the wedges, prove that

$$W(\mu+\tan \alpha)<\mu_1(1-\mu \tan \alpha)(W+W_1),$$

where μ and μ_1 are the coefficients of friction between the two wedges, and between the lower wedge and the table respectively, and α is $\angle CAB$.

90. A smoothly jointed linkage consists of a light rigid bar ABC, jointed at its upper end A to a smooth fixed vertical bar AD and at C to another light rigid bar CD, carrying a ring of negligible weight at D, which can slide on AD. $AB=l$, $AC=a$, $CD=b$, and a weight W is hung at B. Prove that the force to be applied to the ring D in the direction DA in order to keep it at a distance x from A is given by

$$X = Wl(x^2+b^2-a^2)/2ax^2.$$

91. The points O, A, B, C have coordinates $(0, 0)$, $(3, 0)$, $(3, 4)$, $(0, 4)$ respectively. Forces of magnitudes 4, 7, 2, 3, 10 units act along OA, AB, BC, CO, OB respectively in the senses indicated by the order of the letters, and a couple of moment 11 units acts in the sense OABC. Find the magnitude of the resultant

of the five forces and the couple, and show that the equation of its line of action is $3x-2y-10 = 0$. (N.U.)

92. State the principle of Virtual Work.

A bead of mass M can slide on a smooth circular wire hoop of radius a whose plane is vertical, and another bead, of mass m, can slide on a smooth wire which coincides with the vertical diameter of the hoop. The beads are joined by a light inelastic string of length $l(<a)$, and rest in equilibrium in the upper half of the figure. Show that the radius to the bead M makes with the vertical an angle θ (assumed not zero), where

$$\left\{1-\left(\frac{m}{M+m}\right)^2\right\} \cos^2 \theta = 1-l^2/a^2.$$

(It is assumed that the conditions are such as make this angle real.) (O.C.)

93. Four equal uniform rods, each of weight W and length $2a$, are smoothly jointed to form a rhombus ABCD. The rhombus is in equilibrium in a vertical plane with the lower rods, BC and DC, resting on two smooth pegs at the same level, distant $2c$ apart ($c<a$), and with AC vertical. A and C are joined by an elastic string of natural length $2a$ and modulus $2W$. Show that, if $\angle DCB = 2\theta$, then $4a \cos \theta \sin^3 \theta = c$.

Verify that this equation has a root in θ between $0°$ and $45°$, and investigate the stability of the corresponding equilibrium position. (N.U.)

94. A light string of length l passes over a smooth peg and supports a uniform rod of length $2a$. One end of the string is attached to one end of the rod, and the other end of the string is attached to a light ring which slides smoothly on the rod. Show that in the position of equilibrium the rod is inclined at an angle θ to the horizontal, where

$$2a \cos^3 \theta = l \sin \theta.$$

Show further that if in the equilibrium position of the rod the ring is at a distance $\frac{1}{2}a$ from the free end of the rod the length of the string is $\frac{3}{2}a\sqrt{3}$. (N.U.)

95. Explain the Potential Energy method of determining the positions of equilibrium of a material system and testing their stability, paying special attention to systems of one degree of freedom.

A uniform triangular lamina ABC of weight W, with A = B = $45°$ and AB = c, rests with its plane vertical and AC, BC touching two small smooth pegs at the same level and at a distance d apart. Express the potential energy in terms of θ, the angle between the median CG and the vertical; prove that, if $c < 6d$, there are two inclined positions of equilibrium, and test their stability. Show also that the work done by gravity when the lamina moves from an inclined position of equilibrium to the symmetrical one is equal to $Wd(1-c/6d)^2$.   (C.W.B.)

96. Find the potential energy of an elastic string of modulus W and natural length a, when its length is at x.

A uniform rod AB, of weight W and length a, is free to rotate about A, which is fixed. To B is attached one end of an elastic string, of modulus W and natural length a. The rod is hung vertically downwards and the free end of the string is then attached to a fixed point, at a height b vertically above A. Show that the rod is in stable equilibrium if $b < a$.   (O.C.)

97. State how, when the potential energy of a material system in any position is known, the positions of equilibrium and the stability of the equilibrium can be determined.

AB is the horizontal diameter of a smooth circular wire fixed with its plane vertical. A light elastic string whose natural length is AB and whose modulus of elasticity is λ has its ends attached to A and B and passes through a small smooth ring of weight W, which can slide on the lower half on the wire. Show that, if $W > \lambda(1-1/\sqrt{2})$, the ring is in a position of stable equilibrium at the lowest point of wire, and that there are no other positions of equilibrium; but that, if $W < \lambda(1-1/\sqrt{2})$, the equilibrium in this position is unstable and there are two positions of stable equilibrium.   (O.C.)

98. A uniform square board of weight W is supported in a vertical plane on two smooth pegs at the same horizontal level; the distance between the pegs is c and the diagonal AB is of length $b > 4c$. If a weight w is attached to the lowest corner B of the square, prove that, when the diagonal AB is inclined to the horizontal at an angle ϕ, the potential energy of the system is

$$\tfrac{1}{2}W(b \sin \phi + c \cos 2\phi) + \tfrac{1}{2}wc \cos 2\phi.$$

Deduce that the equilibrium position in which AB is vertical is stable if $4cw > W(b-4c)$.   (C.W.B.)

99. Two uniform rods, AB of length $2l$ and weight w and BC of length $2a(a < l)$ and weight W, are freely hinged at B. The end C of the second rod can move freely on a vertical guide and the end A of the first rod is freely pivoted to a fixed point of the guide. In a position in which B is below the levels of C and A the horizontal distance of the centres of mass of the rods from

the guide is x. Show that the potential energy of the system in this position is

$$V = W(a^2-x^2)^{\frac{1}{2}} - (w+2W)(l^2-x^2)^{\frac{1}{2}} + \text{constant}.$$

Show that there is an oblique position of equilibrium if $Wl<(w+2W)a$, and that if this position exists the equilibrium there is unstable.   (O.C.)

100. Three equal uniform rods AB, BC, CD, each of length $2a$, are freely jointed at B and C. The ends A and D are constrained to slide on a smooth wire in the form of a circle centre O and radius b with its plane vertical. If the rods are in equilibrium with the centre rod horizontal and the radii OA, OD in the circle inclined at an angle of 60° with the horizontal, show that

$$b = a\{2+4\sqrt{(3/7)}\}.$$

Show also that this position of equilibrium is unstable.

CHAPTER TEN

VECTORS

10.1. Throughout this book use has been made of vector notation and in some measure of vector methods. We shall now collect together in this chapter all the vector algebra that has been used earlier, and develop it further.

10.2. A *scalar* quantity is one which is completely determined by a single number, its magnitude. It has no intrinsic reference to direction in space. For example, mass, length, time, energy, temperature are all scalars.

A *vector* has magnitude, in the ordinary algebraic sense, as well as direction in space. Two quantities, its magnitude and direction, are needed to specify it completely. For example, force, velocity, acceleration, momentum are all vectors.

The simplest vector is a displacement of a particle, or a displacement of translation of a rigid body, in space. If a particle is displaced from a point P to a point P' the displacement is specified by the vector $\mathbf{PP'}$; its magnitude is PP' and its direction from P to P'.

All vectors can be represented by a line such as PP'. It is often convenient, however, to denote a vector by a single letter such as $\mathbf{a}$, in which case we denote its magnitude by a.

The magnitude of any vector $\mathbf{AB}$ is sometimes called its modulus and is denoted by $|\mathbf{AB}|$.

There is a vector algebra analogous to the ordinary scalar algebra of positive and negative numbers. This we will now develop formally.

10.3. Equality of vectors

Two vectors $\mathbf{a}$ and $\mathbf{b}$ are said to be equal if they have the same magnitude *and* the same direction in space.

We write $\mathbf{a} = \mathbf{b}$, and this implies that $a = b$ and that the vectors are parallel and similarly directed. Strictly this definition of equality refers only to free vectors, and not to vectors localised in a line (such as forces acting on a rigid body).

10.4. Addition and subtraction of vectors

The sum of any two vectors of the same kind represented by **OP** and **PQ** is defined as a vector represented by **OQ**. We write **OP**+**PQ** = **OQ**.

This addition can always be done graphically, that is, by drawing the triangle OPQ (Fig. 10.1).

It will be noted that $OP+PQ \geqslant OQ$.

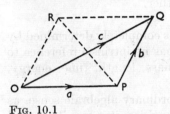

FIG. 10.1

Denoting **OP** by **a**, **PQ** by **b** and **OQ** by **c** we write

$$\mathbf{a}+\mathbf{b} = \mathbf{c}.$$

Complete the parallelogram OPQR. Then from the definition of equality we have

$$\mathbf{OR} = \mathbf{PQ} = \mathbf{b}$$

and

$$\mathbf{RQ} = \mathbf{OP} = \mathbf{a}$$

But,

$$\mathbf{OR}+\mathbf{RQ} = \mathbf{OQ}$$

∴

$$\mathbf{b}+\mathbf{a} = \mathbf{c}.$$

Hence,

$$\mathbf{b}+\mathbf{a} = \mathbf{a}+\mathbf{b}$$

which establishes the *commutative law* of addition for vectors.

The special case when **a** and **b** are such that Q and O coincide is of some importance. In this case OQ is zero, and hence

$$\mathbf{a}+\mathbf{b} = \mathbf{OQ} = \mathbf{0}.$$

A zero vector has zero magnitude; strictly it is represented by a point and has no direction.

If $\mathbf{a}+\mathbf{b} = \mathbf{0}$ we write $\mathbf{b} = -\mathbf{a}$, that is, $-\mathbf{a}$ is a vector of magnitude a, parallel to the vector **a** but in the opposite direction.

Similarly, since

$$OP+PO = 0$$

we write

$$PO = -OP.$$

Quite generally

$$LM = -ML.$$

From Fig. 10.1 we may write

$$RP = RO+OP$$
$$= -OR+OP$$
$$= -b+a$$
$$= a-b$$

Similarly,

$$PR = b-a$$

We note that the two diagonals of a parallelogram represent the sum and difference of the vector represented by the adjacent sides.

The consistency of these rules of addition and subtraction might be illustrated in various ways. We shall show later that some geometrical proofs may be based on them.

10.5. The addition of any number of vectors **a**, **b**, **c**, **d**, . . . may be carried out by a repeated application of the above law of addition. This is indicated in Fig. 10.2.

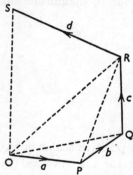

Fig. 10.2

We draw OP, PQ, QR, . . . to represent the vectors **a**, **b**, **c**, . . . respectively.

Then

$$a+b = OQ$$

∴

$$(a+b)+c = OQ+QR$$
$$= OR$$

and so on until the whole addition has been completed.

It is clear that

$$\mathbf{OR} = \mathbf{OP} + \mathbf{PR}$$
$$= \mathbf{a} + (\mathbf{b} + \mathbf{c})$$

and hence

$$(\mathbf{a} + \mathbf{b}) + \mathbf{c} = \mathbf{a} + (\mathbf{b} + \mathbf{c})$$

This shows that the order in which the vectors are added does not affect the sum, and so the brackets are unnecessary. This is referred to as the *associative law* of addition.

We write the sum of the vectors simply as

$$\mathbf{a} + \mathbf{b} + \mathbf{c} + \mathbf{d} + \ldots$$

It also follows from the commutative law that the order in which the vectors are written does not matter.

10.6. In the special case when $\mathbf{a} = \mathbf{b} = \mathbf{c} = \mathbf{d} = \ldots$

we get $\mathbf{a} + \mathbf{b} = \mathbf{a} + \mathbf{a}$, which we write 2**a**.

Also, $\mathbf{a} + \mathbf{b} + \mathbf{c} = \mathbf{a} + \mathbf{a} + \mathbf{a}$, which we write 3**a**.

Indeed, $n\mathbf{a}$, where n is a positive scalar quantity, is defined as a vector in the same direction as **a** but of magnitude na (Fig. 10.3).

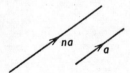

Fig. 10.3

If n is negative, say $-m$ where m is a positive scalar quantity, then $n\mathbf{a} = -m\mathbf{a}$, which is the vector $m\mathbf{a}$ reversed in direction, that is, a vector in the opposite direction to **a** and of magnitude ma.

It follows that a unit vector in the direction of **a** can be written as $\frac{1}{a}\mathbf{a}$. The notation $\hat{\mathbf{a}}$ is sometimes used for this unit vector. Thus

$$\hat{\mathbf{a}} = \frac{1}{a}\mathbf{a} \quad \text{or} \quad \mathbf{a} = a\hat{\mathbf{a}}.$$

10.7. Further, we can show, if **a** and **b** are any two vectors and n is a scalar quantity, that

$$n(\mathbf{a}+\mathbf{b}) = n\mathbf{a}+n\mathbf{b}.$$

If OP (Fig. 10.4) represents **a**, then $\mathrm{OP_1} = n\mathrm{OP}$ represents $n\mathbf{a}$. Also if PQ represents **b** and if $\mathrm{P_1Q_1}$ is drawn parallel to PQ and such that $\mathrm{P_1Q_1} = n\mathrm{PQ}$, then $\mathrm{P_1Q_1}$ represents $n\mathbf{b}$.

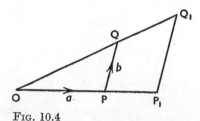

FIG. 10.4

Thus, $\mathrm{OP_1/OP} = \mathrm{P_1Q_1/PQ} = n$, and hence $\mathrm{Q_1}$ lies on OQ produced and is such that $\mathrm{OQ_1/OQ} = n$.

$$\therefore \qquad \mathbf{OQ_1} = n\mathbf{OQ}$$

$$\therefore \qquad \mathbf{OP_1}+\mathbf{P_1Q_1} = n(\mathbf{OP}+\mathbf{PQ})$$

$$\therefore \qquad n\mathbf{a}+n\mathbf{b} = n(\mathbf{a}+\mathbf{b})$$

Figure 10.4 has been drawn corresponding to a positive value of n, but the proof holds for negative values of n too.

10.8. It will be noted that the rules for vectors which govern addition, subtraction, and multiplication by a scalar are identical with the rules governing these operations in ordinary algebra. They are easy to apply to vectors, for no new methods of manipulation have to be learnt. They have been used from time to time as seemed appropriate in these two volumes on statics and dynamics, and numerous examples of their usefulness have been given. They have, however, a wider range of application: in geometry, for instance, vector methods lead to concise and even elegant proofs. A simple example was given in paragraph 1.12 of Volume One; others are given below.

EXAMPLE (i)

Prove that the diagonals of a parallelogram bisect one another

Let S be the point of intersection of the diagonals of the parallelogram OPQR (Fig. 10.5). Let **OP** = **a** and **OR** = **b**.

Then,
$$OS = kOQ, \text{ where } k = OS/OQ$$
$$= k(\mathbf{a} + \mathbf{b})$$

Hence,
$$SP = SO + OP$$
$$= -k(\mathbf{a} + \mathbf{b}) + \mathbf{a}$$
$$= (1 - k)\mathbf{a} - k\mathbf{b}.$$

But,
$$SP = mRP, \text{ where } m = SP/RP$$
$$= m(\mathbf{a} - \mathbf{b}).$$

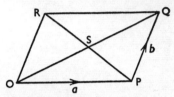

Fig. 10.5

Comparing these two expressions for **SP**, it follows that
$$1 - k = m = k$$

Hence,
$$k = m = \tfrac{1}{2}$$

Hence,
$$OS = \tfrac{1}{2}OQ \text{ and } SP = \tfrac{1}{2}RP,$$

that is, the diagonals bisect one another.

EXAMPLE (ii)

Prove vectorially that if PQ *is drawn parallel to the side* BC *of a triangle* ABC *and cuts* AB, AC *at* P, Q, *then* AP/AB = AQ/AC.

Since PQ, Fig. 10.6, is parallel to BC, we can write

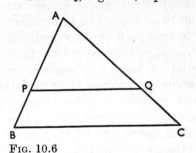

Fig. 10.6

$$PQ = kBC$$

where
$$k = PQ/BC.$$

But
$$PQ = PA + AQ$$

and
$$BC = BA + AC$$

$$\therefore \qquad \mathbf{PA} + \mathbf{AQ} = k\mathbf{BA} + k\mathbf{AC}$$

$$\therefore \qquad \mathbf{PA} - k\mathbf{BA} = k\mathbf{AC} - \mathbf{AQ}$$

$$\therefore \qquad \left(\frac{PA}{BA} - k\right)\mathbf{BA} = \left(k - \frac{AQ}{AC}\right)\mathbf{AC}$$

But **BA** and **AC** are *any* two vectors, and hence this result is only possible if the coefficients $\frac{PA}{BA} - k$ and $k - \frac{AQ}{AC}$ are both zero, that is, $k = PA/BA = AQ/AC$.

Hence $AP/AB = AQ/AC$.

10.9. Components of a vector

Since two vectors can be added or compounded into a single vector, so, conversely, any vector can be split up, or resolved, into two vectors; they are known as its *components*.

Any vector represented by OQ can be resolved into two vectors, **OP** and **PQ**, merely by drawing any triangle of which OQ is one side, or by drawing a parallelogram OPQR of which OQ is a diagonal. This can be done in an infinite number of ways.

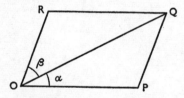

FIG. 10.7

Thus if we require the components of the vector **OQ** which make angles of α and β respectively with OQ we draw the parallelogram OPQR (Fig. 10.7) with $\angle QOP = \alpha$ and $\angle QOR = \beta$.

Then $\qquad\qquad \mathbf{OQ} = \mathbf{OP} + \mathbf{PQ}$

and $\qquad\qquad \dfrac{OP}{\sin \beta} = \dfrac{PQ}{\sin \alpha} = \dfrac{OQ}{\sin (\alpha + \beta)}$

$\therefore \qquad\qquad OP = OQ \sin \beta / \sin (\alpha + \beta)$

and $\qquad\qquad PQ = OQ \sin \alpha / \sin (\alpha + \beta)$.

These are the magnitudes of the components of the vector in the specified directions.

The special case when the components of the vector are perpendicular frequently occurs and should be noticed.

If $\alpha + \beta = 90°$, then

$$OP = OQ \cos \alpha \text{ and } PQ = OQ \sin \alpha.$$

10.10. Since each component of a vector can in turn be resolved into two components, it follows that by repeated application of the addition law a vector can be resolved into any number of components.

Thus a vector **OQ** can be divided into n components by drawing a polygon of $(n+1)$ sides, of which OQ is one side. Moreover, the polygon need not be a plane polygon. There is an infinite number of sets of components of this kind.

10.11. It is sometimes convenient to use components of a vector parallel to certain chosen axes of coordinates.

If P is a point with coordinates (x, y), (Fig. 10.8) the position

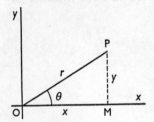

Fig. 10.8

of P relative to the origin may be denoted by the vector **OP**. It is called the position-vector of P and is often denoted by **r**.

If PM is drawn perpendicular to Ox we have

$$OP = OM + MP$$

But **OM** may be written as $x\mathbf{i}$ where $\mathbf{i}$ is a vector of unit magnitude parallel to Ox, and **MP** may be written as $y\mathbf{j}$ where $\mathbf{j}$ is a vector of unit magnitude parallel to Oy.

We have $\mathbf{r} = x\mathbf{i} + y\mathbf{j}.$

If the angle POM $= \theta$ then

$$x = r \cos \theta \text{ and } y = r \sin \theta.$$

For any vector $\mathbf{r}$ the components x and y can be uniquely determined. We note that $r^2 = x^2+y^2$, that is,

$$r = |\,\mathbf{r}\,| = (x^2+y^2)^{\frac{1}{2}}.$$

10.12. More generally, suppose $\mathbf{a}$ and $\mathbf{b}$ are any two non-parallel vectors in a plane. Then any vector $\mathbf{r}$ coplanar with $\mathbf{a}$ and $\mathbf{b}$. can be resolved into components parallel to $\mathbf{a}$ and $\mathbf{b}$ respectively.

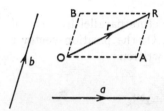

FIG. 10.9

If OR (Fig. 10.9) represents $\mathbf{r}$ we complete the parallelogram $OARB$ with sides parallel to $\mathbf{a}$ and $\mathbf{b}$.

Then,
$$\mathbf{r} = \mathbf{OA}+\mathbf{OB}$$
$$= x\mathbf{a}+y\mathbf{b}$$

where x and y are scalar quantities. In fact, $x = OA/a$ and $y = OB/b$.

Similarly, if $\mathbf{r}_1, \mathbf{r}_2, \mathbf{r}_3, \ldots$ are any set of coplanar vectors they can be resolved into components parallel to the vectors $\mathbf{a}$ and $\mathbf{b}$ as follows:

$$\mathbf{r}_1 = x_1\mathbf{a}+y_1\mathbf{b}$$
$$\mathbf{r}_2 = x_2\mathbf{a}+y_2\mathbf{b}$$
$$\mathbf{r}_3 = x_3\mathbf{a}+y_3\mathbf{b}$$

and so on. We note that their vector-sum is

$$\mathbf{r}_1+\mathbf{r}_2+\mathbf{r}_3+ \ldots = (x_1+x_2+x_3+ \ldots)\,\mathbf{a}+(y_1+y_2+y_3 \ldots)\mathbf{b}.$$

It follows that the components of the sum of any set of vectors are the sums of the components of the separate vectors.

Further, if the vector sum is zero, then

$$x_1+x_2+x_3+ \ldots = 0$$
and
$$y_1+y_2+y_3+ \ldots = 0.$$

The converse of this statement is also true.

10.13. If a vector **r** is not in the plane of the vectors **a** and **b** (or in a plane parallel to this plane) it may be resolved into a component in this plane and a component not in this plane.

It follows that any vector **r** can be resolved into components parallel to any three non-coplanar vectors **a**, **b**, and **c**, and we can write

$$\mathbf{r} = x\mathbf{a} + y\mathbf{b} + z\mathbf{c}$$

where x, y, z are scalar quantities.

In particular, if **i**, **j**, **k** are unit vectors parallel to a set of mutually perpendicular axes Ox, Oy, Oz the position-vector **r** of any point (x, y, z) in space may be written

$$\mathbf{r} = x\mathbf{i} + y\mathbf{j} + z\mathbf{k}.$$

We note that $\qquad r = |\mathbf{r}| = (x^2 + y^2 + z^2)^{\frac{1}{2}}.$

The direction of the vector in space is given by the direction-cosines x/r, y/r and z/r.

10.14. Vector-equation of a line

If OA and OB (Fig. 10.10) represents the vectors **a** and **b**, and P is any point on the line through B parallel to OA, then the position-vector **r** of P with respect to O is given by:

$$\mathbf{r} = \mathbf{OB} + \mathbf{BP}$$
$$= \mathbf{b} + m\mathbf{a}$$

where $m = BP/OA$, a scalar quantity.

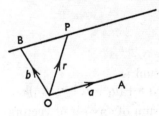

Fig. 10.10

This equation $\mathbf{r} = \mathbf{b} + m\mathbf{a}$ is the vector-equation of the line that passes through the point B (defined by **b**) and is parallel to the vector **a**.

Any particular value of m gives a particular point on the line; m can have any value from minus infinity to plus infinity.

Similarly, if P is any point on the line BA its position vector **r** is given by

$$\mathbf{r} = \mathbf{OB} + \mathbf{BP}$$
$$= \mathbf{b} + m\mathbf{BA}, \text{ where } m = BP/BA$$
$$= \mathbf{b} + m(\mathbf{a} - \mathbf{b})$$
$$= m\mathbf{a} + (1-m)\mathbf{b}.$$

This is the vector-equation of the line through the two points whose position-vectors are **a** and **b**.

The point A corresponds to $m = 1$ and the point B to $m = 0$. Any other point on AB or AB produced corresponds to some particular value of m. The mid-point of AB corresponds to $m = \frac{1}{2}$; for this point $\mathbf{r} = \frac{1}{2}\mathbf{a} + \frac{1}{2}\mathbf{b}$.

10.15. EXAMPLE (i)

ABC *is any triangle, and the bisector of the angle ACB meets AB in* M. *Prove vectorially that* CA/CB = AM/MB.

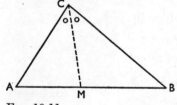

FIG. 10.11

For *any* point M in AB we have (Fig. 10.11)

$$\mathbf{CM} = \mathbf{CA} + \mathbf{AM}$$
$$= \mathbf{CA} + \frac{AM}{AB}\,\mathbf{AB}$$
$$= \mathbf{CA} + \frac{AM}{AB}\,(\mathbf{AC} + \mathbf{CB})$$
$$= \left(1 - \frac{AM}{AB}\right)\mathbf{CA} + \frac{AM}{AB}\,\mathbf{BC}$$
$$= \frac{MB}{AB}\,\mathbf{CA} + \frac{AM}{AB}\,\mathbf{BC}$$

giving the components of **CM** alone CA and CB.

But if CM bisects $\angle$ACB the components of **CM** along CA and CB must be equal in magnitude.

$$\therefore \qquad \frac{MB}{AB}\, CA = \frac{AM}{AB}\, CB$$

$$\therefore \qquad \frac{CA}{CB} = \frac{AM}{MB}$$

EXAMPLE (ii)

Express the vector $\mathbf{r} = 10\mathbf{i} - 3\mathbf{j} - \mathbf{k}$ *as a linear function of the vectors* **a**, **b**, *and* **c**, *given that* $\mathbf{a} = 2\mathbf{i} - \mathbf{j} + 3\mathbf{k}$, $\mathbf{b} = 3\mathbf{i} + 2\mathbf{j} - 4\mathbf{k}$, *and* $\mathbf{c} = -\mathbf{i} + 3\mathbf{j} - 2\mathbf{k}$.

r is a linear function of **a**, **b**, and **c** if it can be written in the form

$$\mathbf{r} = m_1\mathbf{a} + m_2\mathbf{b} + m_3\mathbf{c}$$

where m_1, m_2, m_3 are scalars.

Assuming this relation, we have

$$10\mathbf{i} - 3\mathbf{j} - \mathbf{k} = m_1(2\mathbf{i} - \mathbf{j} + 3\mathbf{k}) + m_2(3\mathbf{i} + 2\mathbf{j} - 4\mathbf{k}) + m_3(-\mathbf{i} + 3\mathbf{j} - 2\mathbf{k}).$$

Hence, equating components, we get

$$2m_1 + 3m_2 - m_3 = 10$$
$$-m_1 + 2m_2 + 3m_3 = -3$$
and $$3m_1 - 4m_2 - 2m_3 = -1.$$

These lead to $m_1 = 1$, $m_2 = 2$ and $m_3 = -2$.

Hence, $\mathbf{r} = \mathbf{a} + 2\mathbf{b} - 2\mathbf{c}$.

EXAMPLE (iii)

Find the foot of the perpendicular from the origin on to the line $\mathbf{r} = 3m\mathbf{i} + 4(1-m)\mathbf{j}$.

The line with the vector-equation

$$\mathbf{r} = 3m\mathbf{i} + 4(1-m)\mathbf{j}$$

is shown in Fig. 10.12. It passes through the point A on the axis of x where $\mathbf{OA} = 3\mathbf{i}$, and the point B on the axis of y where $\mathbf{OB} = 4\mathbf{j}$. A and B correspond to $m = 1$ and $m = 0$ respectively.

If P is the foot of the perpendicular from O on to AB, then $\angle$POA, denoted by θ, equals $\angle$ABO.

But $\qquad$ tan POA = MP/OM = $4(1-m)/3m$

and $\qquad$ tan ABO = OA/OB = 3/4.

Hence, P corresponds to that value of m for which

$$\frac{4(1-m)}{3m} = \tfrac{3}{4}$$

$$\therefore \qquad 16(1-m) = 9k$$

$$\therefore \qquad m = 16/25$$

Hence, the position-vector of P is

$$\frac{48}{25}\mathbf{i} + \frac{36}{25}\mathbf{j}$$

that is, OM $= 48/25$ and MP $= 36/25$.

We note that the length of the perpendicular OP is

$$\left\{\left(\frac{48}{25}\right)^2 + \left(\frac{36}{25}\right)^2\right\}^{\frac{1}{2}},$$

that is, 12/5.

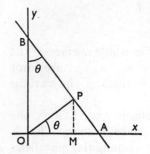

FIG. 10.12

10.16. Areas as vectors

Any plane area has magnitude and orientation in space; the latter can be specified by the direction of the normal to the area. Any plane area can therefore be represented by a vector normal to the area, say $A\mathbf{n}$, where A is the magnitude of the area and $\mathbf{n}$ is unit vector in the direction of the normal. $A\mathbf{n}$ is sometimes written $\mathbf{A}$, it being understood that $\mathbf{A}$ is normal to the area A.

There is an ambiguity as to the positive direction of the normal; it has therefore to be specified carefully in any particular case. Sometimes the positive direction of the normal is taken to be related to positive movement round the boundary of the area by the right-hand screw rule (Fig. 10.13). In turn, movement round the boundary is taken as positive if the area is always on the left-hand side. If, however, the area forms part of a closed surface the *outward* normal is usually taken as positive. For example, the surfaces of a rectangular parallelepiped can be represented (Fig. 10.14) by six vectors normal to the surfaces outwards; they are equal and opposite in pairs.

If a surface S is *not* plane, it may be divided into elements of area dS_1, dS_2, dS_3, . . . each sufficiently small to be regarded as

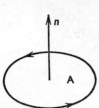

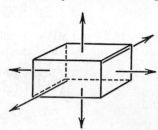

FIG. 10.13   FIG. 10.14

plane. Then the vector **S** representing the whole surface is the vector-sum of $dS_1\mathbf{n}_1$, $dS_2\mathbf{n}_2$, . . . where $\mathbf{n}_1$, $\mathbf{n}_2$, . . . are unit vectors normal to the surface at the positions of the various elements.

For any closed surface this vector-sum is zero, that is, the representative vector of any closed surface is zero. This is obvious in the particular example of a rectangular parallelepiped, as shown in Fig. 10.14. It can be proved generally.

10.17. An area vector can be resolved into components like any other vector. It can be shown that the component of an area vector in any direction equals in magnitude the projection of the area on a plane perpendicular to the given direction.

For if a triangle ABC is projected orthogonally on to a plane its projection is a triangle abc such that

$$\text{area of } \triangle abc = (\text{area of } \triangle ABC)\cos\theta$$

where θ is the angle between the planes of ABC and abc. If $90° < \theta < 180°$ the area of $\triangle abc$ is negative.

This result can be extended to apply to any plane rectilinear figure, for such a figure consists of a set of triangles. Similarly, the result applies to a plane area bounded by any curve, since this area may be regarded as the limit of that of a plane rectilinear figure as the number of sides of the figure increases indefinitely.

Thus, the orthogonal projection of a plane area upon any plane can be represented by the component of the representative vector of the area in a direction normal to the plane of projection.

This result must also be true for any surface whatsoever, since such a surface can be divided into elementary plane surfaces to which the above theorem applies. By vector-summation the theorem follows generally.

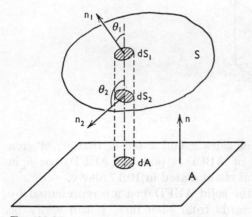

FIG. 10.15

In the case of a closed surface the projection on *any* plane is zero, since as much of the projected area is positive as is negative.

This is clear from Fig. 10.15, where dS_1 is any element of a closed surface S. If the projection of dS_1 on any plane A is denoted by dA, and if n_1 is unit vector normal to the element dS_1, outwards from the surface, then dA equals the component of $dS_1 n_1$ perpendicular to the plane A, that is, in the direction of the unit vector n shown in the figure. Now the cylinder of which dS_1 and dA are the ends cuts the surface S again in some element of area dS_2. If n_2 is unit vector normal to the element dS_2 outwards from the surface, then the component of $dS_2 n_2$ in the direction of the normal to A, that is, in the direction of the unit vector n, is $-dA$.

This is true for all pairs of elements dS_1 and dS_2. Hence, summing for the whole surface, it follows that the representative vector of any closed surface is zero.

10.18. The application of the results discussed above to plane sections of certain solids is obvious.

For instance, a right section of a cube of side x is a square of area x^2. A section ADFE (Fig. 10.16) at an angle θ to the face

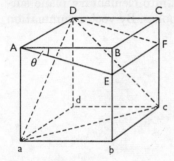

FIG. 10.16

ABCD is a rectangle of sides x and $x \sec \theta$, that is, of area $x^2 \sec \theta$. Hence, area of ABCD = (area of AEFD) $\cos \theta$ in keeping with the general result stated in 10.17 above.

If the six faces of the solid AEFDabcd are represented by vectors, normally outwards from each face, it can easily be shown that their vector sum is zero. This may also be done for the five faces of the wedge ADCBEF.

Similarly, the section Dca of the cube (Fig. 10.16) is an equilateral triangle of side $x\sqrt{2}$, and therefore of area $\frac{1}{2}(x\sqrt{2})^2 \sin 60° = \frac{1}{2}x^2\sqrt{3}$. Its projection on the face abcd of the cube is the triangle dca of area $\frac{1}{2}x^2$. Hence if α is the angle which the section Dca makes with the face abcd we have

$$\tfrac{1}{2}x^2 = (\tfrac{1}{2}x^2\sqrt{3}) \cos \alpha$$

$$\therefore \qquad \cos \alpha = 1/\sqrt{3}.$$

This could also be found by representing the areas of the faces of the tetrahedron Ddca by vectors normally outwards from each face. These have magnitudes $\frac{1}{2}x^2$, $\frac{1}{2}x^2$, $\frac{1}{2}x^2$, and $\frac{1}{2}x^2\sqrt{3}$ respectively, and their vector-sum is zero provided $(\frac{1}{2}x^2\sqrt{3}) \cos \alpha = \frac{1}{2}x^2$, that is, $\cos \alpha = 1/\sqrt{3}$. Conversely, we can prove $\cos \alpha = 1/\sqrt{3}$ geometrically, and then verify that the sum of the vectors representing the four faces is zero.

Again, the right section of a solid cylinder of radius a is a circle of area πa^2. A section at an angle θ to the right section has an area A where

$$A \cos \theta = \pi a^2$$

$$\therefore \quad A = \pi a^2 \sec \theta.$$

This section is, in fact, an ellipse with semi-axes a and $a \sec \theta$, and can be represented by a vector of magnitude $\pi a^2 \sec \theta$ at an angle θ to the axis of the cylinder.

EXAMPLES 10.1

1. ABCDEF is a regular hexagon. Vectors of magnitude 2, $4\sqrt{3}$, 8, $2\sqrt{3}$, and 4 units are directed along AB, AC, AD, AE, AF respectively. Find the magnitude of the sum of the vectors and the inclination of its direction to AB.

2. The position vectors of the points A, B, C, are **a**, **b**, **c**, where
   $\mathbf{a} = a_1\mathbf{i} + a_2\mathbf{j}$, $\mathbf{b} = b_1\mathbf{i} + b_2\mathbf{j}$, $\mathbf{c} = c_1\mathbf{i} + c_2\mathbf{j}$.
   Find the lengths of AB, BC, and CA.

3. (i) If $\mathbf{a} = 2\mathbf{i} + 3\mathbf{j}$, $\mathbf{b} = 3\mathbf{i} - 4\mathbf{j}$ and $\mathbf{c} = -2\mathbf{i} + 5\mathbf{j}$, write down $\mathbf{a} + \mathbf{b} + \mathbf{c}$, $2\mathbf{a} - \mathbf{b} + 3\mathbf{c}$ and $\mathbf{a} + 2\mathbf{b} - 7\mathbf{c}$.
   (ii) Find values of m and n such that $m\mathbf{a} + n\mathbf{b}$ is parallel to **c**.
   (iii) Find the unit vectors $\hat{\mathbf{a}}$, $\hat{\mathbf{b}}$, and $\hat{\mathbf{c}}$.

4. Show that the vectors $a_1\mathbf{i} + a_2\mathbf{j}$ and $b_1\mathbf{i} + b_2\mathbf{j}$ are perpendicular if $a_1b_1 + a_2b_2 = 0$. Find a vector perpendicular to the vector $3\mathbf{i} - 4\mathbf{j}$.

5. Show that the position-vector **r** of the point P which divides the line AB in the ratio $m:n$ is given by $(m+n)\mathbf{r} = n\mathbf{a} + m\mathbf{b}$, where **a** and **b** are the position-vectors of A and B.

6. Prove that the three points, whose position-vectors are **a**, **b**, **c**, are collinear if $m_1\mathbf{a} + m_2\mathbf{b} + m_3\mathbf{c} = 0$ where $m_1 + m_2 + m_3 = 0$.

7. Show that a triangle may be constructed whose sides are parallel and equal to the medians of any given triangle.

8. Prove vectorially that if the medians AD and BE of any triangle ABC intersect at G, then G divides AD and BE in the ratio $2:1$. Show also that the three medians are concurrent.

9. Prove that if the mid-points of the sides of any quadrilateral be joined in order the figure formed is a parallelogram.

10. The diagonals of a given quadrilateral bisect one another; prove that the quadrilateral is a parallelogram.

11. If the points A_1, A_2, ... A_n have position-vectors r_1, r_2, ... r_n with respect to 0, show that the position of the point G defined by the relation

$$(m_1+m_2+ \ldots +m_n)\mathbf{OG} = m_1\mathbf{r}_1+m_2\mathbf{r}_2+ \ldots +m_n\mathbf{r}_n,$$

where m_1, m_2, ... m_n are scalar quantities, is independent of the position of O.

12. Prove vectorially that if a transversal cuts the sides BC, CA, AB of a triangle ABC, in D, E, F respectively, then

$$\frac{BD}{DC} \times \frac{CE}{EA} \times \frac{AF}{FB} = -1.$$

13. Prove vectorially that the internal bisectors of the angles of a triangle are concurrent.

14. Find the point of intersection of the lines of which the vector equations are $\mathbf{r} = k\mathbf{i}+2(1-k)\mathbf{j}$ and $\mathbf{r} = 3(1-m)\,\mathbf{i}-m\mathbf{j}$.

Draw these lines on a diagram, and find the angle between them. Find also the area enclosed by the lines and the axis of x.

15. If A and B have position vectors $\mathbf{a}$ and $\mathbf{b}$, and a parallelogram OACB is constructed with OA and OB as adjacent sides, find the vector equation of the line through C parallel to the diagonal AB. Verify that if this line meets OA produced in D, then $\mathbf{OD} = 2\mathbf{a}$.

16. ABCD is a parallelogram and E is the mid-point of BC. AE cuts the diagonal BD at F. Find the ratio in which F divides BD.

17. Show that the lines of which the vector-equations are

$$\mathbf{r} = k\mathbf{a}+(1-k)\mathbf{b} \text{ and } \mathbf{r} = 2(1+m)\mathbf{a}-(1+2m)\mathbf{b}$$

are coincident.

18. The vectors $\mathbf{a}$, $\mathbf{b}$, $\mathbf{c}$ form the adjacent sides of a parallelepiped. Find the vector-equations of the four diagonals of the parallelepiped and show that these diagonals are concurrent and bisect one another.

19. If the position-vectors of the vertices of a tetrahedron are $\mathbf{r}_1$, $\mathbf{r}_2$, $\mathbf{r}_3$, $\mathbf{r}_4$, show that the position-vector of the centre of mass of the tetrahedron is $\frac{1}{4}(\mathbf{r}_1+\mathbf{r}_2+\mathbf{r}_3+\mathbf{r}_4)$.

20. Show that the three straight lines which join the mid-points of opposite edges of a tetrahedron are concurrent and bisect one another.

21. Find the vector-equation of the line which passes through the points whose coordinates are (1, 2, 3) and (4, −1, 5). Find where the line cuts the (x, y) plane.

22. Show that the four points A, B, C D lie in a plane if their position-vectors $\mathbf{r}_1$, $\mathbf{r}_2$, $\mathbf{r}_3$, $\mathbf{r}_4$ are such that $m_1\mathbf{r}_1 + m_2\mathbf{r}_2 + m_3\mathbf{r}_3 + m_4\mathbf{r}_4 = 0$ and $m_1 + m_2 + m_3 + m_4 = 0$. Find m_1, m_2, m_3, m_4 when: (i) ABCD is a parallelogram; (ii) D is the mid-point of BC; and (iii) D is the centroid of the triangle ABC.

   Show that the equation of the plane through the points whose position-vectors are $\mathbf{r}_1$, $\mathbf{r}_2$, $\mathbf{r}_3$ can be written in the form

   $$\mathbf{r} = m_1\mathbf{r}_1 + m_2\mathbf{r}_2 + (1 - m_1 - m_2)\mathbf{r}_3.$$

23. If the position-vectors of A, B, C are $2\mathbf{i} - \mathbf{j} + 3\mathbf{k}$, $3\mathbf{i} + 2\mathbf{j} - 4\mathbf{k}$ and $-\mathbf{i} + 3\mathbf{j} - 2\mathbf{k}$, find: (i) the equation of the line through A parallel to BC, and (ii) the equation of the plane through A, B and the origin O.

24. If $\mathbf{a} = 2\mathbf{i} - \mathbf{j} + 3\mathbf{k}$, $\mathbf{b} = 3\mathbf{i} + 2\mathbf{j} - 4\mathbf{k}$, and $\mathbf{c} = -5\mathbf{i} - \mathbf{j} + \mathbf{k}$, find the unit vectors $\hat{\mathbf{a}}$, $\hat{\mathbf{b}}$, $\hat{\mathbf{c}}$.

   Show that if three vectors $\mathbf{a}$, $\mathbf{b}$, $\mathbf{c}$ are such that $\mathbf{a} + \mathbf{b} + \mathbf{c} = 0$, then in general $\hat{\mathbf{a}} + \hat{\mathbf{b}} + \hat{\mathbf{c}}$ is not zero. Find the conditions that must be satisfied if $\mathbf{a} + \mathbf{b} + \mathbf{c}$ and $\hat{\mathbf{a}} + \hat{\mathbf{b}} + \hat{\mathbf{c}}$ are both zero.

25. A cube has adjacent sides along the coordinate axes Ox, Oy, Oz. It is divided into two equal parts by a diagonal plane. Write down the vectors which represent the five faces of one of these parts, and verify that their vector-sum is zero.

26. Write down the vectors which represent the four faces of a regular tetrahedron, and verify that their vector sum is zero.

27. A tetrahedron OABC has its vertices A, B, C on Ox, Oy, Oz respectively. Assuming that the sum of the vectors representing the four faces is zero, find the area of the face ABC and the direction of its normal.

28. A regular pentagon ABCDE of side a is projected orthogonally on to a plane through the side AB that makes an angle θ with the plane of ABCDE. If the projection of the triangle ADB is an equilateral triangle, find θ. Find also the area of the projection of the triangle BDC.

29. An equilateral triangle of sides 6 cm has its vertices at heights of 7, 9, 11 cm above a horizontal plane. Find the dimensions of the projection of the triangle on the horizontal and the inclination of the equilateral triangle to the horizontal.

30. A regular hexagon ABCDEF of side 2 cm is projected orthogonally on to the plane through the side AB, which makes an angle of 30° with the plane of the hexagon. Find the lengths of the sides of the projected figure, and the angle between BC and its projection.

31. If the points A, B, C have position vectors $\mathbf{a}$, $\mathbf{b}$, $\mathbf{c}$ from an origin O, show that the equation

$$\mathbf{r} = t\mathbf{a} + (1-t)\mathbf{b},$$

where t is a parameter, represents the straight line AB, and find the equation of BC.

Find the equation of the straight line joining L the mid-point of OA to M the mid-point of BC. Find also the position vector of the point in which the line LM meets the straight line joining the mid-point of OB to the mid-point of AC.   (L.A.)

32. If $\mathbf{u} = 3\mathbf{i} + 4\mathbf{j} + 5\mathbf{k}$ is the position vector of a point P, find the magnitude of $\mathbf{u}$ and calculate to the nearest minute the angles made by $\mathbf{u}$ with the unit vectors, $\mathbf{i}$, $\mathbf{j}$, $\mathbf{k}$. Find also the unit vector in the same direction as $\mathbf{u}$.

Show that the equation

$$\mathbf{r} = 4\,(\mathbf{i} \cos p + \mathbf{j} \sin p),$$

where p is a parameter, represents a circle. Find the position vectors of the points on this circle which are nearest to and farthest from the point P.   (L.A.)

10.19. Differentiation of vector functions

Consider a vector $\mathbf{A}$ which is a function of some independent scalar variable t. In general, both the magnitude and direction of $\mathbf{A}$ will vary with t. If they vary with t in a continuous manner $\mathbf{A}$ is said to be a continuous vector function of t.

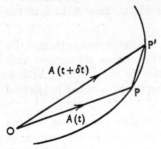

Fig. 10.17

If (Fig. 10.17) we represent $\mathbf{A}(t)$ by the vector $\mathbf{OP}$, then as t varies P will describe some curve.

For example, if $\mathbf{A}$ is the position-vector $\mathbf{r}$ of a particle which is moving in any manner, so that $\mathbf{r}$ is a function of the time t, this curve represents the path of the particle in space.

If (Fig. 10.17) the vector **OP'** represents $\mathbf{A}(t+\delta t)$, that is, the vector **A** when t changes to $t+\delta t$, then

$$\mathbf{PP'} = \mathbf{PO}+\mathbf{OP'}$$
$$= \mathbf{OP'}-\mathbf{OP}$$
$$= \mathbf{A}(t+\delta t)-\mathbf{A}(t)$$

that is, **PP'** represents the increment in the function **A** when t increases to $t+\delta t$. We denote this increment by $\delta\mathbf{A}$.

The limit of $\delta\mathbf{A}/\delta t$ as δt tends to zero is defined as the differential coefficient of **A** with respect to t, and is written $\dfrac{d\mathbf{A}}{dt}$ or $d\mathbf{A}/dt$.

Since **PP'** represents $\delta\mathbf{A}$, the ratio $\delta\mathbf{A}/\delta t$ is represented by a vector parallel to PP' and of magnitude $PP'/\delta t$. Consequently, $d\mathbf{A}/dt$ is represented by a vector in the limiting direction of PP' as P' tends to P, that is, in the direction of the tangent to the curve at P (Fig. 10.17).

In particular, if **A** is the position-vector **r** of a moving particle and if t denotes the time, then $d\mathbf{A}/dt = d\mathbf{r}/dt$ is the velocity of the particle in space, and this is clearly in the direction of the tangent to the path of the particle.

10.20. The differential coefficient $d\mathbf{A}/dt$ can be found from first principles for particular values of the function **A**.

If **A** is a constant vector $d\mathbf{A}/dt$ is zero. Also, if **A** is written as the sum of two vector functions **B** and **C**, then

$$\frac{d\mathbf{A}}{dt} = \frac{d\mathbf{B}}{dt}+\frac{d\mathbf{C}}{dt}.$$

Further, suppose **A** can be written in the form $k\mathbf{B}$, where k is a scalar quantity and **B** is a vector function and both are functions of t. Then

$$\mathbf{A} = k\mathbf{B}$$

and $\qquad \mathbf{A}+\delta\mathbf{A} = (k+\delta k)(\mathbf{B}+\delta\mathbf{B})$

$\therefore \qquad \delta\mathbf{A} = (k+\delta k)(\mathbf{B}+\delta\mathbf{B})-k\mathbf{B}$

$$= \delta k\mathbf{B}+k\delta\mathbf{B}+\delta k\,\delta\mathbf{B}$$

$\therefore \qquad \dfrac{\delta\mathbf{A}}{\delta t} = \dfrac{\delta k}{\delta t}\mathbf{B}+k\dfrac{\delta\mathbf{B}}{\delta t}+\dfrac{\delta k}{\delta t}\,\delta\mathbf{B}$

Hence, as δt tends to zero we get

$$\frac{d\mathbf{A}}{dt} = \frac{dk}{dt}\mathbf{B} + k\frac{d\mathbf{B}}{dt}.$$

In particular, if k is constant

$$\frac{d\mathbf{A}}{dt} = k\frac{d\mathbf{B}}{dt}.$$

Similarly, if $\mathbf{B}$ is a constant vector, then

$$\frac{d\mathbf{A}}{dt} = \frac{dk}{dt}\mathbf{B}.$$

For example, the position-vector $\mathbf{r}$ of a moving particle which is at the point (x, y, z) in space at time t can be written as

$$\mathbf{r} = x\mathbf{i} + y\mathbf{j} + z\mathbf{k}.$$

Differentiating with respect to t we get

$$\frac{d\mathbf{r}}{dt} = \frac{dx}{dt}\mathbf{i} + \frac{dy}{dt}\mathbf{j} + \frac{dz}{dt}\mathbf{k}.$$

This shows that the components of the velocity of the particle parallel to the coordinate axes are

$$\frac{dx}{dt}, \frac{dy}{dt}, \frac{dz}{dt}.$$

Further, since $\dfrac{d\mathbf{r}}{dt}$ is a vector function it can be differentiated again with respect to t. We denote this second differential coefficient by $d^2\mathbf{r}/dt^2$ and we get

$$\frac{d^2\mathbf{r}}{dt^2} = \frac{d^2x}{dt^2}\mathbf{i} + \frac{d^2y}{dt^2}\mathbf{j} + \frac{d^2z}{dt^2}\mathbf{k}.$$

This shows that the components of the acceleration of the particle parallel to the coordinate axes are

$$\frac{d^2x}{dt^2}, \frac{d^2y}{dt^2}, \frac{d^2z}{dt^2}.$$

10.21. Equations of motion

The equation of motion of a particle of mass m subject to a force $\mathbf{F}$ is

$$md^2\mathbf{r}/dt^2 = \mathbf{F} \tag{i}$$

where $\mathbf{r}$ is the position-vector of m at time t.

If $\mathbf{F}$ is constant or a function of t only, this equation can be integrated twice to give in turn $d\mathbf{r}/dt$ and $\mathbf{r}$, from which the characteristics of the motion of the particle can be derived.

For a system of particles $m_1, m_2, \ldots m_n$ subject to forces $\mathbf{F}_1, \mathbf{F}_2, \ldots \mathbf{F}_n$ respectively, as well as to certain internal forces equal and opposite in pairs, we can write down the equation of motion of each particle; on adding all the equations we get

$$\sum m_i \, d\mathbf{r}_i{}^2/dt^2 = \sum \mathbf{F}_i \tag{ii}$$

where $\mathbf{r}_i$ is the position-vector of the mass m_i at time t.

Introducing the position-vector $\mathbf{r}_G$ of the centre of mass G of the system of particles given by

$$M\mathbf{r}_G = \sum m_i \mathbf{r}_i$$

where $M = \sum m_i$, equation (iii) becomes

$$M \, d^2\mathbf{r}_G/dt^2 = \sum \mathbf{F}_i$$

or

$$M \, d\mathbf{v}_G/dt = \sum \mathbf{F}_i \tag{iii}$$

where $\mathbf{v}_G$ is the velocity of the centre of mass of the system at time t (cf. Dynamics, 10.29).

10.22. Many other applications of the differentiation of vectors arise in dynamics. Consider, for example, the motion of a particle in a plane and suppose its position at time t is specified by the polar coordinates (r, θ). The components of its velocity and

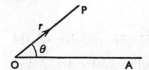

FIG. 10.18

acceleration along and perpendicular to the radius-vector can be found by vector methods (cf. Dynamics, 9.4).

The position-vector of the particle at time t can be written

$$\mathbf{r} = r\hat{\mathbf{r}}$$

where $\hat{\mathbf{r}}$ is unit vector along OP (Fig. 10.18) in the direction of r increasing.

$$\therefore \qquad \frac{d\mathbf{r}}{dt} = \frac{dr}{dt}\hat{\mathbf{r}} + r\frac{d\hat{\mathbf{r}}}{dt}$$

We will evaluate $d\hat{\mathbf{r}}/dt$.

Draw AB (Fig. 10.19) to represent $\hat{\mathbf{r}}(t)$; AB is of unit length and parallel to OP. Similarly, draw AC to represent $\hat{\mathbf{r}}(t+\delta t)$; AC is also of unit length but inclined at an angle $\delta\theta$ to AB. Angle CAB $= \delta\theta$.

FIG. 10.19

Then, BC represents $\delta\hat{\mathbf{r}}$ and is of magnitude $2\sin\frac{1}{2}\delta\theta$ or $\delta\theta$ approximately, and in a direction such that the angle ABC equals $\frac{1}{2}(\pi-\delta\theta)$.

Hence, $d\hat{\mathbf{r}}/dt$ is of magnitude $d\theta/dt$ and is perpendicular to AB, that is, to $\hat{\mathbf{r}}$. We can therefore write

$$\frac{d\hat{\mathbf{r}}}{dt} = \frac{d\theta}{dt}\,\hat{\boldsymbol{\theta}}$$

where $\hat{\boldsymbol{\theta}}$ is unit vector perpendicular to OP in the direction of θ increasing.
Hence,

$$\frac{d\mathbf{r}}{dt} = \frac{dr}{dt}\hat{\mathbf{r}}+r\frac{d\theta}{dt}\hat{\boldsymbol{\theta}}.$$

This shows that the components of the velocity of the particle along and perpendicular to the radius-vector are $\dfrac{dr}{dt}$ and $r\dfrac{d\theta}{dt}$, or $\dot{r}$ and $r\dot{\theta}$ (cf. Dynamics, 9.4).

Differentiating again we get

$$\frac{d^2\mathbf{r}}{dt^2} = \frac{d^2r}{dt^2}\hat{\mathbf{r}}+\frac{dr}{dt}\frac{d\hat{\mathbf{r}}}{dt}+\frac{d}{dt}\left(r\frac{d\theta}{dt}\right)\hat{\boldsymbol{\theta}}+\frac{d\theta}{dt}\frac{d\hat{\boldsymbol{\theta}}}{dt}$$

We must evaluate $d\hat{\boldsymbol{\theta}}/dt$. In the same way as we found $d\hat{\mathbf{r}}/dt$ above we can show that

$$\frac{d\hat{\boldsymbol{\theta}}}{dt} = -\frac{d\theta}{dt}\hat{\mathbf{r}}$$

Hence, we get

$$\frac{d^2\mathbf{r}}{dt^2} = \frac{d^2r}{dt^2}\hat{\mathbf{r}} + \frac{dr}{dt}\times\frac{d\theta}{dt}\hat{\mathbf{\theta}} + \frac{d}{dt}\left(r\frac{d\theta}{dt}\right)\hat{\mathbf{\theta}} - r\left(\frac{d\theta}{dt}\right)^2\hat{\mathbf{r}}$$

$$= \left\{\frac{dr^2}{dt^2} - r\left(\frac{d\theta}{dt}\right)^2\right\}\hat{\mathbf{r}} + \left\{2\frac{dr}{dt}\frac{d\theta}{dt} + r\frac{d^2\theta}{dt^2}\right\}\hat{\mathbf{\theta}}$$

so that the components of the acceleration of the particle along and perpendicular to the radius-vector are

$$\frac{d^2r}{dt^2} - r\left(\frac{d\theta}{dt}\right)^2 \text{ and } 2\frac{dr}{dt}\times\frac{d\theta}{dt} + r\frac{d^2\theta}{dt^2}$$

respectively (cf. Dynamics, 9.4), that is, $\ddot{r} - r\dot{\theta}^2$ and $2\dot{r}\dot{\theta} + r\ddot{\theta}$.

10.23. Some examples in dynamics

EXAMPLE (i)

Three points A, B, and C have position vectors $\mathbf{i}+\mathbf{j}+\mathbf{k}$, $\mathbf{i}+2\mathbf{k}$ *and* $3\mathbf{i}+2\mathbf{j}+3\mathbf{k}$ *respectively, relative to a fixed origin O. A particle P starts from B at time* $t = 0$, *and moves along BC towards C with constant speed 1 unit per sec. Find the position-vector of P after t sec (a) relative to O and (b) relative to A.*

If the angle PAB $= \theta$, *find an expression for* $\cos\theta$ *in terms of t.*

(L.A.)

We have
$$\mathbf{BC} = (3\mathbf{i}+2\mathbf{j}+3\mathbf{k}) - (\mathbf{i}+2\mathbf{k})$$
$$= 2\mathbf{i}+2\mathbf{j}+\mathbf{k}.$$

The velocity of P is therefore in the direction of the vector $2\mathbf{i}+2\mathbf{j}+\mathbf{k}$, and as it has unit magnitude it can be written

$$\tfrac{1}{3}(2\mathbf{i}+2\mathbf{j}+\mathbf{k}).$$

Denoting the position-vector of P relative to O at time t by $\mathbf{r}$, we have

$$\frac{d\mathbf{r}}{dt} = \tfrac{1}{3}(2\mathbf{i}+2\mathbf{j}+\mathbf{k})$$

∴
$$\mathbf{r} = \frac{t}{3}(2\mathbf{i}+2\mathbf{j}+\mathbf{k}) + \mathbf{A}.$$

But
$$\mathbf{r} = \mathbf{i}+2\mathbf{k} \text{ when } t = 0, \text{ and hence}$$

$$\mathbf{r} = \mathbf{i}+2\mathbf{k} + \frac{t}{3}(2\mathbf{i}+2\mathbf{j}+\mathbf{k})$$

that is,
$$\mathbf{OP} = \left(\mathbf{i}+\frac{2t}{3}\right)\mathbf{i} + \frac{2t}{3}\mathbf{j} + \left(2+\frac{t}{3}\right)\mathbf{k}.$$

The position-vector of P relative to A is therefore

$$PA = \frac{2t}{3}\mathbf{i} + \left(\frac{2t}{3}-1\right)\mathbf{j} + \left(1+\frac{t}{3}\right)\mathbf{k}.$$

Now,
$$\cos\theta = \frac{AB^2 + AP^2 - BP^2}{2AB \times AP}$$

where
$$AB = (\mathbf{i}+2\mathbf{k}) - (\mathbf{i}+\mathbf{j}+\mathbf{k}) = -\mathbf{j}+\mathbf{k}$$

and
$$PB = \frac{2t}{3}\mathbf{i} + \frac{2t}{3}\mathbf{j} + \frac{t}{3}\mathbf{k}.$$

Hence,
$$AB^2 = 1+1 = 2$$

and
$$AP^2 = \left(\frac{2t}{3}\right)^2 + \left(\frac{2t}{3}-1\right)^2 + \left(1+\frac{t}{3}\right)^2 = t^2 - \frac{2t}{3} + 2$$

and
$$BP^2 = \frac{t^2}{9}(4+4+1) = t^2.$$

Hence,
$$\cos\theta = \frac{2+t^2-\dfrac{2t}{3}+2-t^2}{2\sqrt{2}\left(t^2-\dfrac{2t}{3}+2\right)}$$

$$= \frac{6-t}{\sqrt{6(3t^2-2t^2+6)}}.$$

EXAMPLE (ii)

Distances being measured in km and speeds in km h^{-1}, a motor boat sets out at 11 a.m. from a position $-6\mathbf{i}-2\mathbf{j}$ relative to a marker buoy, and travels at a steady speed of magnitude $\sqrt{53}$ on a direct course to intercept a ship. The ship maintains a steady velocity vector $3\mathbf{i}+4\mathbf{j}$ and at 12 noon is at a position $3\mathbf{i}-\mathbf{j}$ from the buoy. Find the velocity vector of the motor boat, the time of interception, and the position vector of the point of interception from the buoy. (L.A.)

Denoting the velocity of the motor boat by $u\mathbf{i}+v\mathbf{j}$, then

$$u^2+v^2 = 53.$$

Also the position-vector of the motor boat after time t is

$$-6\mathbf{i}-2\mathbf{j}+(u\mathbf{i}+v\mathbf{j})t. \qquad\qquad \text{(i)}$$

Similarly, the position-vector of the ship after time t is

$$3\mathbf{i}-\mathbf{j}+(3\mathbf{i}+4\mathbf{j})(t-1). \qquad\qquad \text{(ii)}$$

The time of interception is given by the value of t that satisfies the equation

$$-6\mathbf{i}-2\mathbf{j}+(u\mathbf{i}+v\mathbf{j})t = 3\mathbf{i}-\mathbf{j}+(3\mathbf{i}+4\mathbf{j})(t-1)$$

Hence,
$$-6+ut = 3t \qquad\qquad\text{(iii)}$$
$$-2+vt = 4t-5$$

$\therefore\qquad (3t+6)^2+(4t-3)^2 = 53t^2$

$\therefore\qquad 25t^2+12t+45 = 53t^2$

$\therefore\qquad 28t^2-12t-45 = 0$

$\therefore\qquad (14t+15)(2t-3) = 0$

$\therefore\qquad\qquad\qquad t = \tfrac{3}{2}.$

Hence the time of interception is 12.30 p.m.

Also, from (iii) it follows that $u = 7$ and $v = 2$, so that the velocity of the motor boat is $7\mathbf{i}+2\mathbf{j}$.

The position-vector at the time of interception is from (i)

$$-6\mathbf{i}-2\mathbf{j}+(7\mathbf{i}+2\mathbf{j})\tfrac{3}{2} = \tfrac{9}{2}\mathbf{i}+\mathbf{j}$$

or from (ii)

$$3\mathbf{i}-\mathbf{j}+(3\mathbf{i}+4\mathbf{j})\tfrac{1}{2} = \tfrac{9}{2}\mathbf{i}+\mathbf{j}.$$

EXAMPLE (iii)

Two particles are projected simultaneously in the same vertical plane, $\mathbf{i}$ and $\mathbf{j}$ being unit horizontal and vertical vectors in that plane. The first particle is projected from the origin with velocity vector $nV\cos\alpha\,\mathbf{i}+nV\sin\alpha\,\mathbf{j}$, and the second particle is projected from a position $h\mathbf{i}+k\mathbf{j}$ (where $h > 0$, $k > 0$) with velocity-vector

$$-V\cos\beta\,\mathbf{i}+V\sin\beta\,\mathbf{j}.$$

Write down the position-vectors of each of the particles after time t has elapsed.

Show that the particles cannot collide unless $\sin\beta < n\sin\alpha$. and if they do collide, prove that $\sin(\beta+\gamma) = n\sin(\alpha-\gamma)$, where $\tan\gamma = k/h$.

Find the condition imposed on V if the point of collision is above the level of the origin. (L.A.)

The position-vector $\mathbf{r}$ of the first particle after time t is given by

$$\frac{d^2\mathbf{r}}{dt^2} = -g\mathbf{j}$$

$\therefore\qquad \dfrac{d\mathbf{r}}{dt} = -gt\mathbf{j}+(nV\cos\alpha\,\mathbf{i}+nV\sin\alpha\,\mathbf{j})$

and
$\qquad \mathbf{r} = -\tfrac{1}{2}gt^2\mathbf{j}+(nV\cos\alpha\,\mathbf{i}+nV\sin\alpha\,\mathbf{j})t.$

Similarly, the position-vector of the second particle after time t is

$$\mathbf{r} = -\tfrac{1}{2}gt^2\mathbf{j}+(-V\cos\beta\mathbf{i}+V\sin\beta\mathbf{j})t+h\mathbf{i}+k\mathbf{j}.$$

The particles will collide at time T if

$$-\tfrac{1}{2}gT^2\mathbf{j}+(nV\cos\alpha\mathbf{i}+nV\sin\alpha\mathbf{j})T$$
$$= -\tfrac{1}{2}gT^2\mathbf{j}+(-V\cos\beta\mathbf{i}+V\sin\beta\mathbf{j})T+h\mathbf{i}+k\mathbf{j}$$

$\therefore$ $\qquad\qquad (nV\cos\alpha)T = (-V\cos\beta)T+h$

and $\qquad\qquad (nV\sin\alpha)T = (V\sin\beta)T+k$

$\therefore$ $\qquad\qquad \dfrac{n\cos\alpha+\cos\beta}{n\sin\alpha-\sin\beta} = \dfrac{h}{k} = \dfrac{\cos\gamma}{\sin\gamma}$

$\therefore$ $\qquad\qquad n\sin(\alpha-\gamma) = \sin(\beta+\gamma)$

with the proviso that since $k>0$, $n\sin\alpha$ must be greater than $\sin\beta$. If the point of collision is above the level of the origin

$$(nV\sin\alpha)T>\tfrac{1}{2}gT^2$$

that is, $\qquad\qquad nV\sin\alpha>\tfrac{1}{2}gT$

$$nV\sin\alpha>\tfrac{1}{2}gh/V(n\cos\alpha+\cos\beta)$$

$\therefore$ $\qquad\qquad V^2>gh/2n\sin\alpha(n\cos\alpha+\cos\beta)$

or $\qquad\qquad V^2>gk/2n\sin\alpha(n\sin\alpha-\sin\beta).$

EXAMPLE (iv)

The equation of the path of a particle P *is*

$$\mathbf{r} = \mathbf{i}t+\mathbf{k}t^2,$$

where t *is the time. Show that the acceleration of* P *is constant.*

The velocity of another particle Q *relative to* P *is* $(\mathbf{i}-\mathbf{j})$ *and when* $t = 0$, **PQ** $= \mathbf{j}$. *Find the equation of the path of* Q *and the time at which* Q *is nearest to* P. (L.A.)

Since $\qquad\qquad \mathbf{r} = t\mathbf{i}+t^2\mathbf{k}$

$$\frac{d\mathbf{r}}{dt} = \mathbf{i}+2t\mathbf{k}$$

and $\qquad\qquad \dfrac{d^2\mathbf{r}}{dt^2} = 2\mathbf{k}$

$\therefore$ Acceleration of the particle P is constant.

Also the velocity of the particle Q

$$= \mathbf{i}-\mathbf{j}+\mathbf{i}+2t\mathbf{k}$$
$$= 2\mathbf{i}-\mathbf{j}+2t\mathbf{k}$$

$\therefore$ The position-vector of Q is

$$2t\mathbf{i}-t\mathbf{j}+t^2\mathbf{k}+\mathrm{A}$$

where A is a constant-vector.

But at $t = 0$, PQ $= \mathbf{j}$ and hence A $= \mathbf{j}$. Therefore the equation of the path of Q is

$$\mathbf{r} = 2t\mathbf{i}+(1-t)\mathbf{j}+t^2\mathbf{k}.$$

Now, $$\mathbf{PQ} = \mathbf{OQ}-\mathbf{OP}$$
$$= 2t\mathbf{i}+(1-t)\mathbf{j}+t^2\mathbf{k}-t\mathbf{i}-t^2\mathbf{k}$$
$$= t\mathbf{i}+(1-t)\mathbf{j}$$

$\therefore$ PQ is least when $t^2+(1-t)^2$ is least, that is, when $2(t-\frac{1}{2})^2+\frac{1}{2}$ is least, and this occurs when $t = \frac{1}{2}$.

EXAMPLES 10.2

1. If the position-vector of a particle at time t is $\mathbf{r} = (1+t^2)\mathbf{i}+3t\mathbf{j}$, find $d\mathbf{r}/dt$ and $d^2\mathbf{r}/dt^2$. Find also the magnitude and direction of the velocity at time $t = 2$.

2. The position-vector of two particles at time t are

$$(1+t)\mathbf{i}+(1-t^2)\mathbf{j}+(1+t^2)\mathbf{k} \text{ and } (3\mathbf{i}+2\mathbf{j}+\mathbf{k})t.$$

Find when they are travelling at right-angles to one another and their velocities at that instant.

3. If $d^2\mathbf{r}/dt^2 = 2t\mathbf{i}-t^2\mathbf{k}$, find the velocity and position-vector at time t, given that the particle is at rest at time $t = 0$ at the point $\mathbf{r} = 4\mathbf{i}-3\mathbf{j}$.

4. The position-vector of a particle falling freely under gravity is given by $d^2\mathbf{r}/dt^2 = -9\cdot8\mathbf{j}$. Show that if the particle is projected with velocity $100\mathbf{i}$ from the point $\mathbf{r} = 200\mathbf{j}$ at time $t = 0$, it will pass through the point $\mathbf{r} = 400\mathbf{i}+121\cdot6\mathbf{j}$ at time $t = 4$.

5. If $\dfrac{d^2\mathbf{r}}{dt^2} = -n^2\mathbf{r}$, find $\mathbf{r}$ given that $\mathbf{r} = \mathbf{a}$ and $d\mathbf{r}/dt = \mathbf{v}$ when $t = 0$.

6. If $d^2\mathbf{r}/dt^2 = \mathbf{c}$, find $\mathbf{r}$ given that $\mathbf{r} = \mathbf{a}$ and $d\mathbf{r}/dt = \mathbf{b}$ when $t = 0$. If $\mathbf{r}$ is the position-vector of the point P, show that the path of P is a parabola.

7. The position-vector of the point P at time t is

$$a \tan t\mathbf{i}+a \sec t\mathbf{k},$$

where a is a positive constant and $0 \leqslant t < \pi/2$. Show that the velocity and the acceleration of P when $t = 0$ are at right angles to each other.

If A is the point with position-vector $a\mathbf{j}$, obtain the vector equation of the straight line AP at time t. The point Q divides AP internally in the ratio $\cos t:(1-\cos t)$. Show that the acceleration of the point Q is constant in magnitude and is always directed towards a fixed point. (L.A.)

8. The velocity-vectors of the particles P_1, P_2 are

$$u_1\mathbf{i}+v_1\mathbf{j},\ u_2\mathbf{i}+v_2\mathbf{j}$$

respectively. Their relative velocity has the same magnitude as that of the velocity of P_1. If the velocity of one particle is reversed, the magnitude of the relative velocity is doubled. Find the ratio of the speeds of P_1 and P_2 and the sine of the angle between their directions. (L.A.)

9. A particle of mass 3 units is acted on by the forces

$$\mathbf{F}_1 = 2\mathbf{i}+3\mathbf{j},\ \mathbf{F}_2 = 3\mathbf{j}+4\mathbf{k},\ \mathbf{F}_3 = \mathbf{i}+2\mathbf{k},$$

and initially it is at rest at the point $\mathbf{i}-\mathbf{j}-\mathbf{k}$. Find the position and the momentum of the particle after 2 seconds.

10. At time $t = 0$ two particles A and B leave the point $\mathbf{i}+2\mathbf{j}$, the velocity-vector of A being $\mathbf{i}+4\mathbf{j}+\mathbf{k}$ and the velocity of B relative to A being of magnitude $\sqrt{90}$. A third particle C leaves the point $4\mathbf{i}+3\mathbf{k}$ at time $t = 0$ with velocity vector $2\mathbf{i}-\mathbf{j}+3\mathbf{k}$. If the particles travel with constant velocity and if B collides with C, find the initial velocity vector of B. Find also the value of t at the instant when the collision takes place.

If the particles B and C coalesce on collision and then move in a direction perpendicular to the velocity of A, find the ratio of the masses of the particles B and C. (L.A.)

11. Two smooth spheres A and B have masses $2m$ and m respectively, and velocity vectors $3u\mathbf{i}+4u\mathbf{j}$ and $-4u\mathbf{i}+3u\mathbf{j}$ respectively, when they collide with their line of centres parallel to the unit vector $\mathbf{i}$. If the impact causes a loss of energy equal to the original kinetic energy of the sphere B, prove that the coefficient of restitution between the spheres is $\sqrt{(23/98)}$. (L.A.)

12. Define the centre of mass of n particles of masses $m_1, m_2, \ldots, m_n$, placed at points whose position vectors with respect to an origin O are $r_1, r_2, \ldots, r_n$ respectively, and show that it is independent of the choice of origin.

Find the position-vector of the centre of mass of particles of masses 4, 3, 2, 3 units at rest at the points

$$\mathbf{i}+\mathbf{j},\ 2\mathbf{i}-\mathbf{j},\ 2\mathbf{i}+\mathbf{j},\ 2\mathbf{i}+3\mathbf{j}$$

respectively. If each mass is acted upon by a force directed towards the origin and proportional to its distance from the origin, find the direction of the initial acceleration of the centre of mass. (L.A.)

13. Express the velocity **v** of a particle moving in a plane as $v\mathbf{t}$, where **t** is unit vector along the tangent to the path of the particle and find the acceleration. Show that the acceleration has components dv/dt and v^2/ρ along the tangent and normal respectively.

14. A particle moves subject to a central force inversely proportional to the square of the distance r from a fixed point. Show that the velocity v of the particle is given by

$$v^2 = k\left(\frac{2}{r} \pm \frac{1}{a}\right)$$

where k and a are constants.

10.24. Multiplication of vectors

We have earlier defined the product of a scalar quantity and a vector, for example, $m\mathbf{a}$. We shall now consider the product of two vectors. There are two such products; one is called the scalar product, the other the vector product.

10.25. Scalar product of two vectors

The scalar product of two vectors **a** and **b** is defined as the scalar quantity equal to the product of the magnitudes of the two vectors multiplied by the cosine of the angle between the vectors. We write it **a . b**, and it is sometimes referred to as the dot product.

Thus, $\mathbf{a} \cdot \mathbf{b} = ab \cos \theta$, where θ is the angle between **a** and **b**.

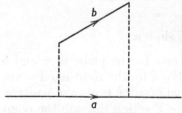

FIG. 10.20

θ lies between 0 and π. The scalar product is negative when $\frac{1}{2}\pi < \theta < \pi$.

This product may be written $a(b \cos \theta)$ and interpreted geometrically (Fig. 10.20) as the magnitude of **a** multiplied by the projection of the vector **b** on the vector **a**. In particular, **â** . **b** equals the projection of the vector **b** on the vector **a**, that is, the component of **b** in the direction of **a**.

How such a product arises in mechanics is easily exemplified. Let **F** denote the force acting on a particle at the point O. If the particle be moved to O' (Fig. 10.21) and the displacement OO'

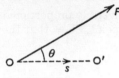

FIG. 10.21

be denoted by the vector **s**, then the work done by the force in this displacement equals

$$OO' \times \text{component of } \mathbf{F} \text{ in the direction } OO'$$
$$= OO' \times F \cos \theta$$
$$= Fs \cos \theta = \mathbf{F} \cdot \mathbf{s}.$$

10.26. Vector product of two vectors

The vector product of two vectors **a** and **b** is defined as the vector of magnitude $ab \sin \theta$ (where θ is the angle between **a** and **b**) and in a direction perpendicular to the plane of **a** and **b**. The direction of this vector is related to the rotation of **a** into **b** in the same way as the motion of translation and the motion of rotation are related in a right-handed screw.

We write it $\mathbf{a} \wedge \mathbf{b}$. It is sometimes written $\mathbf{a} \times \mathbf{b}$ and known as the cross product.

Thus, $\mathbf{a} \wedge \mathbf{b} = ab \sin \theta \, \mathbf{n}$

where **n** is unit vector perpendicular to the plane of **a** and **b** and related to the rotation of **a** into **b** by the right-hand screw rule. θ lies between 0 and π, so that $ab \sin \theta$ is always positive.

As an example, consider a force **F** acting through the point O, as in Fig. 10.22. Suppose P is any point such that its position-vector relative to O is **r**, that is, **OP** = **r**.

The moment of the force F about P $= Fr \sin \theta$, where θ is the angle between **F** and **r**.

This moment is clockwise (as shown in Fig. 10.22) and can

be represented by a vector perpendicular to the plane of the paper downwards, in fact, by the vector-product $\mathbf{F} \wedge \mathbf{r}$. Whatever the position of P, the moment of $\mathbf{F}$ about P can be written as $\mathbf{F} \wedge \mathbf{r}$, where $\mathbf{r}$ is the position-vector of P relative to some point O on the line of action of $\mathbf{F}$. Moreover, the product $\mathbf{F} \wedge \mathbf{r}$ is clearly the same whatever point O on the line of action of $\mathbf{F}$ is chosen.

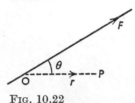

FIG. 10.22

Similarly, if a force $\mathbf{F}$ acts at a point P, of which the position-vector relative to O is $\mathbf{r}$, the moment of $\mathbf{F}$ about O is $\mathbf{r} \wedge \mathbf{F}$, and is the same whatever point P we take on the line of action of $\mathbf{F}$.

10.27. Properties of the scalar product

If $\mathbf{a}$ and $\mathbf{b}$ are perpendicular, $\theta = \frac{1}{2}\pi$ and hence $\mathbf{a} . \mathbf{b} = 0$. Conversely, if $\mathbf{a} . \mathbf{b} = 0$, and neither a nor b is zero, the vectors must be perpendicular.

If $\mathbf{a}$ and $\mathbf{b}$ are parallel and in the same direction, $\theta = 0$ and $\mathbf{a} . \mathbf{b} = ab$. If they are parallel and in opposite directions, $\theta = \pi$ and $\mathbf{a} . \mathbf{b} = -ab$.

It follows that $\mathbf{a} . \mathbf{a} = a^2$ and $\mathbf{\hat{a}} . \mathbf{\hat{a}} = 1$.

If $\mathbf{i}, \mathbf{j}, \mathbf{k}$ are unit vectors parallel to mutually perpendicular axes Ox, Oy, Oz, then.

$$\mathbf{i} . \mathbf{i} = 1 = \mathbf{j} . \mathbf{j} = \mathbf{k} . \mathbf{k}$$
and
$$\mathbf{i} . \mathbf{j} = 0 = \mathbf{j} . \mathbf{k} = \mathbf{k} . \mathbf{i}$$

We can show that a scalar product is independent of the order of the vectors, for by definition

$$\mathbf{b} . \mathbf{a} = ba \cos \theta$$

and hence $\mathbf{b} . \mathbf{a} = \mathbf{a} . \mathbf{b}$, that is, the scalar product is commutative.

We can show that also

$$\mathbf{a} . (\mathbf{b}+\mathbf{c}) = \mathbf{a} . \mathbf{b}+\mathbf{a} . \mathbf{c}$$

Let PQ and QR be drawn (Fig. 10.23) to represent the vectors **b** and **c** respectively.

Then PR represents **b**+**c**.

Suppose AB represents the vector **a**, assumed coplanar with **b** and **c**.

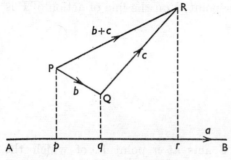

FIG. 10.23

Draw Pp, Qq, Rr perpendicular to the vector **a**.

Then, **a** . (**b**+**c**) = **a**×projection of the vector (**b**+**c**) on **a**

$$= a \times \mathrm{pr}$$
$$= a \times (\mathrm{pq}+\mathrm{qr})$$
$$= a \times \mathrm{pq}+a \times \mathrm{qr}$$
$$= a \times \text{projection of } \mathbf{b} \text{ on } \mathbf{a}+a \times \text{projection of } \mathbf{c} \text{ on } \mathbf{a}$$
$$= \mathbf{a} . \mathbf{b}+\mathbf{a} . \mathbf{c}.$$

This can obviously be extended, so that

$$\mathbf{a} . (\mathbf{b}+\mathbf{c}+\mathbf{d}+\mathbf{e}+ \ldots) = \mathbf{a} . \mathbf{b}+\mathbf{a} . \mathbf{c}+\mathbf{a} . \mathbf{d}+\mathbf{a} . \mathbf{e}+ \ldots$$

Also

$$(\mathbf{a}+\mathbf{b}) . (\mathbf{c}+\mathbf{d}) = (\mathbf{a}+\mathbf{b}) . \mathbf{c}+(\mathbf{a}+\mathbf{b}) . \mathbf{d}$$
$$= \mathbf{a} . \mathbf{c}+\mathbf{b} . \mathbf{c}+\mathbf{a} . \mathbf{d}+\mathbf{b} . \mathbf{d}.$$

10.28. The scalar multiple of a scalar product sometimes arises. For instance, if we form the scalar product of $m\mathbf{a}$ and $n\mathbf{b}$, where m and n are scalar quantities, we get by definition

$$m\mathbf{a} . n\mathbf{b} = (ma)(nb) \cos \theta$$

where θ is the angle between the vectors $m\mathbf{a}$ and $n\mathbf{b}$ and therefore the angle between the vectors **a** and **b**.

Hence,

$$m\mathbf{a} \cdot n\mathbf{b} = mn(ab \cos \theta)$$
$$= mn(\mathbf{a} \cdot \mathbf{b})$$
$$= mn\mathbf{a} \cdot \mathbf{b}$$
$$= \mathbf{a} \cdot mn\mathbf{b}.$$

10.29. EXAMPLE (i)

Use the definition of a scalar product to derive the cosine rule for any triangle.

Consider any triangle ABC. Let the sides BC, CA, AB represent the vectors $\mathbf{a}$, $\mathbf{b}$, $\mathbf{c}$ respectively. Then

$$\mathbf{a} + \mathbf{b} + \mathbf{c} = 0.$$

Hence,

$$\mathbf{a} \cdot \mathbf{a} = (\mathbf{b} + \mathbf{c}) \cdot (\mathbf{b} + \mathbf{c})$$
$$= \mathbf{b} \cdot \mathbf{b} + \mathbf{c} \cdot \mathbf{c} + 2\mathbf{b} \cdot \mathbf{c}$$
$$\therefore \qquad a^2 = b^2 + c^2 + 2bc \cos (180° - A)$$
$$= b^2 + c^2 - 2bc \cos A$$

which establishes the cosine rule.

EXAMPLE (ii)

Express the scalar product of any two vectors $\mathbf{a}$ and $\mathbf{b}$ in terms of the components of the vectors parallel to the coordinate axes.

If $\mathbf{i}$, $\mathbf{j}$, $\mathbf{k}$ are unit vectors parallel to the coordinate axes Ox, Oy, Oz we can write

$$\mathbf{a} = a_1\mathbf{i} + a_2\mathbf{j} + a_3\mathbf{k}$$

and

$$\mathbf{b} = b_1\mathbf{i} + b_2\mathbf{j} + b_3\mathbf{k}.$$

Hence,

$$\mathbf{a} \cdot \mathbf{b} = (a_1\mathbf{i} + a_2\mathbf{j} + a_3\mathbf{k}) \cdot (b_1\mathbf{i} + b_2\mathbf{j} + b_3\mathbf{k})$$
$$= a_1b_1\mathbf{i} \cdot \mathbf{i} + a_2b_1\mathbf{j} \cdot \mathbf{i} + a_3b_1\mathbf{k} \cdot \mathbf{i}$$
$$\qquad + a_1b_2\mathbf{i} \cdot \mathbf{j} + a_2b_2\mathbf{j} \cdot \mathbf{j} + a_3b_2\mathbf{k} \cdot \mathbf{j}$$
$$\qquad + a_1b_3\mathbf{i} \cdot \mathbf{k} + a_2b_3\mathbf{j} \cdot \mathbf{k} + a_3b_3\mathbf{k} \cdot \mathbf{k}$$
$$= a_1b_1 + a_2b_2 + a_3b_3$$

It follows that

$$ab \cos \theta = a_1b_1 + a_2b_2 + a_3b_3$$
$$\therefore \qquad \cos \theta = (a_1b_1 + a_2b_2 + a_3b_3)/ab$$

where

$$a^2 = a_1{}^2 + a_2{}^2 + a_3{}^2 \text{ and } b^2 = b_1{}^2 + b_2{}^2 + b_3{}^3.$$

We note that if the vectors $\mathbf{a}$ and $\mathbf{b}$ are perpendicular $a_1b_1 + a_2b_2 + a_3b_3 = 0$, and conversely. This is a very useful result.

EXAMPLE (iii)

If the position-vectors of the points A, B, *and* C *with respect to* O *are*
$a = 2i-j+3k$, $b = 3i+2j-4k$, $c = -i+3j-2k$, *find unit vector* n
perpendicular to the plane ABC.

Any vector perpendicular to the plane ABC is perpendicular
to the vectors **AB**, **BC**, and **CA**.

Now, $AB = OB - OA = b - a$

                                  $= i+3j-7k$

Also, $BC = OC - OB = c - b$

                                  $= -4i+j+2k$

If $d = d_1i+d_2j+d_3k$ is a vector perpendicular to **AB** and **BC**, we
get from Example (ii) above,

$$d_1+3d_2-7d_3 = 0$$
and
$$-4d_1+d_2+2d_3 = 0.$$

Hence, $2d_1 = d_2 = 2d_3$, so that any vector perpendicular to both
AB and **BC** may be written as

$$d = m(i+2j+k)$$

where m is a scalar quantity.

We can verify that this vector is also perpendicular to

$$CA = a - c = 3i-4j+5k.$$

Hence, unit vector perpendicular to the plane ABC is

$$n = \frac{1}{\sqrt{6}}(i+2j+k).$$

Note. If p is the length of the perpendicular from the origin O to the
plane ABC we have

$$p = a \cdot n$$

$$= (2i-j+3k) \cdot \frac{1}{\sqrt{6}}(i+2j+k)$$

$$= \frac{1}{\sqrt{6}}(2-2+3) = \tfrac{1}{2}\sqrt{6}.$$

Of course, in the same way

$$p = b \cdot n = c \cdot n$$

which again leads to $p = \tfrac{1}{2}\sqrt{6}.$

EXAMPLE (iv)

Show that the vector $\mathbf{r} = \mathrm{a}\, cos\, \phi \mathbf{i} + \mathrm{b}\, sin\, \phi \mathbf{j} + \mathrm{c}\, cos\, \phi \mathbf{k}$ *is perpendicular to the vector* c**i**—a**k**. *Deduce that if* ϕ *is a parameter all the vectors* $\mathbf{r}$ *lie in a plane.*

The scalar product of the two vectors is

$$ac\, \cos \phi - ac\, \cos \phi = 0$$

and hence the vectors are perpendicular whatever the values of ϕ.

It follows that all the vectors $\mathbf{r}$ are perpendicular to a fixed vector c**i**—a**k**, and hence they must all lie in a plane. It is the plane through the pole 0 to which the vector c**i**—a**k** is normal.

10.30. Properties of the vector product

If **a** and **b** are parallel the angle θ between them is zero or π (if they are in opposite directions), and hence $\mathbf{a} \wedge \mathbf{b} = 0$. In particular, $\mathbf{a} \wedge \mathbf{a} = 0$ and $\hat{\mathbf{a}} \wedge \hat{\mathbf{a}} = 0$.

Conversely, if $\mathbf{a} \wedge \mathbf{b} = 0$, and neither a nor b is zero, the vectors **a** and **b** must be parallel.

Similarly, if **a** and **b** are perpendicular, $\mathbf{a} \wedge \mathbf{b} = ab\mathbf{n}$ and **a**, **b**, **n** form a right-handed triad.

If **i**, **j**, **k** are unit vectors parallel to mutually perpendicular axes Ox, Oy, Oz, which form a right-handed system, then

$$\mathbf{j} \wedge \mathbf{k} = \mathbf{i}, \quad \mathbf{k} \wedge \mathbf{i} = \mathbf{j} \quad \text{and} \quad \mathbf{i} \wedge \mathbf{j} = \mathbf{k}.$$

Of course,

$$\mathbf{i} \wedge \mathbf{i} = 0 = \mathbf{j} \wedge \mathbf{j} = \mathbf{k} \wedge \mathbf{k}.$$

We can show that a vector-product is dependent on the order of the vectors, for by definition

$$\mathbf{a} \wedge \mathbf{b} = ab\, \sin \theta\, \mathbf{n}$$

and

$$\mathbf{b} \wedge \mathbf{a} = ba\, \sin \theta\, \mathbf{n}'$$

where, using the right-hand rule, the unit vectors **n** and **n**' are oppositely directed, that is, $\mathbf{n}' = -\mathbf{n}$ (Fig. 10.24).

$$\therefore \qquad \mathbf{b} \wedge \mathbf{a} = -\mathbf{a} \wedge \mathbf{b}.$$

The vector product is *not* commutative, and hence the order of the factors must always be treated with care.

It should be noted, for instance, that $\mathbf{j} \wedge \mathbf{k} = \mathbf{i}$, but

$$\mathbf{k} \wedge \mathbf{j} = -\mathbf{i}.$$

It is sometimes useful to represent the magnitude $ab \sin \theta$ of the vector product $\mathbf{a} \wedge \mathbf{b}$ by the area of the parallelogram (or twice the area of the triangle) of which $\mathbf{a}$ and $\mathbf{b}$ are adjacent sides. In other words, we represent the vector product $\mathbf{a} \wedge \mathbf{b}$ by

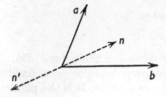

FIG. 10.24

the area vector of the parallelogram of which $\mathbf{a}$ and $\mathbf{b}$ are adjacent sides.

10.31. We can show that

$$(\mathbf{b}+\mathbf{c}) \wedge \mathbf{a} = \mathbf{b} \wedge \mathbf{a} + \mathbf{c} \wedge \mathbf{a}$$

This can be proved by giving the three vector-products a geometrical representation. We will consider first the case when $\mathbf{a}$, $\mathbf{b}$, and $\mathbf{c}$ lie in one plane; the products are accordingly parallel vectors perpendicular to that plane.

Let OB, BC (Fig. 10.25) represent the vectors $\mathbf{b}$ and $\mathbf{c}$ respectively. Then OC represents $\mathbf{b}+\mathbf{c}$.

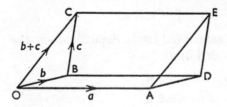

FIG. 10.25

Suppose OA represents the vector $\mathbf{a}$, assumed coplanar with $\mathbf{b}$ and $\mathbf{c}$.

Complete the parallelograms OCEA, OBDA, and BDCE.

Now $(\mathbf{b}+\mathbf{c}) \wedge \mathbf{a} = \mathbf{OC} \wedge \mathbf{OA}$ is a vector perpendicular to the plane of the paper downwards, and of magnitude represented by the parallelogram OCEA.

Also, $\mathbf{b} \wedge \mathbf{a} = \mathbf{OB} \wedge \mathbf{OA}$ is a vector perpendicular to the plane of the paper downwards, and of magnitude represented by the parallelogram OBDA.

Similarly, $\mathbf{c} \wedge \mathbf{a} = \mathbf{BC} \wedge \mathbf{BD}$ is a vector perpendicular to the plane of the paper downwards, and of magnitude represented by the parallelogram BCED.

But, the area of the parallelogram OCEA equals the sum of the areas of the parallelograms OBDA and BCED.

Hence $(\mathbf{b}+\mathbf{c}) \wedge \mathbf{a} = \mathbf{b} \wedge \mathbf{a}+\mathbf{c} \wedge \mathbf{a}$.

This result can be given a statical interpretation. If $\mathbf{b}$ and $\mathbf{c}$ are forces which meet at O and $\mathbf{a}$ is the position-vector of some point P relative to O, then $(\mathbf{b}+\mathbf{c}) \wedge \mathbf{a}$ is the moment of the resultant of the forces about P, and the above theorem is equivalent to Varignon's Theorem, namely, that the sum of the moments of two forces about any point P equals the moment of their resultant about P (cf. 2.13).

10.32. It follows that

$$\begin{aligned} \mathbf{a} \wedge (\mathbf{b}+\mathbf{c}) &= -(\mathbf{b}+\mathbf{c}) \wedge \mathbf{a} \\ &= -\mathbf{b} \wedge \mathbf{a}-\mathbf{c} \wedge \mathbf{a} \\ &= \quad \mathbf{a} \wedge \mathbf{b} +\mathbf{a} \wedge \mathbf{c}. \end{aligned}$$

This result is also true if $\mathbf{a}$ is not coplanar with $\mathbf{b}$ and $\mathbf{c}$.

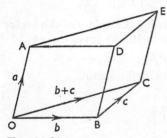

Fig. 10.26

In Fig. 10.26, draw OB to represent $\mathbf{b}$, and BC to represent $\mathbf{c}$. Then OC represents $\mathbf{b}+\mathbf{c}$.

Draw OA to represent $\mathbf{a}$, and draw BD and CE equal and parallel to OA so that $\mathbf{BD} = \mathbf{CE} = \mathbf{a}$. The sum of the area vectors of the faces of the prism OBCEDA is zero, and hence

$$\mathbf{b} \wedge \mathbf{a}+\mathbf{c} \wedge \mathbf{a}-(\mathbf{b}+\mathbf{c}) \wedge \mathbf{a}+\tfrac{1}{2}\mathbf{b} \wedge \mathbf{c}-\tfrac{1}{2}\mathbf{b} \wedge \mathbf{c} = 0.$$

$\therefore$ $\qquad (\mathbf{b}+\mathbf{c}) \wedge \mathbf{a} = \mathbf{b} \wedge \mathbf{a}+\mathbf{c} \wedge \mathbf{a}.$

Similarly, $\qquad \mathbf{a} \wedge (\mathbf{b}+\mathbf{c}) = \mathbf{a} \wedge \mathbf{b}+\mathbf{a} \wedge \mathbf{c}.$

10.33. The scalar multiple of a vector product sometimes arises. For instance, if we form the vector-product of $p\mathbf{a}$ and $q\mathbf{b}$, where p and q are scalar quantities, we get by definition

$$p\mathbf{a} \wedge q\mathbf{b} = (pa)(qb) \sin \theta \, \mathbf{n}$$

where θ is the angle between the vectors $p\mathbf{a}$ and $q\mathbf{b}$ and therefore the angle between the vectors $\mathbf{a}$ and $\mathbf{b}$.

Hence,

$$\begin{aligned}
p\mathbf{a} \wedge q\mathbf{b} &= pq(ab \sin \theta \, \mathbf{n}) \\
&= pq(\mathbf{a} \wedge \mathbf{b}) \\
&= pq\mathbf{a} \wedge \mathbf{b} \\
&= \mathbf{a} \wedge pq\mathbf{b}.
\end{aligned}$$

For example, $5\mathbf{i} \wedge 4\mathbf{j} = 20\mathbf{i} \wedge \mathbf{j} = 20\mathbf{k}$.

10.34. EXAMPLE (i)

Obtain an expression for the vector product of two vectors $\mathbf{a}$ *and* $\mathbf{b}$ *in terms of the components of the vectors parallel to the coordinate axes.*

Writing

$$\mathbf{a} = a_1\mathbf{i} + a_2\mathbf{j} + a_3\mathbf{k}$$

and

$$\mathbf{b} = b_1\mathbf{i} + b_2\mathbf{j} + a_3\mathbf{k}$$

we have

$$\mathbf{a} \wedge \mathbf{b} = (a_1\mathbf{i} + a_2\mathbf{j} + a_3\mathbf{k}) \wedge (b_1\mathbf{i} + b_2\mathbf{j} + b_3\mathbf{k})$$

Expanding the right-hand side and using the values of the vector products of the unit vectors $\mathbf{i}$, $\mathbf{j}$, $\mathbf{k}$ given in 10.30 above, we get

$$\mathbf{a} \wedge \mathbf{b} = (a_2 b_3 - a_3 b_2)\mathbf{i} + (a_3 b_1 - a_1 b_3)\mathbf{j} + (a_1 b_2 - a_2 b_1)\mathbf{k}$$

This expression can be written as the determinant

$$\begin{vmatrix} \mathbf{i} & \mathbf{j} & \mathbf{k} \\ a_1 & a_2 & a_3 \\ b_1 & b_2 & b_3 \end{vmatrix}$$

EXAMPLE (ii)

If the position-vectors of the points A, B, *and* C *with respect to the origin* O *are* $\mathbf{a} = 2\mathbf{i} - \mathbf{j} + 3\mathbf{k}$, $\mathbf{b} = 3\mathbf{i} + 2\mathbf{j} + 2\mathbf{j} - 4\mathbf{k}$, $\mathbf{c} = -\mathbf{i} + 3\mathbf{j} - 2\mathbf{k}$, *find the unit vector* $\mathbf{n}$ *perpendicular to the plane* ABC.

We have solved this problem earlier (10.29, Example (iii)) by using the scalar product; we now solve it using the vector-product.

We have shown that

$$\mathbf{AB} = \mathbf{b} - \mathbf{a} = \mathbf{i} + 3\mathbf{j} - 7\mathbf{k}$$

and

$$\mathbf{BC} = \mathbf{c} - \mathbf{b} = -4\mathbf{i} + \mathbf{j} + 2\mathbf{k}.$$

If we form the vector product of **AB** and **BC** we get, from Example (i) above,

$$\mathbf{AB} \wedge \mathbf{BC} = (6+7)\mathbf{i}+(28-2)\mathbf{j}+(1+12)\mathbf{k} = 13(\mathbf{i}+2\mathbf{j}+\mathbf{k})$$

or

$$\mathbf{AB} \wedge \mathbf{BC} = \begin{vmatrix} \mathbf{i} & \mathbf{j} & \mathbf{k} \\ 1 & 3 & -7 \\ -4 & 1 & 2 \end{vmatrix} = 13(\mathbf{i}+2\mathbf{j}+\mathbf{k})$$

But $\mathbf{AB} \wedge \mathbf{BC}$ is a vector perpendicular to the plane ABC, and hence unit vector **n** perpendicular to this plane is

$$\mathbf{n} = \frac{1}{\sqrt{6}}\,(\mathbf{i}+2\mathbf{j}+\mathbf{k}).$$

The same result would, of course, have been obtained if we had formed the vector product of **BC** and **CA**, or of **CA** and **AB**.

EXAMPLE (iii)

*Show that the equation of the line through the points whose position-vectors are **a** and **b** can be written in the form*

$$(\mathbf{r}-\mathbf{a}) \wedge (\mathbf{b}-\mathbf{a}) = 0.$$

If **r** is the position vector of any point P on the line AB the vector $\mathbf{AP} = \mathbf{r}-\mathbf{a}$ and the vector $\mathbf{AB} = \mathbf{b}-\mathbf{a}$. The angle between these two vectors is zero, and hence

$$(\mathbf{r}-\mathbf{a}) \wedge (\mathbf{b}-\mathbf{a}) = 0.$$

10.35. Moment of a force

We have shown earlier (10.26) that the moment of a force **F** about a point O can be written as $\mathbf{r} \wedge \mathbf{F}$, where **r** is the position vector of any point P on the line of action of **F** relative to O.

If P has coordinates (x, y, z) relative to coordinate axes through O, and if **F** has components X, Y, Z parallel to the axes, the moment of **F** about O is

$$\mathbf{M} = (x\mathbf{i}+y\mathbf{j}+z\mathbf{k}) \wedge (X\mathbf{i}+Y\mathbf{j}+Z\mathbf{k})$$
$$= (yZ-zY)\mathbf{i}+(zX-xZ)\mathbf{j}+(xY-yX)\mathbf{k}$$

We note that

$$\mathbf{i} \cdot \mathbf{M} = yZ-zY$$
$$= \text{moment of the force } \mathbf{F} \text{ about } Ox.$$

Similarly,

$$\mathbf{j} \cdot \mathbf{M} = zX - xZ$$

= moment of the force $\mathbf{F}$ about Oy.

and

$$\mathbf{k} \cdot \mathbf{M} = xY - yX$$

= moment of the force $\mathbf{F}$ about Oz.

Indeed, the moment of the force $\mathbf{F}$ *about any line* OA through O is defined as the component of the vector $\mathbf{M} = \mathbf{r} \wedge \mathbf{F}$ in the direction OA.

10.36. Products of more than two vectors

Consider any three vectors $\mathbf{a}$, $\mathbf{b}$, and $\mathbf{c}$.

Since $\mathbf{b} \wedge \mathbf{c}$ is a vector it can be multiplied by the vector $\mathbf{a}$ to form a scalar product or a vector product. These are sometimes referred to as triple products.

We will first consider the *scalar triple product* $\mathbf{a} \cdot (\mathbf{b} \wedge \mathbf{c})$, or more simply $\mathbf{a} \cdot \mathbf{b} \wedge \mathbf{c}$, for there is no ambiguity when the brackets are removed.

We can represent this product geometrically by drawing (Fig. 10.27) **OA, OB, OC** to represent $\mathbf{a}$, $\mathbf{b}$, $\mathbf{c}$ respectively, and

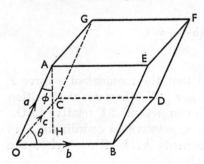

FIG. 10.27

completing the parallelepiped OBDCAEFG. Then the vector-product $\mathbf{b} \wedge \mathbf{c}$ can be represented by the area vector of the parallelogram OBDC, that is, bc sin θ $\mathbf{n}$, where θ is the angle BOC and $\mathbf{n}$ is unit vector normal to the plane of OBDC. If AH is drawn perpendicular to the plane of OBDC, then $\mathbf{n}$ is unit vector parallel to HA.

Hence,

$$\mathbf{a} \cdot \mathbf{b} \wedge \mathbf{c} = bc \sin \theta \, \mathbf{a} \cdot \mathbf{n}$$
$$= (bc \sin \theta) a \cos \phi$$

where ϕ is the angle between $\mathbf{a}$ and the normal to OBDC so that $HA = a \cos \phi$.

$\therefore$ $\mathbf{a} \cdot \mathbf{b} \wedge \mathbf{c} =$ (area of OBDC) $\times$ height of the parallelepiped

               $=$ volume V of the parallelepiped of which $\mathbf{a}$, $\mathbf{b}$, $\mathbf{c}$ are adjacent sides.

Similarly, it can be shown that

$$V = \mathbf{b} \cdot \mathbf{c} \wedge \mathbf{a} = \mathbf{c} \cdot \mathbf{a} \wedge \mathbf{b}$$

Hence,

$$\mathbf{a} \cdot \mathbf{b} \wedge \mathbf{c} = \mathbf{b} \cdot \mathbf{c} \wedge \mathbf{a} = \mathbf{c} \cdot \mathbf{a} \wedge \mathbf{b}$$

the cyclic order a, b, c being maintained.

This can also be written

$$\mathbf{b} \wedge \mathbf{c} \cdot \mathbf{a} = \mathbf{c} \wedge \mathbf{a} \cdot \mathbf{b} = \mathbf{a} \wedge \mathbf{b} \cdot \mathbf{c}.$$

We note that if the vectors $\mathbf{a}$, $\mathbf{b}$, $\mathbf{c}$ are coplanar, then $V = 0$, and hence $\mathbf{a} \cdot \mathbf{b} \wedge \mathbf{c}$ and all the other scalar triple products vanish. Conversely, if $\mathbf{a} \cdot \mathbf{b} \wedge \mathbf{c} = 0$ the three vectors $\mathbf{a}$, $\mathbf{b}$, $\mathbf{c}$ are coplanar (assuming none of the vectors is zero).

We note also that if two of the three vectors are equal $V = 0$, and hence the scalar triple product is zero.

The scalar triple products involving $\mathbf{i}$, $\mathbf{j}$, $\mathbf{k}$ all vanish except $\mathbf{i} \cdot \mathbf{j} \wedge \mathbf{k} = 1 = -\mathbf{i} \cdot \mathbf{k} \wedge \mathbf{j}$.

In terms of the components of the vectors parallel to the coordinate axes we can show that

$$\mathbf{a} \cdot \mathbf{b} \wedge \mathbf{c} = V = \begin{vmatrix} a_1 & a_2 & a_3 \\ b_1 & b_2 & b_3 \\ c_1 & c_2 & c_3 \end{vmatrix}$$

One example of this triple product may be given here. The moment of a force $\mathbf{F}$ acting at P about any line OA through O is defined as the component of the vector $\mathbf{r} \wedge \mathbf{F}$ in the direction of OA, that is, $\mathbf{l} \cdot \mathbf{r} \wedge \mathbf{F}$, where $\mathbf{l}$ is unit vector in the direction the of line OA, and $\mathbf{r} = \mathbf{OP}$ (cf. 10.35).

10.37. The vector triple product

As suggested above, the vector product $\mathbf{b} \wedge \mathbf{c}$ can be multiplied vectorially by the vector $\mathbf{a}$ to form the triple product $(\mathbf{b} \wedge \mathbf{c}) \wedge \mathbf{a}$, which equals $-\mathbf{a} \wedge (\mathbf{b} \wedge \mathbf{c})$.

It should be noted immediately that the brackets must be retained. For $\mathbf{a} \wedge (\mathbf{b} \wedge \mathbf{c})$ is a vector perpendicular to the vectors $\mathbf{a}$ and $\mathbf{b} \wedge \mathbf{c}$, while $\mathbf{b} \wedge \mathbf{c}$ is perpendicular to the vectors $\mathbf{b}$ and $\mathbf{c}$. Hence $\mathbf{a} \wedge (\mathbf{b} \wedge \mathbf{c})$ must lie in the plane of $\mathbf{b}$ and $\mathbf{c}$, and so is expressible in the form $p_1 \mathbf{b} + q_1 \mathbf{c}$, where p_1 and q_1 are scalars.

Similarly, $(\mathbf{a} \wedge \mathbf{b}) \wedge \mathbf{c}$ is a vector in the plane of $\mathbf{a}$ and $\mathbf{b}$, and so is expressible in the form $p_2\mathbf{a} + q_2\mathbf{b}$, where p_2 and q_2 are scalars.

Hence, in general, $\mathbf{a} \wedge (\mathbf{b} \wedge \mathbf{c}) \neq (\mathbf{a} \wedge \mathbf{b}) \wedge \mathbf{c}$.
We will show that

$$\mathbf{a} \wedge (\mathbf{b} \wedge \mathbf{c}) = (\mathbf{a} \cdot \mathbf{c})\mathbf{b} - (\mathbf{a} \cdot \mathbf{b})\mathbf{c}$$

Similarly,

$$(\mathbf{a} \wedge \mathbf{b}) \wedge \mathbf{c} = -\mathbf{c} \wedge (\mathbf{a} \wedge \mathbf{b})$$
$$= -\{(\mathbf{c} \cdot \mathbf{b})\mathbf{a} - (\mathbf{c} \cdot \mathbf{a})\mathbf{b}\}$$
$$= (\mathbf{a} \cdot \mathbf{c})\mathbf{b} - (\mathbf{b} \cdot \mathbf{c})\mathbf{a}$$

Method (i): As shown above we can write

$$\mathbf{a} \wedge (\mathbf{b} \wedge \mathbf{c}) = p_1\mathbf{b} + q_1\mathbf{c} \tag{i}$$

where p_1 and q_1 are scalar quantities.

If we multiply throughout scalarly by $\mathbf{a}$ we get

$$\mathbf{a} \cdot \mathbf{a} \wedge (\mathbf{b} \wedge \mathbf{c}) = p_1\mathbf{a} \cdot \mathbf{b} + q_1\mathbf{a} \cdot \mathbf{c}$$

But the expression on the left-hand side is zero, since two of the vectors are equal, as proved in 10.29, and hence

$$\frac{p_1}{\mathbf{a} \cdot \mathbf{c}} = -\frac{q_1}{\mathbf{a} \cdot \mathbf{b}} = \lambda$$

where λ is some scalar quantity which is independent of $\mathbf{a}$, $\mathbf{b}$, and $\mathbf{c}$.

Hence, $\mathbf{a} \wedge (\mathbf{b} \wedge \mathbf{c}) = \lambda\{(\mathbf{a} \cdot \mathbf{c})\mathbf{b} - (\mathbf{a} \cdot \mathbf{b})\mathbf{c}\}$ $\qquad$ (ii)

Putting $\mathbf{a} = \mathbf{i}, \mathbf{b} = \mathbf{i}$, and $\mathbf{c} = \mathbf{j}$, we get

$$\mathbf{i} \wedge (\mathbf{i} \wedge \mathbf{j}) = \lambda\{(\mathbf{i} \cdot \mathbf{j})\mathbf{i} - (\mathbf{i} \cdot \mathbf{i})\mathbf{j}\}$$

$\therefore \qquad\qquad \mathbf{i} \wedge \mathbf{k} = \lambda\{-\mathbf{j}\}$

$\therefore \qquad\qquad\qquad \lambda = 1.$

Alternatively, we can show that $\lambda = 1$ by equating the components, parallel to the axis of x, of the vectors represented by the two sides of equation (ii).

Hence,

$$\mathbf{a} \wedge (\mathbf{b} \wedge \mathbf{c}) = (\mathbf{a} \cdot \mathbf{c})\mathbf{b} - (\mathbf{a} \cdot \mathbf{b})\mathbf{c}.$$

Method (ii): Since $\mathbf{a}$, $\mathbf{b}$, and $\mathbf{c}$ are non-coplanar vectors, we can write

$$\mathbf{a} = \lambda_1\mathbf{b} + \lambda_2\mathbf{c} + \lambda_3\mathbf{b} \wedge \mathbf{c} \qquad \text{(iii)}$$

where λ_1, λ_2, λ_3 are scalar quantities.

Multiplying throughout vectorially by $\mathbf{b} \wedge \mathbf{c}$, we get

$$\mathbf{a} \wedge (\mathbf{b} \wedge \mathbf{c}) = \lambda_1\mathbf{b} \wedge (\mathbf{b} \wedge \mathbf{c}) + \lambda_2\mathbf{c} \wedge (\mathbf{b} \wedge \mathbf{c}) + \lambda_3(\mathbf{b} \wedge \mathbf{c}) \wedge (\mathbf{b} \wedge \mathbf{c})$$

$$= \lambda_1\mathbf{b} \wedge (\mathbf{b} \wedge \mathbf{c}) + \lambda_2\mathbf{c} \wedge (\mathbf{b} \wedge \mathbf{c}) \qquad \text{(iv)}$$

We will now evaluate $\mathbf{b} \wedge (\mathbf{b} \wedge \mathbf{c})$ and $\mathbf{c} \wedge (\mathbf{b} \wedge \mathbf{c})$.

Consider $\mathbf{c} \wedge (\mathbf{b} \wedge \mathbf{c})$. It can be written as

$$\mathbf{c} \wedge bc \sin \theta \, \mathbf{n}$$

where (Fig. 10.28) θ is the angle between $\mathbf{b}$ and $\mathbf{c}$, and $\mathbf{n}$ is unit vector perpendicular to the plane of the paper downwards.

$$\therefore \qquad \mathbf{c} \wedge (\mathbf{b} \wedge \mathbf{c}) = bc^2 \sin \theta \, \mathbf{m}$$

where $\mathbf{m}$ is unit vector perpendicular to $\mathbf{c}$ in the plane of the

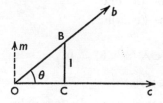

Fig. 10.28

paper, as shown in Fig. 10.28. Now, from triangle OBC, where BC is drawn of unit length, we have

$$\mathbf{m} = \mathbf{CO} + \mathbf{OB}$$

But $OC = \cot \theta$ and $OB = \operatorname{cosec} \theta$.

$$\therefore \qquad \mathbf{m} = -\frac{\cot \theta}{c}\mathbf{c} + \frac{\operatorname{cosec} \theta}{b}\mathbf{b}$$

Hence,

$$\mathbf{c} \wedge (\mathbf{b} \wedge \mathbf{c}) = -bc \cos \theta \ \mathbf{c} + c^2 \mathbf{b}$$
$$= -(\mathbf{b} \cdot \mathbf{c})\mathbf{c} + (\mathbf{c} \cdot \mathbf{c})\mathbf{b}$$

From this we can deduce the value of $\mathbf{b} \wedge (\mathbf{b} \wedge \mathbf{c})$ for

$$\mathbf{b} \wedge (\mathbf{b} \wedge \mathbf{c}) = -\mathbf{b} \wedge (\mathbf{c} \wedge \mathbf{b})$$
$$= (\mathbf{c} \cdot \mathbf{b})\mathbf{b} - (\mathbf{b} \cdot \mathbf{b})\mathbf{c}$$

by interchanging $\mathbf{b}$ and $\mathbf{c}$ in the above result.

Hence, equation (iv) becomes

$$\mathbf{a} \wedge (\mathbf{b} \wedge \mathbf{c}) = \{\lambda_1(\mathbf{c} \cdot \mathbf{b}) + \lambda_2(\mathbf{c} \cdot \mathbf{c})\}\mathbf{b} - \{\lambda_1(\mathbf{b} \cdot \mathbf{b}) + \lambda_2(\mathbf{b} \cdot \mathbf{c})\}\mathbf{c}$$

But from equation (iii) on multiplying scalarly by $\mathbf{b}$ and $\mathbf{c}$ respectively we get

$$\mathbf{a} \cdot \mathbf{b} = \lambda_1 \mathbf{b} \cdot \mathbf{b} + \lambda_2 \mathbf{c} \cdot \mathbf{b}$$

and

$$\mathbf{a} \cdot \mathbf{c} = \lambda_1 \mathbf{b} \cdot \mathbf{c} + \lambda_2 \mathbf{c} \cdot \mathbf{c}$$

Hence,

$$\mathbf{a} \wedge (\mathbf{b} \wedge \mathbf{c}) = (\mathbf{a} \cdot \mathbf{c})\mathbf{b} - (\mathbf{a} \cdot \mathbf{b})\mathbf{c}.$$

Many other methods of proof have been devised. They have various degrees of elegance.

It should be noted that $\mathbf{a} \wedge (\mathbf{b} \wedge \mathbf{c})$ does not retain its value on cyclic permutation of the vectors $\mathbf{a}$, $\mathbf{b}$, and $\mathbf{c}$. In fact,

$$\mathbf{a} \wedge (\mathbf{b} \wedge \mathbf{c}) = (\mathbf{a} \cdot \mathbf{c})\mathbf{b} - (\mathbf{a} \cdot \mathbf{b})\mathbf{c}$$

and

$$\mathbf{b} \wedge (\mathbf{c} \wedge \mathbf{a}) = (\mathbf{b} \cdot \mathbf{a})\mathbf{c} - (\mathbf{b} \cdot \mathbf{c})\mathbf{a}$$

and

$$\mathbf{c} \wedge (\mathbf{a} \wedge \mathbf{b}) = (\mathbf{c} \cdot \mathbf{b})\mathbf{a} - (\mathbf{c} \cdot \mathbf{a})\mathbf{b}$$

Adding

$$\mathbf{a} \wedge (\mathbf{b} \wedge \mathbf{c}) + \mathbf{b} \wedge (\mathbf{c} \wedge \mathbf{a}) + \mathbf{c} \wedge (\mathbf{a} \wedge \mathbf{b}) = 0.$$

10.38. Products of more than three vectors

If $\mathbf{a}, \mathbf{b}, \mathbf{c}, \mathbf{d}, \dots$ are any set of vectors various products involving four or more vectors can be formed. For example, $\mathbf{a} \wedge (\mathbf{b} \wedge \mathbf{c})$ being a vector can be multiplied scalarly or vectorially by $\mathbf{d}$ to give $\mathbf{a} \wedge (\mathbf{b} \wedge \mathbf{c}) \cdot \mathbf{d}$ or $\mathbf{a} \wedge (\mathbf{b} \wedge \mathbf{c}) \wedge \mathbf{d}$.

The process can clearly be continued indefinitely.

10.39. EXAMPLE (i)

If the position vectors of the points A, B, C are $\mathbf{a}$, $\mathbf{b}$, $\mathbf{c}$ respectively, show that the equation of the plane ABC can be written in the form $\mathbf{r} \cdot \mathbf{N} = \mathbf{a} \cdot \mathbf{b} \wedge \mathbf{c}$, where $\mathbf{N} = \mathbf{b} \wedge \mathbf{c} + \mathbf{c} \wedge \mathbf{a} + \mathbf{a} \wedge \mathbf{b}$.

If $\mathbf{r}$ is the position-vector of any point P in the plane ABC, the vector $\mathbf{AP} = \mathbf{r} - \mathbf{a}$. Hence if $\mathbf{n}$ is unit vector normal to the plane ABC we have

$$(\mathbf{r} - \mathbf{a}) \cdot \mathbf{n} = 0.$$

We note that $\mathbf{AB} = \mathbf{b} - \mathbf{a}$ and $\mathbf{BC} = \mathbf{c} - \mathbf{b}$ and hence

$$\mathbf{AB} \wedge \mathbf{BC} = (\mathbf{b} - \mathbf{a}) \wedge (\mathbf{c} - \mathbf{b})$$
$$= \mathbf{b} \wedge \mathbf{c} - \mathbf{a} \wedge \mathbf{c} + \mathbf{a} \wedge \mathbf{b}$$
$$= \mathbf{b} \wedge \mathbf{c} + \mathbf{c} \wedge \mathbf{a} + \mathbf{a} \wedge \mathbf{b}.$$

Hence the vector $\mathbf{N}$ is normal to the plane ABC and consequently

$$(\mathbf{r} - \mathbf{a}) \cdot \mathbf{N} = 0$$
$$\therefore \quad\quad \mathbf{r} \cdot \mathbf{N} = \mathbf{a} \cdot \mathbf{N}$$
$$= \mathbf{a} \cdot \mathbf{b} \wedge \mathbf{c}$$

the other two terms on the right-hand side being zero.

EXAMPLE (ii)

Find an expression for $(\mathbf{a} \wedge \mathbf{b}) \cdot (\mathbf{c} \wedge \mathbf{d})$ *where* $\mathbf{a}, \mathbf{b}, \mathbf{c}, \mathbf{d}$ *are any four vectors.*

We can write

$$(\mathbf{a} \wedge \mathbf{b}) \cdot (\mathbf{c} \wedge \mathbf{d}) = \mathbf{e} \cdot (\mathbf{c} \wedge \mathbf{d}) \text{ where } \mathbf{e} = \mathbf{a} \wedge \mathbf{b}$$
$$= \mathbf{c} \cdot \mathbf{d} \wedge \mathbf{e}$$
$$= \mathbf{c} \cdot \mathbf{d} \wedge (\mathbf{a} \wedge \mathbf{b})$$
$$= \mathbf{c} \cdot [(\mathbf{d} \cdot \mathbf{b})\mathbf{a} - (\mathbf{d} \cdot \mathbf{a})\mathbf{b}]$$
$$\therefore \quad (\mathbf{a} \wedge \mathbf{b}) \cdot (\mathbf{c} \wedge \mathbf{d}) = (\mathbf{d} \cdot \mathbf{b})(\mathbf{c} \cdot \mathbf{a}) - (\mathbf{d} \cdot \mathbf{a})(\mathbf{c} \cdot \mathbf{b})$$

Note. If we put $\mathbf{c} = \mathbf{a}$ and $\mathbf{d} = \mathbf{b}$, we get

$$(\mathbf{a} \wedge \mathbf{b}) \cdot (\mathbf{a} \wedge \mathbf{b}) = (\mathbf{b} \cdot \mathbf{b})(\mathbf{a} \cdot \mathbf{a}) - (\mathbf{b} \cdot \mathbf{a})(\mathbf{a} \cdot \mathbf{b})$$
$$= a^2 b^2 - (\mathbf{a} \cdot \mathbf{b})^2$$

Writing this relation in terms of the components a_1, a_2, a_3 and b_1, b_2, b_3 of the vectors $\mathbf{a}$ and $\mathbf{b}$, we get

$$(a_2 b_3 - a_3 b_2)^2 + (a_3 b_1 - a_1 b_3)^2 + (a_1 b_2 - a_2 b_1)^2$$
$$= (a_1^2 + a_2^2 + a_3^2)(b_1^2 + b_2^2 + b_3^2) - (a_1 b_1 + a_2 b_2 + a_3 b_3)^2$$

which is known as Lagrange's identity.

It follows if θ is the angle between $\mathbf{a}$ and $\mathbf{b}$ that

$$\sin^2 \theta = \{(a_2 b_3 - a_3 b_2)^2 + (a_3 b_1 - a_1 b_3)^2 + (a_1 b_2 - a_2 b_1)^2\}/$$
$$(a_1^2 + a_2^2 + a_3^2)(b_1^2 + b_2^2 + b_3^2)$$

10.40. Motion of a rigid body

The equations of motion of a *rigid body* can be derived from the principles of linear and angular momentum (cf. Dynamics, 10.29). If a body of mass M is moving in two dimensions subject to a set of forces $F_1, F_2, \ldots F_n$ acting at the points $r_1, r_2, \ldots r_n$, the equations of motion are

$$M \, dv_G/dt = \sum F_i$$

and

$$dH'/dt = \sum r_i{}' \wedge F$$

where v_G is the velocity of the centre of mass G of the body, H' is the angular momentum of the body about the axis of rotation through G, and r'_i is the radius vector of the point of application of the force F_i relative to G, that is, $r'_i = r_i - v_G$, where r_G is the position-vector of G.

For a more general motion of a rigid body the above equations still apply, but the angular momentum has the value derived in 10.41 below.

10.41. Angular Momentum of a rigid body

If a rigid body is rotating about an axis through a fixed point O we can denote the angular velocity ω by means of a vector $\boldsymbol{\omega}$ along the axis of rotation.

The velocity of any particle P of the body can be written as

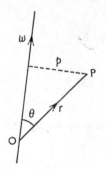

FIG. 10.29

$\omega p n$, where p is the perpendicular distance of P from the axis of rotation and n is unit vector normal to the plane of the paper downwards (Fig. 10.29), if $\boldsymbol{\omega}$ is in the direction shown.

$\therefore \qquad$ velocity of P $= \omega r \sin \theta \, n = \boldsymbol{\omega} \wedge r$

where **r** is the position-vector of P with respect to the point O on the axis of rotation.

If m is the mass of the particle P its linear momentum is $m\boldsymbol{\omega} \wedge \mathbf{r}$, and its moment of momentum about O is $\mathbf{r} \wedge (m\boldsymbol{\omega} \wedge \mathbf{r})$.

Hence, the angular momentum **H** of the body about O is $\sum \mathbf{r} \wedge (m\boldsymbol{\omega} \wedge \mathbf{r})$, the summation being throughout the body. Hence,

$$\mathbf{H} = \sum \mathbf{r} \wedge (m\boldsymbol{\omega} \wedge \mathbf{r})$$
$$= \sum m\{(\mathbf{r} \cdot \mathbf{r})\boldsymbol{\omega} - (\mathbf{r} \cdot \boldsymbol{\omega})\mathbf{r}\}$$
$$= \boldsymbol{\omega}\sum m\mathbf{r} \cdot \mathbf{r} - \sum m(\mathbf{r} \cdot \boldsymbol{\omega})\mathbf{r}$$

For a lamina rotating about an axis perpendicular to its plane $\mathbf{r} \cdot \boldsymbol{\omega}$ is everywhere zero, and hence $\mathbf{H} = \boldsymbol{\omega}\sum mr^2 = I\boldsymbol{\omega}$, where I is the moment of inertia of the lamina about the axis of rotation (cf. Dynamics, 10.19).

Quite generally, however, if we denote the components of **H** parallel to the coordinate axes through O by H_1, H_2, H_3 and the components of ω by ω_1, ω_2, ω_3 and the components of **r** by x, y, z we get

$$H_1 = \omega_1\sum mr^2 - \sum m(x\omega_1 + y\omega_2 + z\omega_3)x$$
$$= \omega_1\sum m(y^2 + z^2) - \omega_2\sum mxy - \omega_3\sum mzx$$
$$H_2 = -\omega_1\sum mxy + \omega_2\sum m(z^2 + x^2) - \omega_3\sum myz$$
$$H_3 = -\omega_1\sum mzx - \omega_2\sum myz + \omega_3\sum m(x^2 + y^2)$$

These may be written

$$H_1 = A\omega_1 - F\omega_2 - E\omega_3$$
$$H_2 = -F\omega_1 + B\omega_2 - D\omega_3$$
$$H_3 = -E\omega_1 - D\omega_2 + C\omega_3$$

where A, B, C are the moments of inertia of the body about Ox, Oy, Oz and D, E, F are the products of inertia of the body with respect to the coordinate planes taken in pairs.

The kinetic energy T of the rigid body is given by

$$2T = \sum m(\boldsymbol{\omega} \wedge \mathbf{r}) \cdot (\boldsymbol{\omega} \wedge \mathbf{r})$$
$$= A\omega_1^2 + B\omega_2^2 + C\omega_3^2 - 2D\omega_2\omega_3 - 2E\omega_3\omega_1 - 2F\omega_1\omega_2$$

For a body rotating about the axis of z, $\omega_1 = \omega_2 = 0$ and $\omega_3 = \omega$, so that $2T = C\omega^2$, or $T = \frac{1}{2}C\omega^2$ in keeping with the usual expression for the kinetic energy of a body rotating about a fixed axis (cf. Dynamics, 10.2).

EXAMPLES 10.3

1. If $\mathbf{a} = 4\mathbf{i} + 2\mathbf{j}$, $\mathbf{b} = 3\mathbf{i} - \mathbf{j}$, find $\mathbf{a} \cdot \mathbf{b}$ and $\mathbf{a} \wedge \mathbf{b}$. Evaluate $\mathbf{a} \wedge \mathbf{b} \cdot \mathbf{c}$ when (i) $\mathbf{c} = 2\mathbf{j} + \mathbf{k}$ and (ii) $\mathbf{c} = 3\mathbf{i} - \mathbf{j} + 2\mathbf{k}$.

2. If $\mathbf{a} = 3\mathbf{i} + 2\mathbf{j}$, $\mathbf{b} = 4\mathbf{i} - \mathbf{j}$ and $\mathbf{c} = 2\mathbf{i} - 5\mathbf{j}$, find $\mathbf{a} \wedge \mathbf{b}$, $\mathbf{a} \wedge \mathbf{c}$, $(\mathbf{a} \wedge \mathbf{b}) \cdot (\mathbf{a} \wedge \mathbf{c})$, and $(\mathbf{a} \wedge \mathbf{b}) \wedge (\mathbf{a} \wedge \mathbf{c})$.

3. If ABCD is any parallelogram form the scalar product of the two vectors represented by AC and BD; deduce that the diagonals of a rhombus are at right angles.

4. If A, B, C, D are four points such that $AB^2 + CD^2 = AC^2 + BD^2$, prove that BC is perpendicular to AD.

5. Prove that $(\mathbf{a} - \mathbf{b}) \wedge (\mathbf{a} + \mathbf{b}) = 2\mathbf{a} \wedge \mathbf{b}$, and interpret the result geometrically.

6. Prove that the vector area of the triangle for which the position-vectors of the vertices are $\mathbf{a}$, $\mathbf{b}$, and $\mathbf{c}$ may be written as

$$\tfrac{1}{2}(\mathbf{b} \wedge \mathbf{c} + \mathbf{c} \wedge \mathbf{a} + \mathbf{a} \wedge \mathbf{b}).$$

7. Prove that the area of the triangle formed by joining the mid-point of one of the non-parallel sides of a trapezium to the extremities of the opposite side is one-half that of the trapezium.

8. Show that the angle in a semicircle is a right angle.

9. If P, Q, R, S are the four points $(3, -2)$, $(2, 1)$, $(4, -1)$, and $(-2, 1)$ in the (x, y) plane, express the vectors $\mathbf{PQ}$ and $\mathbf{RS}$ in the form $x\mathbf{i} + y\mathbf{j}$, and find the products $\mathbf{PQ} \cdot \mathbf{RS}$ and $\mathbf{PQ} \wedge \mathbf{RS}$.

10. Find the unit vector $\mathbf{n}$ that is perpendicular to the two vectors $\mathbf{a} = 2\mathbf{i} + 3\mathbf{j}$ and $\mathbf{b} = \mathbf{k} - 2\mathbf{i}$, so that $\mathbf{a}$, $\mathbf{b}$, $\mathbf{n}$ is a right-handed triad.

11. If $\mathbf{a} = 3\mathbf{i} + 2\mathbf{j} + \mathbf{k}$, $\mathbf{b} = 4\mathbf{i} - \mathbf{j} - 2\mathbf{k}$, and $\mathbf{c} = 2\mathbf{i} + 3\mathbf{j} + 2\mathbf{k}$, verify that $\mathbf{a} \cdot (\mathbf{b} + \mathbf{c}) = \mathbf{a} \cdot \mathbf{b} + \mathbf{a} \cdot \mathbf{c}$ and that $\mathbf{a} \wedge (\mathbf{b} + \mathbf{c}) = \mathbf{a} \wedge \mathbf{b} + \mathbf{a} \wedge \mathbf{c}$.

12. If $\mathbf{a}$, $\mathbf{b}$, $\mathbf{c}$ have the values given in question 11 above, find $\mathbf{b} \wedge \mathbf{c}$, $\mathbf{a} \cdot \mathbf{b} \wedge \mathbf{c}$, $\mathbf{a} \wedge (\mathbf{b} \wedge \mathbf{c})$, and $(\mathbf{a} \wedge \mathbf{b}) \wedge \mathbf{c}$.

13. Find the unit vector perpendicular to each of the vectors $2\mathbf{i} - \mathbf{j} + \mathbf{k}$ and $3\mathbf{i} + 4\mathbf{j} - \mathbf{k}$. Calculate the sine of the angle between these two vectors.

14. Use vectors to find the area of the triangle of which the vertices are the points $(3, 4, 2)$, $(1, 0, 5)$, and $(-1, -2, 3)$.

15. If $\mathbf{a} = a_1\mathbf{i} + a_2\mathbf{j} + a_3\mathbf{k}$, show that $a_1 = \mathbf{a} \cdot \mathbf{i}$, $a_2 = \mathbf{a} \cdot \mathbf{j}$, and $a_3 = \mathbf{a} \cdot \mathbf{k}$.

16. A force $\mathbf{F} = 2\mathbf{i} - 3\mathbf{j} - 6\mathbf{k}$ passes through the point whose position-vector is $3\mathbf{i} - 2\mathbf{j} + 5\mathbf{k}$. Find the moment of $\mathbf{F}$ about the point

whose position-vector is $6\mathbf{i}+\mathbf{j}-7\mathbf{k}$. Find also the moment of F about a line through this point parallel to the vector $4\mathbf{i}-\mathbf{j}+3\mathbf{k}$.

17. Show that the equation of the line through a given point, whose position-vector is $\mathbf{a}$, and parallel to a given vector $\mathbf{b}$ can be written as $(\mathbf{r}-\mathbf{a}) \wedge \mathbf{b} = 0$.

    Deduce the Cartesian equivalents.

18. Show that the equation of the line through the point whose position-vector is $\mathbf{a}$ and perpendicular to the vectors $\mathbf{b}$ and $\mathbf{c}$ can be written as $(\mathbf{r}-\mathbf{a}) \wedge (\mathbf{b} \wedge \mathbf{c}) = 0$.

    Interpret the equation $(\mathbf{r}-\mathbf{a}) \cdot \mathbf{b} \wedge \mathbf{c} = 0$.

19. (i) Show that any vector $\mathbf{R}$ may be expressed in the form $\lambda_1\mathbf{a}+\lambda_2\mathbf{b}+\lambda_3\mathbf{c}$, where $\mathbf{a}$, $\mathbf{b}$, $\mathbf{c}$ is any set of three non-coplanar vectors and λ_1, λ_2, λ_3 are appropriate scalars. Show that $\lambda_1 = \mathbf{R} \cdot \mathbf{b} \wedge \mathbf{c}/\mathbf{a} \cdot \mathbf{b} \wedge \mathbf{c}$.

    (ii) if $\mathbf{a}$, $\mathbf{b}$, $\mathbf{c}$ are any three non-coplanar vectors, show that $\mathbf{a}$ can be expressed in the form $k_1\mathbf{b}+k_2\mathbf{c}+k_3\mathbf{b} \wedge \mathbf{c}$. Show that $k_3 = \mathbf{a} \cdot \mathbf{b} \wedge \mathbf{c}/(\mathbf{b} \wedge \mathbf{c}) \cdot (\mathbf{b} \wedge \mathbf{c})$.

20. Show that $\mathbf{a} \cdot \mathbf{b} \wedge \mathbf{c}$ equals the volume of the parallelepiped of which $\mathbf{a}$, $\mathbf{b}$, $\mathbf{c}$ are adjacent sides. Find the volume of the pyramid of which the vertices are the origin and the points $(2, 3, 4)$, $(4, 5, 7)$, and $(7, 6, 2)$.

21. If $\mathbf{a}$ and $\mathbf{b}$ are any two vectors, show that the components of $\mathbf{b}$ parallel and perpendicular to $\mathbf{a}$ can be written as

$$\frac{\mathbf{a} \cdot \mathbf{b}}{\mathbf{a} \cdot \mathbf{a}}\mathbf{a} \quad \text{and} \quad \frac{\mathbf{a} \wedge (\mathbf{b} \wedge \mathbf{a})}{\mathbf{a} \cdot \mathbf{a}} \quad \text{respectively.}$$

22. Show that the equation of a plane can be expressed in the form $(\mathbf{r}-\mathbf{p}) \cdot \mathbf{p} = 0$, where $\mathbf{r}$ is the position-vector of any point in the plane and $\mathbf{p}$ is the position-vector of the foot of the perpendicular from the origin to the plane.

    Find the length of the perpendicular from the origin to the plane $3x+2y+4z = 14$.

23. Find the vector product $\mathbf{AB} \wedge \mathbf{CD}$, given that A, B, C, D are the points $(5, 7, -2)$, $(2, 8, -3)$, $(-3, 3, 6)$, and $(0, 1, 2)$ respectively. Deduce the shortest distance between the lines AB and CD.

24. Show that $\mathbf{a} \cdot \mathbf{b} \wedge \mathbf{c} = \mathbf{a} \wedge \mathbf{b} \cdot \mathbf{c}$, and verify when $\mathbf{a} = 3\mathbf{i}+2\mathbf{j}+\mathbf{k}$, $\mathbf{b} = 4\mathbf{i}-\mathbf{j}-2\mathbf{k}$, and $\mathbf{c} = 2\mathbf{i}+3\mathbf{j}+2\mathbf{k}$.

25. If $\mathbf{a}$, $\mathbf{b}$, $\mathbf{c}$ $\mathbf{d}$ are any four vectors, show that

$$(\mathbf{a} \wedge \mathbf{b}) \wedge (\mathbf{c} \wedge \mathbf{d}) = (\mathbf{a} \cdot \mathbf{c} \wedge d)\mathbf{b}-(\mathbf{b} \cdot \mathbf{c} \wedge d)\mathbf{a}$$

$$= (\mathbf{a} \cdot \mathbf{b} \wedge d)\mathbf{c}-(\mathbf{a} \cdot \mathbf{b} \wedge \mathbf{c})\mathbf{d}.$$

Deduce that if $\mathbf{a}$, $\mathbf{b}$, $\mathbf{c}$ are any three non-coplanar vectors any other vector $\mathbf{d}$ can be written as a linear function of $\mathbf{a}$, $\mathbf{b}$, and $\mathbf{c}$.

26. Prove from first principles that

$$\frac{d}{dt} (\mathbf{a} \cdot \mathbf{b}) = \frac{d\mathbf{a}}{dt} \cdot \mathbf{b} + \mathbf{a} \cdot \frac{d\mathbf{b}}{dt}$$

By differentiating $\hat{\mathbf{r}} \cdot \hat{\mathbf{r}} = 1$ with respect to t, show that $\hat{\mathbf{r}}$ and $d\hat{\mathbf{r}}/dt$ are perpendicular.

27. Prove from first principles that

$$\frac{d}{dt}(\mathbf{a} \wedge \mathbf{b}) = \frac{d\mathbf{a}}{dt} \wedge \mathbf{b} + \mathbf{a} \wedge \frac{d\mathbf{b}}{dt}$$

$$\frac{d}{dt}(\mathbf{a} \cdot \mathbf{b} \wedge \mathbf{c}) = \frac{d\mathbf{a}}{dt} \cdot \mathbf{b} \wedge \mathbf{c} + \mathbf{a} \cdot \frac{d\mathbf{b}}{dt} \wedge \mathbf{c} + \mathbf{a} \cdot \mathbf{b} \wedge \frac{d\mathbf{c}}{dt}$$

28. Show that $|\mathbf{a}|$ is constant if $\mathbf{a} \cdot \dfrac{d\mathbf{a}}{dt} = 0$, and that $\mathbf{a}$ is in a fixed direction if $\mathbf{a} \wedge \dfrac{d\mathbf{a}}{dt} = 0$.

29. (i) If $\mathbf{a} \wedge \dfrac{d\mathbf{b}}{dt} = \mathbf{b} \wedge \dfrac{d\mathbf{a}}{dt}$, show that $\mathbf{a} \wedge \mathbf{b}$ is constant.

    (ii) Show that $\dfrac{d}{dt}\left(\mathbf{a} \wedge \dfrac{d\mathbf{a}}{dt} \right) = \mathbf{a} \wedge \dfrac{d^2\mathbf{a}}{dt^2}$.

30. If $\mathbf{r}$, the position-vector of any point P in a plane, is given at time t by $\mathbf{r} = \mathbf{a} \cos nt + \mathbf{b} \sin nt$, where $\mathbf{a}$ and $\mathbf{b}$ are given vectors in the plane, show that the path of P is an ellipse.

    Show that the axes of the ellipse are given by the values of t that satisfy the equation

$$\tan 2nt = 2\mathbf{a} \cdot \mathbf{b}/(a^2 - b^2).$$

Solve this equation when $\mathbf{a} = 3\mathbf{i} - \mathbf{j}$ and $\mathbf{b} = 4\mathbf{j}$.

31. If $\mathbf{r} = \mathbf{a} \cos nt + \mathbf{b} \sin nt$ where $\mathbf{a}$ and $\mathbf{b}$ are constant vectors, prove that $d^2\mathbf{r}/dt^2 = -n^2\mathbf{r}$.

    Show also that $\dfrac{d}{dt}\left(\mathbf{r} \wedge \dfrac{d\mathbf{r}}{dt} \right) = 0$.

32. If the vector-displacement of a particle of mass m at time t is $\mathbf{r}$, show that the angular momentum of the particle about the origin O can be written as $\mathbf{H} = \mathbf{r} \wedge m \, d\mathbf{r}/dt$. Deduce that $d\mathbf{H}/dt = \mathbf{r} \wedge \mathbf{F}$, where $\mathbf{F}$ is the vector-sum of the forces acting on m.

33. A, B, C are three fixed points and P is a variable point. Show that if $PA^2 + PB^2 + PC^2$ is a minimum, P must be the centroid of the triangle ABC.

34. A particle of mass m is moving under the action of a single force. Its position-vector with respect to the origin at time t is

$$\mathbf{r} = a \sin pt\mathbf{i} + 2a \cos pt\mathbf{j} + a \cos pt\mathbf{k}.$$

Show that the particle is moving in a plane and find an expression for the force acting on it.

Another particle is describing simple harmonic motion with period $2\pi/p$ between the points $\pm a\mathbf{j}$, and when $t = 0$ its acceleration is $-ap^2\mathbf{j}$. Show that the magnitude of the relative velocity of the particles is greatest when they are closest together.

(L.A.)

35. If α is a constant such that $0 < \alpha < \frac{1}{2}\pi$, and ϕ is a parameter, prove that the curves

$$\mathbf{r} = b \cos \phi\mathbf{i} + b \sin \phi\mathbf{j} + b \cos \phi \tan \alpha\mathbf{k},$$
$$\mathbf{r} = b \cos \phi\mathbf{i} + b \sin \phi\mathbf{j} + (2h - b \cos \phi \tan \alpha)\mathbf{k}$$

are both ellipses of eccentricity $\sin \alpha$.

Show that, provided $h^2 < b^2 \tan^2\alpha$, the ellipses intersect in two points distant $2(b^2 - h^2 \cot^2 \alpha)^{1/2}$ apart. (L.A.)

36. A rigid body is rotating about the z-axis uniformly at 20 rad s^{-1}. Find the velocity and acceleration of the point of the body whose position-vector is $4\mathbf{i} + 3\mathbf{j} - 12\mathbf{k}$.

37. A rigid body is rotating uniformly at 5 rev s^{-1} about an axis through the origin parallel to the vector $2\mathbf{i} + 2\mathbf{j} + \mathbf{k}$. Find the velocity and acceleration of the point whose coordinates are $(3, 2, 1)$.

38. A solid cube of mass M and side $2a$ is acted upon by a force $5\mathbf{k} - 5\mathbf{i}$ at the point $(a, 0, 0)$, a force $5\mathbf{i} - 5\mathbf{j}$ at the point $(0, a, 0)$, and a force $5\mathbf{j} - 5\mathbf{k}$ at the point $(0, 0, a)$, the origin of coordinates being at the centre of the cube. Describe the motion at time t assuming the cube is initially at rest, and find the angular velocity at time t.

39. A rigid body rotates at time t with angular speed $\omega = 2t^3 - 3t$ about a fixed axis, taken as the axis of z. If the position-vector of a point A of the body at time $t = z$ is $\mathbf{r} = 3\mathbf{i} - 4\mathbf{j}$, find the velocity and acceleration of A at that instant.

40. Examine the motion of a uniform sphere moving over a rough horizontal plane. Show that in general the path of the centre of the sphere is a parabola as long as slipping occurs and is a straight line when the sphere is rolling.

ANSWERS TO THE EXAMPLES

EXAMPLES 1.1 (p. 9)

1. (i) 15; (ii) 12; (iii) 13; (iv) 10;
   (v) $\cos^{-1}(-\frac{2}{5})$; (vi) $\sqrt{58}$.
3. (i) 120°; (ii) 60°.
4. $Q = \sqrt{3}P$.
5. 15 N, 9 N, $\cos^{-1}(-\frac{3}{5})$.
6. 10 g N on each peg.
7. $5\sqrt{3}$ g N on the upper peg;
   5 g N on the lower peg.
8. $\cos^{-1}(\frac{7}{15})$.
10. 120°.
11. 4 N, $\cos^{-1}(-\frac{3}{5})$.
12. $\sqrt{76}$ N.
13. 10·3 N.
14. $4\sqrt{3}$ N, at 90° to the 4 N force.
16. 20 N, $\tan^{-1}\frac{3}{4}$.
17. $5\sqrt{2}$; $3\sqrt{10}$ and $\sqrt{10}$ N.
18. $\cos^{-1}\frac{1}{4}$ to P.
19. 8·8 N; 43° with 3 N.
20. 353·5 N; 234·5 N.
21. 8·7, 22·0 g N.

EXAMPLES 1.2 (p. 19)

1. 16 g N and 12 g N.
2. 72 g N and 54 g N.
3. 24 g N and 10 g N.
5. 7·8 g N and 6·6 g N.
6. 4·8 g N and 1·4 g N.
7. a. $5\sqrt{3}$ g N.
   b. $10\sqrt{3}$ g N.
8. 8·1 g N and 5·5 g N.
9. $M = 3$; $3\sqrt{3}$ g N.
10. 9·03 g N in AC; 7·1 g N in BC.
12. (i) $7\frac{1}{2}$ g N, 6 g N;
    (ii) 60 g N, 36 g N;
    (iii) $2·89 \times 10^3$ g N, $2\frac{1}{2} \times 10^3$ g N.
13. 5 g N, $5\sqrt{3}$ g N.
14. At right angles to the first
    string; $2·5\sqrt{3}$ g N and 2·5 g N.
15. 30·1 g N.
16. $5\sqrt{2}(\sqrt{3}-1)$ g N and
    $10(\sqrt{3}-1)$ g N.
17. 3·6 g N; 67° 58′, 44° 3′.
18. $5\sqrt{2}$; 5 g N.
19. 63·9 N; 23°; 80 g N; 54°.
21. $5\sqrt{2}$, $5\sqrt{2}$, 10, $5(\sqrt{2}-1)$.
23. 10 g.

EXAMPLES 1.3 (p. 28)

1. 7·7 N, 29°.
2. 10 N at $\tan^{-1}(\frac{3}{4})$ with the
   4 N force.
3. 7 N, between the 13 and 10 N
   forces at $\tan^{-1}(5\sqrt{3}/11)$ with
   the former.
4. 9·198 N.
5. 64·5 N, 13° 57′ S. of W.
6. 20 N at 60° to AB.
7. 228·8 N, 49° 39′ S. of E.
8. 23·9 g N on A; 10 g N on B and
   C; $10\sqrt{3}$ g N on D.
9. $\sqrt{3}F$; $\tan^{-1}(3-2\sqrt{2})$.
10. 112·8 N, $\tan^{-1}3(\sqrt{2}-)1$ S. of
    E.
11. 12·17 N, $25\frac{1}{4}$°.
12. 6·3 N at $\tan^{-1}(7\sqrt{3}/3)$, with
    the 2 N force.
13. $\sqrt{281}$ N at $\tan^{-1}(\frac{16}{5})$ with
    AB.

14. 8 N along GA.
15. 30 N, 34 N.
16. 10 N, 53° 8′; 10 N, 36° 52′ E. of N.
17. 2·91 N, 290° 6′.

18. $2\sqrt{2}$ N parallel to DB through E.
19. $\sqrt{3}$ N perpendicular to BC, 7·5 cm from B.

EXAMPLES 1.4 (p. 33)

1. 8 N; 15 N.
2. $12\sqrt{5}$; $4\sqrt{2}$.
3. $31\frac{1}{4}$ N; $P = 25$ N.
4. 0·0726 g N.

5. 13 N; $\tan^{-1} \frac{12}{5}$ to BA.
6. 16 N opposite to 5 N.
7. 3·77 N at 142°.
9. $\frac{1}{2}\sqrt{3}$, $-\frac{1}{2}$.

EXAMPLES 1.5 (p. 42)

1. 10 g N, 8·96 g N.
2. 0·25.
3. 4·85 g N.
5. $\frac{1}{8}$.
6. $10\sqrt{3}/3$ g N.
7. (i) 8·8 g N; (ii) 15·2 g N.
8. $1/\sqrt{39}$.
9. 12 g N, 213 g N.
10. 73·57 g N.
13. $\tan^{-1}(\frac{1}{4})$ with the plane; 56 g N.
14. $\sin \alpha = \sin \beta + \mu \cos \beta$, where μ is the coefficient of friction, and α the inclination of the smooth plane.

16. (i) 0·4; (ii) 171·6 N.
17. $P = W\dfrac{\sin \alpha + \mu \cos \alpha}{\cos \beta + \mu \sin \beta}$;

$Wl \cos \beta \dfrac{\sin \alpha + \mu \cos \alpha}{\cos \beta + \mu \sin \beta}$;

608 J.
18. $\cos \lambda = \dfrac{PW}{Q\sqrt{(P^2 + W^2)}}$.
19. 52° 30′, 7° 30′, $W/\sqrt{2}$ at $7\frac{1}{2}°$ to plane.
20. $2R\alpha$, where α is in radians.
21. 0·3535; 1·096 g N.
22. 10 g N, 14·33 g N.

EXAMPLES 2.1 (p. 52)

1. 110 N, 14 cm from A; 30 N, $51\frac{1}{3}$ cm, from A beyond B.
2. 21 N, 24 cm from A; 3 N, 168 cm from A beyond B.
3. 4 N, 24 cm from A on BA produced.
4. 12 N, 10 cm from the 8 N force.

5. 75 N, 25 N, the 75 N force being 1 m from the resultant.
11. 2 units; cuts diagonal BD distance $\frac{1}{10}$ (diagonal) from centre.
12. If D is on CA such that CD = 2 cm resultant bisects BD.

EXAMPLES 2.2 (p. 59)

1. 32 N cm, 64 N cm, 16 N cm about A, B, C respectively.
2. 180 N cm, 180 N cm.
3. $5\sqrt{3}$ N m.

4. 18 N m, zero.
5. 5 N m, 3 N m.
6. $21\sqrt{3}$ N cm, $21\sqrt{3}$ N cm.

EXAMPLES 2.3 (p. 62)

1. 7 g N on the support nearer the 3 kg; 6 g N on the other.
2. 20 g N at A; 40 g N on the the other.
3. 62·5 g N on the nearer; 37·5 g N on the other.
4. 37·5 kg.
5. $7\frac{1}{3}$ g N at A; $10\frac{2}{3}$ g N at B.
6. 0·865 m from the end from which distances are measured.
7. 5 units at 14 m from A on DA produced.
8. 2·1 m from A.
9. 0·6 m.
10. 7·35 g N at C; 4·65 g N at D; 51·5 g N cm.
11. $\frac{9}{34}$ m on either side of the centre.
13. Not nearer than 17·5 cm to either string.
14. A couple of moment 37 units, 10 g N — 3·7 m from O.

16. $4\frac{2}{3}$ g N on the prop further from the 2 kg; $7\frac{1}{3}$ g N on the nearer one.
17. 26 g N and 14 g N.
18. 12·5 kg.
19. $p+q$.
20. AC = 0·6 m, DB = 0·3 m.
21. 202 N, 1·78 m.
22. At B, $\dfrac{48+9M_1-3M_2}{7}g$,

    At C, $\dfrac{36+10M_2-2M_1}{7}g$;

    each equal to 6 kg.
23. 0·46 m.
24. 60°.
25. 25 g N at 2·4 m from the end from which distances are measured; 15 g N at this end, 10 g N at the other end.

EXAMPLES 2.4 (p. 72)

1. $\frac{1}{8}$.
3. Mark off distances a, $2a$, etc., from M towards A, a being the distance of the mid-point of BD from C.

5. x must not be less than 7 kg.
    $x = 7\sqrt{2}$ kg.
7. 14 g N; $\frac{3}{14}$ m from centre.
9. $\frac{1}{2}$ cm; $W_0 = 4W_1$.

EXAMPLES 3.1 (p. 85)

1. 45°.
2. $\tan^{-1}\left(\frac{3}{2}\right)$; $1\frac{1}{4}$ times weight of rod at $\tan^{-1}\left(\frac{4}{3}\right)$ to the horizontal.
4. $W \sin 15°$; $W \cos 15°$, at 15° to the vertical.
8. $4\frac{3}{8}$ g N.
9. 5 g $\tan 40° = 4·2$ g N nearly. 10·8 g N at 67° 14′ to horizontal.

10. $\frac{1}{6}W$; $W\sqrt{37}/6$; at $\tan^{-1}$ 6 to the horizontal.
11. $W\sqrt{3}/3$; $2W\sqrt{3}/3$; at 60° to horizontal.
12. $2W\sqrt{3}/3$; $W\sqrt{3}/3$.
15. $7\frac{1}{4}$ g N.
16. 77·1 g N; 163·4 g N, 63° 39′ to the horizontal.
18. 0·46 m each.

19. ACB $= 90°$, and B is below C. Least pressure $= W \sin \alpha$.
21. $3\frac{1}{8}$ g N.
22. $20 \cdot 5$ g N, $4 \cdot 5$ g N.
24. Perpendicular to AB; $\frac{1}{2}w\sqrt{(4-3\sin^2\theta)}$.
25. 10 N; $5\sqrt{3}$ N.

26. $10\sqrt{3}$ g N on 30° plane; 10 g N on 60° plane; at 30° to horizontal.
27. $W/(1+8\cos^2\alpha)^{\frac{1}{2}}$, $2W/(1+8\cos^2\alpha)^{\frac{1}{2}}$.
28. $9w/4$, $15w/4$; $\frac{1}{4}(5\sqrt{5}-9)w$, $7w/2$.

EXAMPLES 3.2 (p. 100)

1. 33 g N; $30\cdot5$ g N, at $\tan^{-1}\frac{25}{56}$ to the horizontal.
2. $1\cdot5$ g N; $9\cdot1$ g N nearly, at $\tan^{-1}6$ to the horizontal.
4. $11\cdot4$ g N.
6. $\frac{1}{8}$ of the length of the beam from the end on the 30° plane.
8. $5\cdot25$ g N, 7 g N, vertically.
9. $R_1 = \frac{67}{56}W$; $R_2 = \frac{45}{56}W$.
10. $\sin^{-1}\left(\dfrac{m+n}{m}\sin\theta\right)$; $0\cdot73$.

11. Perpendicular to the line joining the centre of roller to edge of kerb.
12. Perpendicular to the line joining the centre of roller to edge of step, $1:2$. There is a tendency to slip at the step, but not at the ground.
14. 1476 N, 75° S. of E.; 4400 N.
15. $(8-\sqrt{3})$ g N.

EXAMPLES 3.3 (p. 110)

1. $W\sqrt{3}/6$ horizontal at the lowest joint; $W\sqrt{39}/6$ at $\tan^{-1}2\sqrt{3}$ to the horizontal at each of the upper joints.
2. $2W$ in AC; $W/2$ horizontal at B and D.
3. Tension in string $= 2W$. Reaction at B and D $= W\sqrt{3}/6$ horizontal.
4. W; W vertical at A and C; W horizontal at B.
5. $\frac{3}{4}W$.
6. Horizontal and vertical components of reactions are at B, $\dfrac{ab(a+b)}{2(a^2+b^2)}Mg$, $\dfrac{a^3-b^3}{2(a^2+b^2)}Mg$; at C, $\dfrac{ab(a+b)}{2(a^2+b^2)}Mg$, $\dfrac{a^3+2a^2b+b^3}{2(a^2+b^2)}Mg$; at A, $\dfrac{ab(a+b)}{2(a^2+b^2)}Mg$, $\dfrac{a^3+2ab^2+b^3}{2(a^2+b^2)}Mg$.

7. $W\left(1+\dfrac{b}{2a}\right)$; $W\left(2-\dfrac{b}{2a}\right)$; $\dfrac{c(a+b)}{2a\sqrt{(a^2-c^2)}}W$.
9. $2W$; the horizontal and vertical components of the reaction at the hinge are $2W$ and $\frac{5}{3}W$.
12. Horizontal and vertical components of the reaction at B or E are $\dfrac{W}{4\cos18}$ and $W\left(\dfrac{1}{4\sin18}-\frac{1}{2}\right)$.
13. $w\sqrt{7}/2$ at A and D; $w\sqrt{19}/2$ at B and C.
14. $0\cdot05$ m on side nearer B. The horizontal and vertical components of the stresses are: at A, $0\cdot84$, and $1\cdot63$ g N; at B, $0\cdot84$, and $1\cdot87$ g N; at C, $0\cdot84$, and $0\cdot37$ g N.
15. $W\cot\alpha$; $W\operatorname{cosec}\alpha$.

16. $\frac{1}{2}W(\tan \alpha +(a/c) \operatorname{cosec}^2 \alpha)$.
17. 2 g N, 2 g N.
20. 200 g N horizontal and vertical at both D and B.
23. 11·74 g N.
24. 0·6 m from B; 2·5 g N.

25. 2·5 g N horizontal.
26. $\cos^{-1} \{M/(2M+m)\}$.
29. (i) 1·875 g N;
    (ii) 5 g N. vertical;
    (iii) 1·875 g N.

EXAMPLES 3.4 (p. 120)

1. 0·74a from A on the side opposite to B.
2. At C.
3. 0·75 m.
4. 6·05 at 70° 42′ to AB on the side opposite to the square.
5. Divides BC in the ratio 1:2.
6. 24$\sqrt{2}$ N.
7. 30 N.
8. 57·6.
9. 2·37 P, it cuts BC at $\frac{7}{9}$BC from B, and makes an angle $\tan^{-1} \sqrt{15}/5$ with BC.
10. 1 from C to D, 1 from C to B, 1 from A to D.
    (i) 2 along BC; (ii) a couple of moment twice the side of the square.

11. A force of 21 from B to A; 54·5 nearly.
13. $5\sqrt{2}/4$ N; $5\sqrt{2}$ n.
14. 2 N from D to A.
15. 13, $\sqrt{173}$.
16. 2P parallel to AB bisecting AD.
18. 1 : 0·27 : 1 : 1·55.
19. 2 N parallel to DA, cutting BA produced at a distance 3AB from A.
20. 8 cm.
21. When P = Q they are equivalent to a couple.
22. $22y = 54x + 55$; $58y = 6x + 145$.
23. 7·6 N; 71·25 cm.

EXAMPLES 3.5 (p. 127)

8. Another circle.
9. 6DG where D is the mid-point of AB, and G the point of trisection of BC which is nearer to C.
10. The line bisecting AC at right angles, where C is the mid-point of AB.

11. The centre is at G, the mid-point of the line joining the mid-points of opposite sides of the quadrilateral; the radius is PG.
12. CF = AE.
16. $\sqrt{10}$ N at $\tan^{-1} \frac{1}{3}$ with AB, cutting AB produced at 1·36 AB from A.
21. Through B.
22. $l = 3-m$; 2 : 1.

EXAMPLES 3.6 (p. 134)

1. $\frac{1}{2}Wa$.
2. Vertical force 7·5 N; couple 18 N m.

5. Reactions on BC at B: vertical force 2W, horizontal force 2W, couple 5Wa; and at C vertical force W, horizontal force 2W, couple Wa where W = weight of AB, 2a = length of AB.

6. Two equal forces $12P/5$ forming a couple.
8. $k = -4$.

EXAMPLES 3.7 (p. 143)

1. $3\frac{3}{4}$; 5.
2. $(21\sqrt{3}/2)a$; where a is a side of the hexagon; 6 parallel to the force of 5.
3. $6F$, $\tan^{-1}(5\sqrt{5}-9/8)$ to AB; $3\cdot17$ of AB beyond A, $0\cdot27$ of BC beyond C.
5. $G/a\sqrt{3}$ where G^2 equals $G_1{}^2+G_2{}^2+G_3{}^2-G_2G_3-G_3G_1$ $-G_1G_2$.
6. $N-xY+yX = 0$.
7. 50; $0\cdot047$ m on BA produced, $0\cdot094$ m between A and D.
8. 1 N; $\sqrt{3}x-y-\sqrt{3}a = 0$.
10. $\sqrt{74}$ N at $\tan^{-1}(-\frac{7}{5})$ to AB; 36 units counter-clockwise.
11. $19\cdot05$ units at $1°\ 36'$ with the horizontal sides of square; $23\cdot11$ units anticlockwise.

10. kb parallel to AD, $\frac{3}{2}a$ from A; couple kab.

12. 6N in direction CO; $90\cdot9$ N cm.
13. 10, $5\sqrt{2}$, $5\sqrt{2}$ N.
14. $4\sqrt{5}/5$ N along BC, $16\sqrt{5}/15$ N along AC, $8\sqrt{5}/15$ N along AB.
15. 6 N inwards at mid-point of BC; 8 N outwards at mid-point of AC.
17. (i) $\sqrt{10}$ N at $\tan^{-1}3$ with AB; 6 N m anticlockwise.
(ii) $\sqrt{160}$ N at B; $\sqrt{90}$ N at C.
20. $1\cdot59$ N m.
22. $\sqrt{194}$; $5x+13y-17a = 0$.
24. $2P$, perpendicular to BD at a distance $\frac{1}{4}$ BD from D.

EXAMPLES 3.8 (p. 149)

1. $120\cdot7$ g N.
2. $\frac{40}{3}+6\sqrt{3}$, $\frac{40}{3}+10\sqrt{3}$, $\frac{220}{3}-16\sqrt{3}$ N at A, B, C respectively.

4. 20 g N.
5. $16\frac{2}{3}$ g N; 100 g N.
6. $14\cdot93$ g N.

REVISION EXAMPLES A (pp. 150–157)

1. 6, $3\sqrt{3}$.
3. $P\cos(\pi-\theta)$, $P\sin(\pi-\theta)$.
7. $\tan^{-1}\{(2-\sqrt{3})/\sqrt{3}\}$.
8. $(\sqrt{3}-1)/2$; $\tan^{-1}\sqrt{3}/(2\sqrt{3}-1)$.
11. 4 : 3.
12. $w/2\sqrt{3}$.
13. $W+wd/4b$; $W+w(1-d/4b)$.
14. $5W/2 \cos \pi/5$.
17. $\frac{1}{2}W$, $\frac{3}{2}W$.
18. 553 g N.
19. Ties: DA 117 g N, AC 224 g N. Struts: CD 168 g N, CB 280 g N.

22. kb parallel to AD, $\frac{3}{2}a$ from A; kab.
23. $(P+Q)\cos \alpha$, $(P-Q)\sin \alpha$.
24. $\sqrt{2}$.
25. $5M/12a$.
26. 1 N parallel to AB; $1\cdot2\sqrt{3}$ m; $1\cdot2\sqrt{3}$ N m; 1 N parallel to BA.
27. $10a$ in the sense DCBA; $3\sqrt{2}$ from C to A.
28. $10\sqrt{(4-\sqrt{3})}$ N at $\tan^{-1}(2-\sqrt{3})/\sqrt{3}$ with BA; 10 N m.
29. $(4\cos 36°-1)(6\cos 36°+1)$.

30. $4, -8\sqrt{3}; 18a\sqrt{3}$.

31. $\{4(M_1-M_2)^2+(M_2+M_3)^2\}^{\frac{1}{2}}$
$\div 2a$;

$\tan^{-1}\dfrac{M_2+M_3}{2(M_1-M_2)}; \dfrac{2aM_2}{M_2+M_3}$.

32. $M_1+M_2-M_3$.

33. (a) $\sqrt{5}: 2\sqrt{2}: \sqrt{5}$.
    (b) $\sqrt{5}: \sqrt{2}: \sqrt{5}$.

34. 15 kg; 1·43, 9·82, 18·75 g N.

35. $-5, 3, -4; -5, 7, -6$.

37. $mnW/(m^2+n^2)^{\frac{1}{2}}; \tan^{-1}(n/m)$.

39. $\frac{1}{2}W(\sqrt{2}+1)$ horizontally.

EXAMPLES 4.1 (p. 165)

1. Between the 5 and 4 N forces; 0·08 m from the former.
2. 9.7×10^3 g N at the left-hand end; 9.3×10^3 g N at the other end.
3. Between the 7 and 4 N forces, at 0·24 m from the former.
4. 1·08 m from the end.
5. 7·2 g N at end from which distances are measured, 7·8 g N at the other end.
6. 10·5 g N, and 8·5 g N.
7. 2·66 m from the 2 N force.

8. 4·8 m from the first 10-Mg load.
9. 6·3 m from the end nearer to the 4 kg; 8·2 g N on the end nearer the 4 kg, and 13·8 gN on the other end.
10. 5·46 m.
11. 44·375 g N at A; 105·625 g N at the support near B.
13. 2·5, 5·5, 2·8, 3·9 g N.
14. 6·2 N at 76° 6′ to AB, cutting AB between A and B at $\frac{6}{7}$ cm from A.
15. 5·9 N at 13° with 3 N force.

EXAMPLES 4.2 (p. 176)

1. 14·5 N; BC and CD, +12·2; AB and AD, +8·45.
2. 90 g N; 64 g N at 100° 36′ with DB.
4. AB, $-w$; BC, $-w$; CD, $-0.62w$; DE, $-0.38w$; EA, $-0.62w$; BE, $+0.6w$; BD, $+0.62w$.
5. 2·19 g N; 4·8 g N.
6. P = 25 g N; AB, +43·3; AC, -50; BC, +86·5.
7. AB and DC, -0.38; AD and BC, -0.76; BD, +0·66.
8. AB and AD, +18·75; BC and CD, +25; AE, -22.5; BE and ED, -31.25.
9. AB, -15; BC, -9; CD, -5.5; AC, +9·25 N.
10. BD, +10; BC and AD, -8.3; AB and CD, -5.5.
11. YD and YC, -1.73; AD and BC, -1.5; XD and XC, +0·86; XA and XB, -1.15; AB, -0.86.

12. Each force = 10, AB, BC, CD, DE, +20; AC and CE, -11.55; AE, +5·77.
13. 21·21 in each supporting string; 17·32 in AB and AC; 6·34 in BC.
14. AE, +358; AD, -320; EC, +134; ED, +223; CD, -150; CF, +268; FD, +56; FB, +324; DB, -145. Reactions, 160 at A, 290 at B.
15. Reactions, 27·7 at A, 22·3 at B; AC, -28.3; AD, -7.7; CD, +20; AB, -20; BD, +3·6; DE, +17·3; BE, -15; BF, -20; EF, +17·3.
16. Reactions, 27·5 at A, 22·5 at B; AC, +55; AF, -48; CD, +25; CF, +30; DF, -25; DE, +25; EF, +20; EB, +45; FB, -39.0.
17. Reaction at D = 22·4; at A = 20; AB, -20; BD, +14; BC, 14; DC, +10.

18. AB, $+11 \cdot 5$; BD, $+34 \cdot 5$; DE, $+11 \cdot 5$; DF, $+23 \cdot 0$; CD, $+11 \cdot 5$; EF, $-11 \cdot 5$; CE, $-11 \cdot 5$; AC, -23; CD, $-11 \cdot 5$.

19. OB, -60; AB, $+42 \cdot 3$; AC, $+30$; BD, -20; BC, $-14 \cdot 2$; CD, $+14 \cdot 2$; CE, $+10$; DE, $-14 \cdot 2$. Reaction at A $= 66 \cdot 9$.

20. AF, -40; AB, -20; BC, $+20$; BF, $+28 \cdot 3$; FE, -20; CE, $+14 \cdot 1$; CD, $+10$; ED, $-14 \cdot 1$; CF, -20.

21. Reaction at A $= 126$; AC, $+56 \cdot 5$; BC, -120; AD, $+80$; CD, $-56 \cdot 5$; CE, -40; DE, $+28 \cdot 2$; DF, $+20$; EF, $-28 \cdot 2$.

22. AB, $-0 \cdot 5W$; AC, $+1 \cdot 12W$; BC, $+W$; EB, $-1 \cdot 12W$; CD, $+3 \cdot 3W$; EC, $-1 \cdot 4W$.

23. TB, $1 \cdot 03$; TC, $1 \cdot 10$; TD, $1 \cdot 25$; TE, $1 \cdot 4$.

24. AC, $-4 \cdot 62$; BC, $+2 \cdot 3$; BD, $+5 \cdot 63$; CD, -4.

25. AB, $+18$; BF, $+18$; BG, $-11 \cdot 3$; FG, -18; CG, $+15$.

26. AF and BG zero; AD, $+0 \cdot 66$; DF, $-0 \cdot 94$; DE, $+0 \cdot 66$; FE, $-0 \cdot 37$; ED, $+0 \cdot 37$; FG, $-0 \cdot 5$; EC, $+0 \cdot 34$; CG, $-0 \cdot 47$; BC, $+0 \cdot 34$.

27. Reaction, $0 \cdot 6$ at A, $0 \cdot 72$ at B; AD, $-1 \cdot 1$; AC, $+1 \cdot 32$; DE, -1; CD, $-0 \cdot 20$; EC, $-0 \cdot 20$; BC, $+0 \cdot 5$; EB, $-1 \cdot 1$.

28. CE, -2; EF, zero; AD, $+14$; DC, $+10$; AE, $-12 \cdot 1$; DE, $+4$; EB, $-8 \cdot 7$; CF, $+10$; BF, $+10$.

29. $33°$, $+2 \cdot 6$.

30. AC, $+4 \cdot 18$; AE, $+14 \cdot 26$; ED, $+15 \cdot 7$; CD, $-11 \cdot 83$; BC, $-16 \cdot 01$; CE, $-3 \cdot 53$.

31. AC, $-43 \cdot 2$; BC, $+30$; BD, $+28 \cdot 2$; CD, -20; CE, $-14 \cdot 1$, DE, $+10$.

32. AB, $+20$; AC, $+7 \cdot 0$; BC, -10; BD, $+14 \cdot 1$; CE, $-7 \cdot 0$; DE, $+10$; CD, zero.

33. Supporting forces, 25 at A, 75 at B; AB, $+14 \cdot 5$; AD, $-28 \cdot 9$; BD, $+50$; CD, $-57 \cdot 8$; BC, $-115 \cdot 6$.

34. 5 g N.

35. $25 \cdot 2$ units on D, $29 \cdot 8$ units on F, AD, $+26$; DE, $-5 \cdot 5$; AE, $-16 \cdot 2$; AB, $+22 \cdot 4$; BE, $+12 \cdot 5$; BC, $+22 \cdot 4$; EC, $-15 \cdot 5$, EF, $-6 \cdot 1$; CF, $+30 \cdot 5$.

36. AC and BE, $+7 \cdot 1$; AD and BD, $+20$; CD and DE, $+5$; AB, $-22 \cdot 3$.

EXAMPLES 4.3 (p. 191)

1. CD, $+3 \cdot 5W$; GD, $+0 \cdot 56W$; GH, $-3 \cdot 75W$.

2. AB, $+W$; AF, $+0 \cdot 52W$; AD, $-1 \cdot 5W$; BF, $+0 \cdot 57W$; DF, $+0 \cdot 57W$; DE, $-1 \cdot 15W$; where $2W$ is the weight supported at B.

3. AB, $+28$; AE, -25; BE, $-13 \cdot 5$; ED, $-15 \cdot 0$.

4. AE, $+12 \cdot 3$; AF, $-8 \cdot 9$; EC, $+11 \cdot 3$; EF, $+1 \cdot 7$; FC, $-4 \cdot 8$; FG, $-4 \cdot 4$; CG, $-9 \cdot 5$; CD, $+14 \cdot 7$; GD, $+3 \cdot 6$; BD, $+15 \cdot 8$; BG, $-13 \cdot 6$. Reaction at B, $6 \cdot 3$ vertical. Reactions at A, $5 \cdot 1$ vertical, 2 horizontal.

EXAMPLES 5.1 (p. 204)

1. $\tan^{-1}(\frac{3}{2})$.
3. $\frac{2}{5}$ of the weight of the ladder.
11. $3\frac{1}{3} g$ N.

12. Zero; $\frac{1}{6}W$. The friction will be limiting at both wall and ground.
13. $\sqrt{3}W/6$; $50 g$ N.
15. The thirteenth rung.

EXAMPLES 5.2 (p. 208)

2. $2\tan^{-1}(\frac{1}{16})$.
4. $a/2h < \mu$.
9. (i) by sliding; (ii) by tilting.

11. $\mu \not< \frac{1}{2}(w/W)$; $\sqrt{2}(a/b)$.
12. $\frac{2}{3}$, $7a/2$.

EXAMPLES 5.3 (p. 211)

1. $\frac{30}{59}$; the longer rod will slip.
2. $\frac{1}{2}$.

4. At A, $\dfrac{W(3+\cos 2\alpha)}{4\cos\alpha}$;

at C, $\dfrac{W(3\cos 2\alpha+1)}{4\cos\alpha}$;

at A, $5\sqrt{3}/7$.

EXAMPLES 5.4 (p. 215)

1. $\dfrac{\mu}{\cos\alpha+\mu\sin\alpha}$.
2. $P = W\sqrt{(\mu^2-n^2)}$,

   $\tan^{-1}\dfrac{n}{\sqrt{(\mu^2-n^2)}}$.

6. $W_1/W = \dfrac{\mu}{\cot 2\theta-\mu}$.
7. $\frac{1}{7}$.
8. $\dfrac{m(\sin\alpha-\mu\cos\alpha)}{(M+m)(\mu\sin\alpha+\cos\alpha)}$.
17. $P = W/\sqrt{(2-1)}$; not.

18. $\mu = 1/(3\mu'+4)$.
22. $\frac{1}{2}$.
24. $P/\sqrt{2}$; $\mu \not< P/(P+W)$.
25. $(Wa/r)\sin^2\theta$.
26. $(l-x)w+W$.
28. $\frac{1}{2}W\sec(\alpha-\lambda)$.
30. (i) $W\cos 2\lambda\operatorname{cosec}\lambda$, 2λ.
33. $\frac{1}{2}$; $\frac{1}{4}W\sqrt{13}$ at $\tan^{-1}\frac{2}{3}$ to the horizontal.
36. At A.

EXAMPLES 5.5 (p. 230)

2. $\tan^{-1}\left(\dfrac{a\sin\theta}{a\cos\theta+h}\right)$.
3. The smaller of $\dfrac{\mu}{\mu'}W$ and $\dfrac{a}{\mu'h}W$.
8. $\dfrac{4-\sqrt{7}}{3}$; $\dfrac{4-\sqrt{7}}{9}$.
9. $\dfrac{W}{2\sin\beta(\mu\cos\alpha-\sin\alpha)}$.
10. $X = \dfrac{Wa\sin\lambda}{b-a\sin\lambda}$.

12. $R\lambda\pm\dfrac{m}{M+m}l$ on the two sides of C. The $+$ sign on the side from which m is suspended.
22. $60°$.

29. $\dfrac{W}{g}\left(\dfrac{a'g-fh}{a+a'}\right)$ on back, and

$\dfrac{W}{g}\left(\dfrac{ag+fh}{a+a'}\right)$ on front.

When $a = a'$ the maximum retarding force for back-wheel braking is $\dfrac{\mu a W}{2a+\mu h}$, and for front-wheel braking $\dfrac{\mu a W}{2a-\mu h}$. The first is less than $\frac{1}{2}\mu W$, i.e. less than $\frac{1}{2}$ the full braking force.

43. $w+\dfrac{nw}{4}\left(2-\dfrac{x}{l}\right)$ and

$w+\dfrac{nw}{4}\left(2+\dfrac{x}{l}\right)$,

when $2l$ is the length and w the weight of each ladder.

EXAMPLES 6.1 (p. 250)

1. $32\cdot875\,g$ N.
2. $9\cdot5\,g$ N.

4. $25\,g$ N, $25\,g$ N, $50\,g$ N, $100\,g$ N, $200\,g$ N. Difference is the effort, $25\,g$ N.

EXAMPLES 6.2 (p. 258)

1. $10\,g$ N.
2. $7\frac{1}{2}\,g$ N.

3. $50\,g$ N.
4. 50; $16\frac{2}{3}\,g$ N.

EXAMPLES 6.3 (p. 260)

1. 4 strings; 2 units.
2. 3 units, M.A. = 6.
3. 5; 24 N.
4. 85·4 units, 31·74 per cent, 38·09 per cent, 40·82 per cent.
5. $W = 15P+11w$.
6. 672 N.
7. $7\,g$ N.
8. $129\,g$ N, $129\,g$ N.
9. $50\cdot8\,g$ N.
10. $P/Q = \frac{1}{2}$.
11. 187·5 N.

12. 0·68.
13. The side wound on the 30 cm drum.
14. $\frac{9}{8}W$.
15. $\frac{1}{11}$.
16. 0·23.
17. 4000π N.
18. 1·88 cm.
19. 0·10.
20. $Wb/4a$; $b/4an$.
21. $40\,g$ N, $196\,g$ N.
22. $Pc = \frac{1}{2}W(a-b)$; 864 N.

EXAMPLES 7.1 (p. 269)

1. 3·1 cm.
2. 2·5 m.
3. 2·828 cm from BC; $\frac{1}{3}$ cm from AD.
4. 5·6 cm; 4 cm.
5. $\sqrt{171}/2$ cm.
6. 21·9 cm nearly.
7. $2\frac{2}{3}$ m.

9. Half the length of a side on the line joining the 15 kg and 3 kg masses.
10. 4 cm, towards the 5 kg mass on the line joining this mass to the 2 kg mass.
11. 3·15 cm.
12. 8·4 m; 3 m.

EXAMPLES 7.2 (p. 278)

1. $9\frac{1}{3}$ cm.
2. $\frac{6}{35}$ cm on diameter through hole.
3. $3\sqrt{2}/17$ cm from centre on diameter bisecting line joining centres of holes.
4. $6\frac{2}{3}$ cm; $3\frac{1}{9}$ cm.
5. $5\frac{2}{7}$ cm; $6\frac{2}{7}$ cm.
7. On AC at 6·07 cm from A.
9. $\tan^{-1}(\frac{16}{33})$.
11. $\frac{16}{45}$ of its length up the median.
12. 46·4 cm.
13. 15·05 cm, 15·07 cm.
15. 72 cm.
18. $55\sqrt{3}/18$ cm.
20. 2·0 cm.
21. 3·12 m.
22. In each case $\frac{1}{4}$ of the way up the line joining the vertex to the centre of gravity of the base.
24. $26\sqrt{2}/9$ cm up the median drawn to the side of 8 cm.
28. 13: 54.
31. $\tan^{-1}[\frac{3}{10}(5+\sqrt{5})]$.
33. At the point (2, 3) taking the outer edges as axes.
34. $\frac{185}{376}$ of an edge.
35. $\frac{11}{56}\sqrt{\frac{2}{3}}a$ from the base.
36. $\frac{1}{3}b\frac{3a+c}{2a+c}; \frac{3a^2+3ac+c^2}{3(2a+c)}$.

37. 3·15 m from the free end of the first plank, and at the height of the upper surface of the second plank.
38. 0·52a from the top.
39. At a distance $2mca^2/Mb^2$ from the axis.
   $$\frac{m}{M} \not> \frac{b^2(a-b)}{2a^2c}.$$
41. 3 cm from BC.
43. (i) At the intersection of the medians of ABC.
   (ii) 1·9 mm.
45. $\frac{110}{29}$ cm; $\frac{126}{29}$ cm.
46. AC = 3 cm.
48. $\frac{4}{3}$ cm; $\frac{5}{3}$ cm; 3 cm.
49. of $\frac{513}{1070}$ the height from base. At $\tan^{-1}(\frac{2675}{6134})$ to the vertical.
50. $\dfrac{5\sqrt{3}-\sqrt{15}}{12}a$ from the base.
51. $c+\dfrac{5a^2b}{8(3ab+3ac+2bc)}$ from the base.
   $\dfrac{3}{2}a+\dfrac{3a^2b(2-\sqrt{3})}{8(3ab+3ac+2bc)}$ from the end to which the lid of length a is attached.
52. $\dfrac{a}{n}\cos\dfrac{\pi}{2n}\left[\sin\dfrac{\pi}{2n}+\sin\dfrac{3\pi}{2n}+\ldots\right]$
   where a is the radius of the circumcircle.

EXAMPLES 7.3 (p. 286)

3. $\dfrac{(a^2+ab+b^2)\sqrt{[4c^2-(b-a)^2]}}{6c(a+b)}$ from DA;
   $\dfrac{4ac^2+2bc^2-a^3+b^3}{6c(a+b)}$ from the other axis.

4. 0·50, 2·69.
6. $\frac{5}{3}$ cm from each diagonal.

EXAMPLES 7.4 (p. 297)

2. 24 cm from C.

3. $\dfrac{(a^2+ab+b^2)\sqrt{[4c^2-(b-a)^2]}}{6c(a+b)}$ from DA;

$\dfrac{4ac^2+2bc^2-a^3+b^3}{6c(a+b)}$ from the other axis.

5. $\frac{1}{2}(3-\sqrt{5})$.

6. $\dfrac{a^2+ax+x^2}{3(a+x)}$ and $\dfrac{a(2a+x)}{3(a+x)}$, where $x = AE$. AE not less than $\frac{1}{2}(\sqrt{3}-1)a$.

7. With BC or DE in contact with the plane; with BD or EC in contact with it if DE is above BC.

10. Yes.

11. Not less than $3:1$.

12. 1·59.

13. $(n-1)c/2$ from the centre of lowest block and at a vertical height of $na/2$.

14. $r\sqrt{3}$.

15. 53·9 cm; 29°.

EXAMPLES 7.5 (p. 309)

2. $\sqrt{2}:1$.

7. $3a/2\pi$ from the centre along the middle radius, a being the radius of the disc.

9. $\frac{13}{14}r$.

10. $\sin^{-1}(\frac{3}{4}\pi \sin \theta)$.

11. At $a/(2\pi-3\sqrt{3})$ from the centre.

12. $\frac{1}{4}W \tan \alpha$.

13. 810 kg, 6·3 m from the butt.

14. $(\pi^2-1)^{-\frac{1}{2}}$.

16. $\sin^{-1}\frac{1}{3}$.

17. 6, 7 and 8.

19. $\dfrac{2}{3}\dfrac{6a^2+3ax-x^2}{\pi(2a+x)+4(a+x)}$ from the centre of the circle.

20. In the line joining the point where the planes meet the sphere at a distance $3(4\sqrt{3}+3)/104$ of the radius from the centre.

21. (i) $38:7$. (ii) $2a\sqrt{2}/3(4-\pi)$.

22. 62° 29′.

23. $a/4(2\pi-1)$, $29a/12(2\pi-1)$.

24. $3Wc/l$, $3\sqrt{2}Wc/l$, $W(1-3c/l)$.

25. $\tan^{-1}\frac{3}{4}$.

REVISION EXAMPLES B (pp. 313–325)

1. $\frac{1}{4}\sqrt{3}$.

4. μW.

5. $W/2\mu \sin \alpha (\sin \alpha - \mu \cos \alpha)$; $W/2\mu \sin \alpha (\mu \sin \alpha + \cos \alpha)$.

7. $(W-w)(1-\mu)/\sqrt{2}\mu$; $(W-w)(1+\mu)/\sqrt{2}\mu$.

8. AB, compression $= 35$ g N; BC, compression $= 88$ g N; CA, tension $= 173$ g N; CD, compression $= 135$ g N; force at A $= 153$ g N; at 60° to AD.

9. $\sqrt{3}$ (A), $\sqrt{7}$ (B); AD (tension) $2\sqrt{3}$, CD (compression) 4, AC (compression) 2, BC (compression) $2\sqrt{3}$, AB (tension) 1.

10. AB, CD compression $= 10\sqrt{3}$; BC, DA, tension $= 20\sqrt{3}$; AC, tension $= 40$.

11. Tension in string $= w/\sqrt{3}$; OA, thrust $= 2w/\sqrt{3}$; OB, tension $= \frac{1}{2}w$; OC, thrust $= w/2\sqrt{3}$; OD, thrust $= \frac{1}{2}w$; OE, tension $= 2w/\sqrt{3}$.

12. Stress in rods $= W/\sqrt{3}$, $W(1+1/\sqrt{3})$, $W(\sqrt{3}+1)$; tension $= W(\sqrt{6}+\sqrt{2})/2$; BC, CD in tension.

13. AD, AE tension $= P/\sqrt{2}$; BD, BE, tension $= 3P/\sqrt{2}$; CD, CE thrust $= P\sqrt{5}$; CB, thrust $= 2P$.

14. $16° 6'$; BC (thrust) 8.3 g N, CA (tension) 11.1 g N, AB (tension) 33.3 g N.

15. DC 10, DE 10, EC 8.7, EF 15, CB 15, EB 0, BA 30, FA 30, FB 26.0, in g N to 1 dp.

16. 6, 8, $2\sqrt{3}$ tension, $2/\sqrt{3}$ compression, $7/\sqrt{3}$ compression.

17. $2\sin^{-1}(\frac{1}{3})$.

18. On AC, distant $\pi ab^2\sqrt{2}/2(4a^2-3\pi b^2)$ from O.

19. $(\frac{7}{4}, \frac{19}{4})$, $(\frac{47}{28}, \frac{153}{28})$.

20. $\tan^{-1}(4/3\pi)$.

24. $(2a^2+ab)/(3a+3b)$, $(a^2+ab+b^2)/(3a+3b)$.

27. $\tan^{-1}(\frac{24}{13})$.

32. 15.9 N.

34. 4; 3.45; 0.86. $\{W+(2^n-1)w\}/2^n$.

35. $\frac{1}{4}(W+w)$; $4W/5(W+w)$.

36. 68 N; $49/54\pi$, $49/51\pi$, $49/50\pi$.

37. $M \geqslant Wa$.

39. Tension: AB $3\sqrt{3}$, BC $3\sqrt{3}$, BF $3\sqrt{3}$, CD $\sqrt{3}$, CE 2. Compression: CF 2, FE $2\sqrt{3}$, ED 2.

   $\sqrt{19}$ at $\tan^{-1}\left(\dfrac{7}{3\sqrt{3}}\right)$ to FE downwards.

40. Tension: PQ 2 g N, QR 1.79 g N, SP 1.73 g N, QS 4 g N. Compression RS 3.46 g N.

41. 7.67 at $32°$ to vertical; 16.5. Tension: AB 7.55, BC 7.55, CD 11.55. Compression: DE 5.77, EA 7.75, EB 7.75.

42. Tension: 4.7 units; Reaction: 5.0 units. Thrusts: PQ 2.7, QR 3.1, RS 2.8, ST 3.8, RT 2.2, QT 3.0, PT 6.6.

43. BA, BC, BD.

44. $\frac{1}{8}Wa\sin\theta$.

45. $\frac{1}{3}\sqrt{2}$.

47. $\mu=1$.

48. $\sqrt{2}-1$, $3-2\sqrt{2}$.

49. $W-P$, $W+P$.

EXAMPLES 8.1 (p. 336)

1. B.M. increases from zero at A to 12 g at C, and then to 20 g at D, where it is a maximum; it then decreases to zero at B.

2. Shear is 16 g up from first support to centre and 20 g down from centre to second support. Bending moment increases uniformly from zero at first support to 80 g at centre, and then decreases uniformly to zero at the second support.

3. If x is measured from free end B.M. $= 7\frac{1}{2}-x(0\leqslant x\leqslant 5)$ and $5-\frac{1}{2}x(5\leqslant x\leqslant 10)$ $S.F.=1$ and $\frac{1}{2}$.

4. (i) $1-x/10$ and $x/10$ unit; (ii) $-x/10$ if $x<2$, $-(1+x/10)$ if $x>2$; (iii) $-4x/5$ is $x<2$, $x/5-2$ if $x>2$; (iv) $x=2$.

5. Shear A to C $+1.3$, C to D $+0.3$, D to B -1.7; B.M. A to C $-1.3x$, C to D $-3-0.3x$, D to B $-17+1.7x$.

6. 13.6 at A and 13.3 at B; -4.4 and -93.1.

7. Shearing force $= -1.05$. Bending moment $= -43.4$.

8. Shearing force $= -1.91$. Bending moment $= -25.6$.

9. Between C and D at $\frac{102}{113}$ m from D.

10. 13 and 15 units.

EXAMPLES 8.2 (p. 342)

1. Shearing force $= -6.3$ g.
Bending moment $= -5.635$ g.

2. Maximum at 1.2 m from A, where its value is 1.8. The value at B is 1.6, and it decreases uniformly from there to zero at C.

3. Shear $+2$ from A to C, zero C to D, and -4 from D to B. B.M. increases from 0 at A to 8 at C, then constant to D, and decreases to zero at B. In second case shear curve is a straight line from $+3$ at A to -3 at B, and B.M. curve is a parabola with maximum value 9 at mid-point.

4. $\dfrac{W}{2}\left(\dfrac{l}{4}-a\right), +\dfrac{W}{2l}\left(\dfrac{l}{2}-a\right)^2$;

   $a = l\left(1-\dfrac{\sqrt{2}}{2}\right)$.

5. 35 cm from end nearer 4 kg mass; 202.5 g.

6. Shear A to C, $+W$ to $+2W$; C to D, $-\frac{9}{4}W$ to $-\frac{5}{4}W$; D to B $-\frac{1}{4}W$ to $-\frac{3}{4}W$. B.M. values at x from A.

   A to C, $W\left(x+\dfrac{x^2}{2a}\right)$;

   C to D, $W\left(\dfrac{x^2}{2a}-\dfrac{13}{4}x+\dfrac{17}{4}a\right)$;

   D to B, $W\left(\dfrac{x^2}{2a}-\dfrac{9}{4}x+\dfrac{9}{4}a\right)$.

7. If x is measured from C the value of the shearing force and bending moment for sections CD, DA, etc., are $-(x+2)$; $\frac{1}{2}x^2+2x$;

   $-(x+4)$; $\frac{1}{2}x^2+4x-6$;
   $-(x-18)$; $\frac{1}{2}x^2-18x+126$;
   $-(x-17)$; $\frac{1}{2}x^2-17x+118$;
   $-(x-13)$; $\frac{1}{2}x^2-13x+82$;
   $-(x-12)$; $\frac{1}{2}x^2-12x+72$.
   All multiplied by W.

8. (a) S.F. $= -W\left(\dfrac{1}{2}+\dfrac{x}{16}\right)$
   $$(0<x<4)$$
   $= -W\left(-\dfrac{3}{4}+\dfrac{x}{16}\right)$
   $$(4<x<12)$$
   $= -W\left(-1+\dfrac{x}{16}\right)$
   $$(12<x<16)$$
   B.M. $= \left(\dfrac{1}{2}x+\dfrac{x^2}{32}\right)W$
   $= \left(5-\dfrac{3}{4}x+\dfrac{x^2}{32}\right)W$
   $= \left(8-x+\dfrac{x^2}{32}\right)W$

   (b) S.F. $= -Wx/16$
   $$(0<x<4)$$
   $= \left(\dfrac{3}{4}-\dfrac{x}{16}\right)W$
   $$(4<x<8)$$
   $M = Wx^2/32 (0<x<4)$
   $= \left(3-\dfrac{3}{4}x+\dfrac{x^2}{32}\right)W$
   $$(4<x<8)$$

9. $\frac{85}{36}W$. Diminished.

10. Shearing force, $-xw+\frac{1}{2}aw$ for AB. For BC, $-xw+\frac{1}{2}aw$ from B to D, $-xw+2aw$ from D to C. Bending moment, $\frac{1}{2}(x^2-ax)w$ for AB. For BC, $\frac{1}{2}(x^2-ax)w$ from B to D, $\frac{1}{2}(2a-x)^2w$ from D to C. x measured from end of rod.

11. 9 units on each support. Shearing force curve consists of three straight lines, $y = \pm 3$ between ends and supports and $y = -x+9$ between supports. Bending moment curves are
    $y = 3x,$
    $y = \frac{1}{2}(x-9)^2-9,$
    $y = 3(18-x).$

12. Shearing force varies from $+0\cdot8w$ at A to $-9\cdot2w$ at B, then $+10w$ to $+4w$ at C. Bending moment,

$\left(\dfrac{1}{2}x^2-0\cdot8x\right)w$ from A to B,

$\left(\dfrac{1}{2}x^2-20x+192\right)w$ from B to C.

13. S.F. $= +(81-9x)$ g
$\qquad (0<x<3)$
$\quad = +(45-9x)$ g
$\qquad (3<x<10).$
B.M. $= (-81x+9x^2/2)$ g
$\qquad (0<x<3)$
$\quad = -9(12-x)(2+x)$ g $/2$
$\qquad (3<x<12).$

14. S.F. $= +(9-3x)$ g
$\qquad (0<x<8)$
$\quad = (-3x+36)$ g
$\qquad (8<x<12).$
B.M. $= 3x(x-6)$ g $/2$
$\qquad (0<x<8)$
$\quad = 3(12-x)^2$ g $/2$
$\qquad (8<x<12).$

16. (a) 27, 9
$\quad$ (b) $4\cdot5$, 0.

17. S.F. $= W\left(\dfrac{3}{4}-\dfrac{x}{l}\right)\left(0<x<l\right)$
$\qquad = -\tfrac{1}{4}W \quad (l<x<2l)$
B.M. $= Wx(2x-3l)/4l$
$\qquad (0<x<l)$
$\qquad = \tfrac{1}{4}W(x-2l)$
$\qquad (l<x<2l)$

18. S.F. $= -1 \quad (0<x<6)$
$\qquad = 14\cdot5-x \ (6<x<13\cdot5)$
$\qquad = 12\cdot5-x \ (13\cdot5<x<21)$
$\qquad = 2 \ (21<x<24)$
B.M. $= x \ (0<x<6)$
$\qquad = 75-14\cdot5x+\tfrac{1}{2}x^2$
$\qquad (6<x<13\cdot5)$
$\qquad = 48-12\cdot5x+\tfrac{1}{2}x^2$
$\qquad (13\cdot5<x<21)$
$\qquad = 2(24-x)$
$\qquad (21<x<24)$

19. (a) 0, $-18\tfrac{1}{2}$, -34, $-46\tfrac{1}{2}$, -42, $-34\tfrac{1}{2}$, -24, $-10\tfrac{1}{2}$, 6, 0.
$\quad$ (b) 7 units at 6 m from end, 3 units at other end, and uniform load of $\tfrac{3}{4}$ unit per m over 16 m length. Other support 16 m from end.

20. $S_p = S_q+wb+W$;
$M_p = M_q+\tfrac{1}{3}Wb+\tfrac{1}{2}wb^2.$
$2kh^3/q\sqrt{3}.$

EXAMPLES 8.3 (p. 357)

1. $2:1.$
5. Weight of 27 Mg. The portions of chain starting from the middle link are inclined at angles $\tan^{-1}$ $(\tfrac{1}{3})$, $\tan^{-1}$ $(\tfrac{2}{3})$, and $45°$ to the horizontal.
6. $2\cdot4$ m above roadway. Slopes of links starting from lowest are $\tan^{-1}$ $(\tfrac{1}{5})$, $\tan^{-1}$ $(\tfrac{2}{5})$, $\tan^{-1}$ $(\tfrac{3}{5})$, $\tan^{-1}$ $(\tfrac{4}{5})$, $\tan^{-1}$ (1), $\tan^{-1}$ $(\tfrac{6}{5})$.

8. Middle, $\dfrac{an(n-1)}{2k}W.$
$\quad$ End, $\dfrac{n\sqrt{(4k^2+a^2(n-1)^2)}}{2k}W.$

9. $3\cdot21$ g N; $2\cdot69$ g N; $3\cdot84$ m; $3\cdot23$ m.

EXAMPLES 8.4 (p. 371)

1. $\dfrac{4l}{3n^2}.$

7. $50a$ and $14a$.
$\quad 48a \ \sinh^{-1}\ (\tfrac{7}{24}).$

10. $l\sqrt{10}$.
11. 49 m.
13. $a/\sqrt{3}$.
14. 5·7 m, 10·3 m.

16. 0·62.
22. Weight of $120/\sqrt{61}$ Mg.
27. $2\cdot4\times10^6$ N.
28. 2·2.

EXAMPLES 9.1 (p. 388)

3. $W(\frac{1}{2}l \cot \theta - r \sec^2 \theta)$.
7. $2W$.
9. $\sqrt{3}W$.
14. $2\sqrt{3}W$.

19. $\dfrac{a\sqrt{3}}{9\sqrt{(3l^2-a^2)}}W$.

28. $\dfrac{\sqrt{17}}{2}W$ at $\tan^{-1} 4$ to the horizontal.

EXAMPLES 9.2 (p. 399)

1. $\tan^{-1}(\frac{1}{8})$; stable.
9. $Mga(1-\cos\theta)$
   $-2\sqrt{2}mga(1-\cos\frac{1}{2}\theta)$.
11. $Mga(1-\cos\theta)$
    $-4mga\sin\frac{1}{2}\alpha(1-\cos\frac{1}{2}\theta)$.
15. BC horizontal, unstable; AC horizontal, unstable; AB in contact with one plane, stable.

19. $3b>5a$.
20. $\dfrac{El}{2}\left(\dfrac{1-\cos\theta}{\cos\theta}\right)^2 - \dfrac{wl}{2}\tan\theta$.
26. $\sin^2\theta = \frac{1}{12}$; unstable.
28. Unstable.
29. Stable.

EXAMPLES 9.3 (pp. 423–444)

14. In DR, $Wa/2b \cos\theta$;
    in AP, $W/2\sin\theta$;
    in AQ, $\dfrac{W}{2}\left(\dfrac{1}{\sin\theta}-\dfrac{a}{b\cos\theta}\right)$;
    (i) $\tan\theta>b/a$;
    (ii) $\tan\theta<b/a$;
    (iii) $\tan\theta = b/a$.
16. $\dfrac{W(2a\sin\theta-c\,\mathrm{cosec}^2\theta)}{a\cos\theta}$,
    B and D move outwards.
21. 18·7 g N.
26. $\dfrac{1}{10}a\left(3-\dfrac{20}{\sqrt{101}}\right)$.
27. $E^2r^2 = 2W\,Elr\sin\alpha\cos\theta$
    $+W^2(\mu^2\cos^2\alpha-\sin^2\alpha)$,
    taking the hole as origin and line of greatest slope as initial line.
31. $\dfrac{w(a+a\mu+c\mu\sin\theta)}{a(1+\mu)}$.

32. $\dfrac{W}{\sqrt{6}}$ and $\dfrac{W}{3\sqrt{6}}$.
43. $\dfrac{\sqrt{6}}{2}W$ in AB, AC, AD.
    $\dfrac{\sqrt{6}}{6}W$ in BC, CD, DB.
52. $\dfrac{1}{4}T\dfrac{AB}{XY}$ in each. Tension in CD, thrusts in the others.
61. At a depth
    $l\cot\alpha\left(1+\dfrac{W\cot\alpha}{\lambda\pi}\right)$
    below the vertex.
63. The line joining the centre of the circle to B makes an angle
    $\cos^{-1}\left(\dfrac{3a^2-4l^2}{3a^2}\right)^{\frac{1}{2}}$,
    with the upward vertical in the upper position, and with the downward vertical in the lower position.

68. (i) on back

$$\tfrac{1}{2}W\left(\frac{2h}{a}\sin\alpha+\cos\alpha\right),$$

on front

$$\tfrac{1}{2}W\left(\cos\alpha-\frac{2h}{a}\sin\alpha\right).$$

(ii) the same values interchanged; 0·38.

69. $\tfrac{3}{2}W$.

72. 2·68 g N on B and C; 4·64 g N on A.

73. About a point $\tfrac{1}{2}\sqrt{2}$ times the length of the rod from the end to which the force is applied.

74. $W+(l-x)w$.

79. (a) 22° 01′, (b) 20° 33′.

81. $12W/25$, $36W/25$, $\tan^{-1}(17/156)$ with horizontal.

84. $\tfrac{1}{2}W(1+3\tan\phi)/(1+\tan\phi)$.

87. $Wl(2\mu\sin\theta-\cos\theta)$.

91. $4\sqrt{13}$.

95. $\tfrac{1}{6}W(2c\cos\theta-3d\cos2\theta)$. Stable.

EXAMPLES 10.1 (p. 461)

1. 20 units at 60° to AB.
2. $AB^2=(a_1-a_2)^2+(b_1-b_2)^2$.
3. (i) $3i+4j$, $-5i+25j$, $22i-40j$.
   (ii) 7, -16. (iii) $(2i+3j)/\sqrt{13}$, $(3i-4j)/5$, $(-2i+5j)/\sqrt{29}$.
4. $4i+3j$.
14. $\tan^{-1}7$, $-4/7$.
15. $r=(1+k)a+(1-k)b$.
16. $2:1$.
21. $r=(1+3m)i+(2-3m)j+(3+2m)k$; $(-3\tfrac{1}{2},6\tfrac{1}{2})$.

23. (i) $(2-4m)i+(m-1)j+(3+2m)k$, (ii) $(2m_1+3m_2)i+(2m_2-m_1)j+(3m_1-4m_2)k$.
24. $(2i-j+3k)/\sqrt{14}$; $a=b=c$.
27. $\tfrac{1}{2}(b^2c^2+c^2a^2+a^2b^2)^{\frac{1}{2}}$; $bc:ca:ab$.
28. 55° 45′, 0·27a^2.
29. $4\sqrt{2}$, $4\sqrt{2}$, $2\sqrt{5}$ cm, 41° 29′.
30. 2, $\tfrac{1}{2}\sqrt{13}$ cm, 25° 38′.
31. $r=\tfrac{1}{2}ta+\tfrac{1}{2}(1-t)(b+c)$; $r=\tfrac{1}{4}(a+b+c)$.
32. $\pm(12i+16j)/5$.

EXAMPLES 10.2 (p. 473)

1. $2ti+3j$; $2i$. 5 units at $\tan^{-1}\tfrac{3}{4}$ to axis of x.
2. $t=\tfrac{3}{2}$. $i-3j+3k$, $3i+2j+k$.
3. $t^2i-\tfrac{1}{3}t^3k$; $(4+\tfrac{1}{3}t^3)i-3j-\tfrac{1}{12}t^4k$.
5. $a\cos nt+\dfrac{1}{n}v\sin nt$.
6. $r=a+tb+\tfrac{1}{2}t^2c$.

7. $r=ma\tan ti+(1-m)aj+ma\sec tk$.
8. $\sqrt{2}:\sqrt{3}$; $\sqrt{5}/2\sqrt{2}$.
9. $3i+3j+6k$; $6i+12j+12k$.
10. $5i-3j+6k$, $t=1$; $1:1$.
12. $\tfrac{5}{3}i+j$, in the direction of the vector $-7i-4j$.

EXAMPLES 10.3 (p. 494)

1. 10; $-10k$. -10, -20.
2. $-12k$, $-19k$, 228, 0.
9. $-i+3j$, $-6i+2j$; 12, 16k.
10. $(3i-2j+6k)/7$.

12. $4i-4j+10k$; 14; $24i-26j-20k$; $53i+28j+11k$.

13. $(-3\mathbf{i}+5\mathbf{j}+11\mathbf{k})/\sqrt{155}$;
    $\sqrt{(155/156)}$.
14. $\sqrt{78}$.
16. $54\mathbf{i}+6\mathbf{j}+15\mathbf{k}$; $255/\sqrt{26}$.
20. $2\frac{1}{2}$.
22. $14/\sqrt{29}$.
23. $-6\mathbf{i}-15\mathbf{j}+3\mathbf{k}$; $44/\sqrt{30}$.
30. $\tan nt = +\frac{1}{2}, -2$.
34. $-mp^2\mathbf{r}$.

36. $-60\mathbf{i}+80\mathbf{j}$; $-1600\mathbf{i}-1200\mathbf{j}$.
37. $\dfrac{10\pi}{3}(\mathbf{j}-2\mathbf{k})$;
    $\dfrac{10\pi^2}{9}(-5\mathbf{i}+4\mathbf{j}+2\mathbf{k})$.
38. $w_0 - \dfrac{15}{Ma}(\mathbf{i}+\mathbf{j}+\mathbf{k})t$.
39. $-40\mathbf{i}+30\mathbf{j}$; $-216\mathbf{i}-337\mathbf{j}$.